METHODS IN MOLECULAR BIOLOGY

Series Editor
John M. Walker
School of Life and Medical Sciences
University of Hertfordshire
Hatfield, Hertfordshire, UK

For further volumes:
http://www.springer.com/series/7651

For over 35 years, biological scientists have come to rely on the research protocols and methodologies in the critically acclaimed *Methods in Molecular Biology series.* The series was the first to introduce the step-by-step protocols approach that has become the standard in all biomedical protocol publishing. Each protocol is provided in readily-reproducible step-by step fashion, opening with an introductory overview, a list of the materials and reagents needed to complete the experiment, and followed by a detailed procedure that is supported with a helpful notes section offering tips and tricks of the trade as well as troubleshooting advice. These hallmark features were introduced by series editor Dr. John Walker and constitute the key ingredient in each and every volume of the *Methods in Molecular Biology series.* Tested and trusted, comprehensive and reliable, all protocols from the series are indexed in PubMed.

HIV Protocols

Fourth Edition

Edited by

Vinayaka R. Prasad

Department of Microbiology & Immunology, Albert Einstein College of Medicine, Bronx, NY, USA

Ganjam V. Kalpana

Department of Genetics, Albert Einstein College of Medicine, Bronx, NY, USA

Humana Press

Editors
Vinayaka R. Prasad
Department of Microbiology
& Immunology
Albert Einstein College of Medicine
Bronx, NY, USA

Ganjam V. Kalpana
Department of Genetics
Albert Einstein College of Medicine
Bronx, NY, USA

ISSN 1064-3745 ISSN 1940-6029 (electronic)
Methods in Molecular Biology
ISBN 978-1-0716-3861-3 ISBN 978-1-0716-3862-0 (eBook)
https://doi.org/10.1007/978-1-0716-3862-0

Cover Caption: High resolution fluorescence microscope image of reactivated HIV-1 latent cells obtained after single molecule RNA-FISH showing transcription site (white arrows) and single molecules of gag RNA (yellow arrow heads). Blue, DAPI; Magenta, HIV-1 gag RNA. See chapter 4 by Pathak et al.

This Humana imprint is published by the registered company Springer Science+Business Media, LLC, part of Springer Nature.
The registered company address is: 1 New York Plaza, New York, NY 10004, U.S.A.

Paper in this product is recyclable.

Preface

During the past decade and a half, we edited two versions of Springer *HIV Protocols*—second and third editions, in 2009 and 2016, respectively. Thus, it is incredible being given an opportunity to put together another update in the form of *HIV Protocols Fourth Edition*. These three editions will have little overlap and thus can serve as a useful three-edition collection for use by every preclinical HIV-1 researcher.

HIV epidemic is still a health burden, and it has not been possible to eliminate HIV-1 due to its persistence. Thus, the research to develop new methodologies to study persistent virus, define new drug targets, develop new drugs, and make new effort to deal with the pathogenic outcomes is as essential today as it was prior to the start of COVID19 epidemic. We are beginning to understand new intricacies in the precise order of events in viral replication cycle due to several advanced live cell imaging and other techniques; we are learning of new types of modifications to the viral RNA that impact HIV biology, new types of intracellular compartments (membraneless organelles) all of which can help in combating the epidemic in novel ways. In addition, novel methods to: (a) study HIV latency reactivation via single-molecule RNA detection; (b) study T cell responses; (c) generate new and efficacious anti-HIV CAR T cells; (d) perform highly sensitive analysis of mucosal HIV infection; (e) prepare and analyze 3D brain organoids to study neuro AIDS; or (f) study the transfer of antibodies across the blood-brain barrier all further enrich the tool set currently available to understand HIV-1 epidemic and its persistence in different anatomical sites including CNS.

As is typical to all MiMB series of books, the chapters in this book are complete with a Notes section. As the notes come from the expert practitioners of the technique, they include many useful tips, tricks, and details that are rarely found in experimental protocols found elsewhere. Authors of each chapter made a special effort to come up with highly useful tips that are crucial for the protocols to work.

We would like to thank Springer for the opportunity to edit this book, the series editor Dr. John Walker for his continuous support and expert guidance on how to generate this compilation, and Patrick Marton for his constant support. We extend our sincere gratitude to all authors for their valuable contributions to this collection and for their prompt and timely communication. We wish to specially express our gratitude to the members of our two research groups (Emilie Ernst, Dina Mofed, Anthony Morante, Swati Haldar and Rajiv Pathak) who helped with preparing the subject index. Special Thanks are due to Emilie who contributed above and beyond the call of duty.

Bronx, NY, USA | *Vinayaka R. Prasad*
Ganjam V. Kalpana

Contents

Part VII HIV-1 Preclinical Studies

Contributors

DAVID AJASIN • *Department of Neurobiology, The University of Texas Medical Branch (UTMB), Galveston, TX, USA*
KIM ANTHONY-GONDA • *Caring Cross, Gaithersburg, MD, USA*
VELPANDI AYYAVOO • *Department of Infectious Diseases and Microbiology, School of Public Health, University of Pittsburgh, Pittsburgh, PA, USA*
ALBERTO BOSQUE • *Department of Microbiology, Immunology and Tropical Medicine, George Washington University, Washington, DC, USA*
HRIMKAR BUCH • *HIV-AIDS Laboratory, Molecular Biology and Genetics Unit, Jawaharlal Nehru Centre for Advanced Scientific Research, Jakkur, Bangalore, Karnataka, India*
RYAN C. BURDICK • *Viral Mutation Section, HIV Dynamics and Replication Program, Center for Cancer Research, National Cancer Institute at Frederick, Frederick, MD, USA*
EDWARD M. CAMPBELL • *Department of Microbiology and Immunology, Stritch School of Medicine, Loyola University, Maywood, Chicago, IL, USA*
SUTANUKA CHAKRABORTY • *Department of Biosciences, Sri Sathya Sai Institute of Higher Learning, Anantapur, Andhra Pradesh, India*
AJIT CHANDE • *Molecular Virology Laboratory, Department of Biological Sciences, Indian Institute of Science Education and Research, Bhopal, Madhya Pradesh, India*
DAVID G. COURTNEY • *Wellcome-Wolfson Institute for Experimental Medicine, Queen's University Belfast, Belfast, UK*
HANNAH L. COUTTS • *Wellcome-Wolfson Institute for Experimental Medicine, Queen's University Belfast, Belfast, UK*
ADARSH DHARAN • *Department of Biology, Indiana University, Bloomington, IN, USA; Department of Microbiology and Immunology, Stritch School of Medicine, Loyola University, Maywood, Chicago, IL, USA*
FRANCESCA DI NUNZIO • *Unit of Advanced Molecular Virology, Institut Pasteur, Université Paris Cité, Paris, France*
FELIPE DIAZ-GRIFFERO • *Department of Microbiology and Immunology, Albert Einstein College of Medicine, New York, NY, USA*
RAYANE DIBSY • *CNRS, University of Montpellier, Institut de Recherche en Infectiologie de Montpellier – IRIM, UMR9004, Montpellier, France*
ROBERTA S. DOS REIS • *Department of Infectious Diseases and Microbiology, School of Public Health, University of Pittsburgh, Pittsburgh, PA, USA*
BORO DROPULIĆ • *Caring Cross, Gaithersburg, MD, USA*
ALICE DUCHON • *Viral Recombination Section, HIV Dynamics and Replication Program, Center for Cancer Research, National Cancer Institute at Frederick, Frederick, MD, USA*
GODFREY DZHIVHUHO • *Myles H. Thaler Center for AIDS and Human Retrovirus Research, University of Virginia, Charlottesville, VA, USA; Department of Microbiology, Immunology and Cancer Biology, University of Virginia, Charlottesville, VA, USA*
CAROLINA ELISCOVICH • *Department of Medicine (Hepatology), and Developmental and Molecular Biology, Albert Einstein College of Medicine, Bronx, NY, USA*
ELISEO A. EUGENIN • *Department of Neurobiology, The University of Texas Medical Branch (UTMB), Galveston, TX, USA*

CYRIL FAVARD • *CNRS, University of Montpellier, Institut de Recherche en Infectiologie de Montpellier – IRIM, UMR9004, Montpellier, France*
STEPHANIE GAMBUT • *Department of Microbial Pathogens & Immunity, Rush University Medical Center, Chicago, IL, USA*
FRANZISKA K. GEIS • *Columbia University Medical Center, New York, NY, USA*
STEPHEN P. GOFF • *Columbia University Medical Center, New York, NY, USA*
HARRIS GOLDSTEIN • *Department of Pediatrics, Albert Einstein College of Medicine, Bronx, NY, USA; Departments of Microbiology & Immunology, Albert Einstein College of Medicine, Bronx, NY, USA*
MARIE-LOUISE HAMMARSKJOLD • *Myles H. Thaler Center for AIDS and Human Retrovirus Research, University of Virginia, Charlottesville, VA, USA; Department of Microbiology, Immunology and Cancer Biology, University of Virginia, Charlottesville, VA, USA*
ETHAN HONEYCUTT • *Department of Molecular Biology and Microbiology, School of Medicine, Case Western Reserve University, Cleveland, OH, USA*
THOMAS J. HOPE • *Department of Biomedical Engineering, Northwestern University, Chicago, IL, USA; Department of Cell and Development Biology, Northwestern University, Chicago, IL, USA; Department of Obstetrics and Gynecology, Northwestern University, Chicago, IL, USA; Feinberg School of Medicine, Northwestern University, Chicago, IL, USA; Chemistry of Life Processes Institute, Northwestern University, Chicago, IL, USA*
JING HUANG • *Myles H. Thaler Center for AIDS and Human Retrovirus Research, University of Virginia, Charlottesville, VA, USA; Department of Microbiology, Immunology and Cancer Biology, University of Virginia, Charlottesville, VA, USA*
WEI-SHAU HU • *Viral Recombination Section, HIV Dynamics and Replication Program, Center for Cancer Research, National Cancer Institute, Frederick, MD, USA*
KAUSHIK INAMDAR • *CNRS, University of Montpellier, Institut de Recherche en Infectiologie de Montpellier – IRIM, UMR9004, Montpellier, France*
DANA F. INDIHAR • *Division of Hematology/Oncology, Department of Medicine, Heersink School of Medicine, University of Alabama at Birmingham, Birmingham, AL, USA*
PATRICK E. H. JACKSON • *Myles H. Thaler Center for AIDS and Human Retrovirus Research, University of Virginia, Charlottesville, VA, USA; Division of Infectious Diseases and International Health, University of Virginia, Charlottesville, VA, USA*
JENNIFER J. JONES • *Division of Hematology/Oncology, Department of Medicine, Heersink School of Medicine, University of Alabama at Birmingham, Birmingham, AL, USA*
GANJAM V. KALPANA • *Department of Genetics, Albert Einstein College of Medicine, Bronx, NY, USA*
JOHN C. KAPPES • *Division of Hematology/Oncology, Department of Medicine, Heersink School of Medicine, University of Alabama at Birmingham, Birmingham, AL, USA; Birmingham VA Health Care System, Research Service, Birmingham, AL, USA*
JONATHAN KARN • *Department of Molecular Biology and Microbiology, School of Medicine, Case Western Reserve University, Cleveland, OH, USA*
FREDRICK KIZITO • *Department of Molecular Biology and Microbiology, School of Medicine, Case Western Reserve University, Cleveland, OH, USA*
YUEKUN LANG • *Department of Microbiology and Immunology, Albert Einstein College of Medicine, Bronx, NY, USA; Department of Veterinary Pathobiology, College of Veterinary Medicine, Columbia, MO, USA*
CALLIE LEVINGER • *Department of Microbiology, Immunology and Tropical Medicine, George Washington University, Washington, DC, USA*

Charlotte Luchsinger • *Department of Microbiology and Immunology, Albert Einstein College of Medicine, Bronx, NY, USA*
Holden T. Maecker • *Institute for Immunity, Transplantation, and Infection, Stanford University School of Medicine, Stanford, CA, USA*
João I. Mamede • *Department of Microbial Pathogens & Immunity, Rush University Medical Center, Chicago, IL, USA*
Tarun Mishra • *Department of Microbiology and Immunology, Carver College of Medicine, University of Iowa, Iowa City, IA, USA*
Dina Mofed • *Department of Microbiology and Immunology, Albert Einstein College of Medicine, Bronx, NY, USA*
Delphine Muriaux • *CNRS, University of Montpellier, Institut de Recherche en Infectiologie de Montpellier – IRIM, UMR9004, Montpellier, France*
J. Natalie Howard • *Department of Microbiology, Immunology and Tropical Medicine, George Washington University, Washington, DC, USA*
Christina Ochsenbauer • *Division of Hematology/Oncology, Department of Medicine, Heersink School of Medicine, University of Alabama at Birmingham, Birmingham, AL, USA*
Rimas J. Orentas • *Caring Cross, Gaithersburg, MD, USA*
Arun Panchapakesan • *Molecular Biology Laboratory, Y R Gaitonde Centre for AIDS Research and Education (YRG CARE), Chennai, Tamil Nadu, India*
Rajiv Pathak • *Department of Genetics and Department of Microbiology and Immunology, Albert Einstein College of Medicine, Bronx, NY, USA*
Vinay K. Pathak • *Viral Mutation Section, HIV Dynamics and Replication Program, Center for Cancer Research, National Cancer Institute at Frederick, Frederick, MD, USA*
Stacia Phillips • *Department of Microbiology and Immunology, Carver College of Medicine, University of Iowa, Iowa City, IA, USA*
Vinayaka R. Prasad • *Department of Microbiology and Immunology, Albert Einstein College of Medicine, Bronx, NY, USA*
Brendan Prideaux • *Department of Neurobiology, The University of Texas Medical Branch (UTMB), Galveston, TX, USA*
Pavitra Ramdas • *Molecular Virology Laboratory, Department of Biological Sciences, Indian Institute of Science Education and Research, Bhopal, Madhya Pradesh, India*
Udaykumar Ranga • *HIV-AIDS Laboratory, Molecular Biology and Genetics Unit, Jawaharlal Nehru Centre for Advanced Scientific Research, Jakkur, Bangalore, Karnataka, India*
Vasudev R. Rao • *Department of Microbiology and Immunology, Albert Einstein College of Medicine, Bronx, NY, USA; Division of AIDS Research, National Institute of Mental Health, National Institutes of Health, Bethesda, MD, USA*
David Rekosh • *Myles H. Thaler Center for AIDS and Human Retrovirus Research, University of Virginia, Charlottesville, VA, USA; Department of Microbiology, Immunology and Cancer Biology, University of Virginia, Charlottesville, VA, USA*
Anjelica F. Reyes • *Department of Microbial Pathogens & Immunity, Rush University Medical Center, Chicago, IL, USA*
Jeffrey R. Schneider • *Department of Microbial Pathogens and Immunity, Rush University Medical Center, Chicago, IL, USA*
Viviana Scoca • *Unit of Advanced Molecular Virology, Institut Pasteur, Université Paris Cité, Paris, France*

SATHISH SELVAM • *Department of Infectious Diseases and Microbiology, School of Public Health, University of Pittsburgh, Pittsburgh, PA, USA*

ROBERT H. SINGER • *Department of Cell Biology and Neuroscience, Albert Einstein College of Medicine, Bronx, NY, USA*

HANG SU • *Departments of Microbiology & Immunology, Albert Einstein College of Medicine, Bronx, NY, USA*

SHOBITH SURESH • *HIV-AIDS Laboratory, Molecular Biology and Genetics Unit, Jawaharlal Nehru Centre for Advanced Scientific Research, Jakkur, Bangalore, Karnataka, India*

THOMAS SWEET • *Department of Nutrition, School of Medicine, Case Western Reserve University, Cleveland, OH, USA*

SILVANA VALDEBENITO • *Department of Neurobiology, The University of Texas Medical Branch (UTMB), Galveston, TX, USA*

MARC C. E. WAGNER • *Department of Infectious Diseases and Microbiology, School of Public Health, University of Pittsburgh, Pittsburgh, PA, USA*

SAMANTHA J. WELNINSKI • *Department of Microbial Pathogens and Immunity, Rush University Medical Center, Chicago, IL, USA*

LI WU • *Department of Microbiology and Immunology, Carver College of Medicine, University of Iowa, Iowa City, IA, USA*

Part I

Visualization of HIV-1 RNA and Proteins

Chapter 1

Infectious Virus Tracking by Fluorescent Live Cell Imaging in Primary Cells

Stephanie Gambut, Thomas J. Hope, and João I. Mamede

Abstract

To successfully infect a cell, HIV-1 has to overcome several host barriers while exploiting cellular cofactors. HIV-1 infection is highly inefficient with the great majority of viral particles not being able to successfully integrate into the target cell genome. Nonproductive HIV-1 particles are degraded or accumulated in cellular compartments. Thus, it becomes hard to distinguish between viral behaviors that lead to effectively infecting the cell from the ones that do not by using traditional methods. Here, we describe the infectious virus tracking method that detects and quantifies individual fluorescent viral particles over time and links viral particle behavior to its infectivity. This method employs live-cell imaging at ultra-low MOIs to detect the outcome of infection for every HIV-1 particle.

Key words Live cell imaging, HIV uncoating, HIV fusion, Wide-field microscopy, HIV early steps of infection, HIV viral core

1 Introduction

Many biological processes are naturally stochastic [1–4]. When so, biological molecules such as proteins or macromolecular assemblies from a defined clonal genetic background do not follow deterministic outcomes in the same controlled experimental setting. The evolutionary arms race between hosts and pathogens often leads to interactions between host proteins and pathogens that have a number of different outcomes. In the particular case of HIV-1, it is well defined that for each infectious event there are a majority of viral particles that do not result in infection. In addition, there is a nongenetically driven loss of infectivity due to incomplete fusion, lack of completion of reverse transcription, incomplete crossing of the nuclear envelope, and inability to fully integrate into the host cell genome as a provirus. Classical biochemistry and functional assays provide averaged pool behaviors of viral particles, thus obfuscating the normal infectious behavior of a viral particle. Live-cell

Vinayaka R. Prasad and Ganjam V. Kalpana (eds.), *HIV Protocols*, Methods in Molecular Biology, vol. 2807,
https://doi.org/10.1007/978-1-0716-3862-0_1,

fluorescence imaging has been able to shed light on the steps that HIV completes from binding to the cell receptors and fusion, microtubule interaction, capsid integrity loss, reverse transcription, nuclear translocation, integration, and transcription from a provirus [5–13]. However, many of these methods focus on a small time scale and specific processes due to the experimental limitations of phototoxicity and photobleaching. Thus, methods that identify single particle behavior of viral particles for the whole replication cycle are of very high utility to differentiate between particles that infect a cell, from the ones that do not.

Here, we present a fluorescence live-cell imaging method designed to identify the viral behavior that leads to productive HIV replication in its infectious context and natural target cells. Our methodology tracks fully fluorescently labeled viruses and cells over long periods of time at low multiplicity of infection. The overall objective of the infectious virus tracking (IVT) method is to challenge cells with less than one viral particle per cell, on average, which allows us to distinguish between infectious behaviors of HIV-1 during its early steps of infection and behaviors that are nonproductive [7]. Here, we will use dual-labeled fluorescent viral particles prepared by transfecting two different HIV constructs: HIV-iGFP [14] and HIV-CA-mRuby3 (Fig. 1a). These allow one to identify virus–cell membrane fusion, loss of capsid integrity (HIV-iGFP), and the direct loss of CA with mRuby3 inserted in the cyclophilin A-binding loop (HIV-CAmRuby3) (Fig. 1b, c).

2 Materials

2.1 Cell Culture Materials

1. HEK-293T cells (ATCC). Culture with DMEM supplemented with 10% FBS and 1 U/mL of penicillin, 1 μg/mL streptomycin, and 3.38 mg/mL glutamine.
2. HOS-Lamin cells. Culture with FluoroBrite DMEM supplemented with 10% FBS and 1 U/mL of penicillin, 1 μg/mL streptomycin, and 3.38 mg/mL glutamine.
3. Monocyte-derived macrophages (MDMs). Culture with RPMI supplemented with 10% FBS and 1 U/mL of penicillin, 1 μg/mL streptomycin, 3.38 mg/mL glutamine, and 50 ng/mL of M-CSF.
4. FluoroBrite DMEM.
5. Dulbecco's modified Eagle's medium (DMEM).
6. Roswell Park Memorial Institute (RPMI) 1640 Media (no red phenol).
7. Fetal bovine serum (FBS).
8. Penicillin–streptomycin–glutamine.

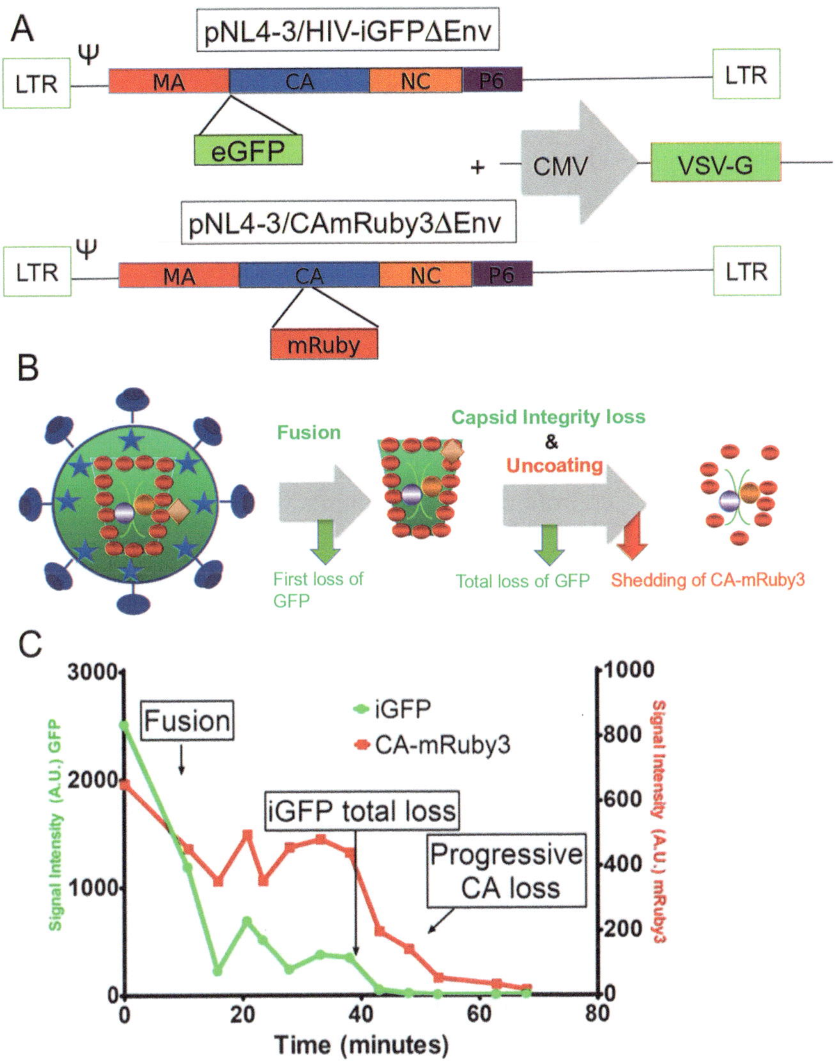

Fig. 1 Schematic representation of HIV-iGFP and HIV-CA-mRuby3. (**a**) HIV-iGFP and HIV-CA-mRuby3 are pNL4-3-based HIVΔEnv constructs allowing for pseudotyping with VSV-G or different HIV Env. The integration into the target genome results in the expression of GFP or mRuby3 at the same ratio as used in the transfection mix (*see* Subheading 3.1). (**b**) The dual-labeled viruses contain the fluid-phase GFP trapped inside of virions, both inside and outside of the viral core. After fusion, there is a first loss of GFP intensity associated with a viral particle and after capsid integrity is lost there is a second loss of GFP from the viral core. CA-mRuby3 is fused into the CypA-binding loop and reports CA content of the viral core. (**c**) Example of a particle that fused into the cytoplasm, lost integrity ~30 min post-fusion (total loss of detectable GFP), and started to shedding CA (reduction of mRuby3 signal over time) from the viral core

9. MEM Non-Essential Amino Acids (MEM-NEAA).
10. Macrophage colony-stimulating factor (M-CSF).
11. 100-mm diameter culture dishes.
12. 150-mm diameter culture dishes.

13. 0.45-μm PVDF filters.
14. 20-mL syringe Luer-Lok.
15. Polypropylene cryovials.

2.2 Monocyte-Derived Macrophage Isolation

1. Heparin-containing EDTA collection tubes, lavender cap.
2. EasySep Direct Human Monocyte Isolation Kit (STEMCELL).
3. EasyEights™ Magnet (STEMCELL).
4. 14-mL polystyrene tubes.
5. 5-mL polystyrene tube.

2.3 Virus Production

1. PEI 1 mg/mL in water.
2. Vector pCMV-VSV-g coding for vesicular stomatitis virus glycoprotein (Addgene item #8454).
3. pHIV-iGFP-ΔEnv which has a GFP between the MA and CA protein cleavage sites in gag reading frame (HIV reagent program HRP-12455).
4. pHIV-CARU-ΔEnv has a coding mRuby3 in the CA protein site based on a pNL4-3 HIV-1 plasmid.
5. pSIV3+ is a packaging vector for SIVmac containing all accessory genes (HIV reagent program ARP-13456).

2.4 Live Cell Imaging

1. The protocol was optimized for Nikon wide-field Ti-e2 equipped with an LED light source and a back-illuminated sCMOS camera (PCO Edge BI4.2) using emission filters for DAPI/FITC/TRITC/Cy5 with a plan Apo γ 60X Nikon Lens.
2. DeltaT heated lid with perfusion (Bioptechs).
3. DeltaT culture dishes 0.17-mm thick (Bioptechs).
4. Heating system with humidity set at 37 °C.
5. CO_2 system injecting 5% CO_2.
6. Light-duty tissue wipes.
7. Polybrene at 10 mg/mL (hexadimethrine bromide).
8. L-Ascorbic acid at 250 mM.
9. Ammonium chloride at 4 M.
10. NucSpot Live 650 + verapamil (Biotium).

2.5 Virus on Glass and Infectivity Assays

1. Coverslip glass #1.5.
2. Fibronectin.
3. Fixing solution composed of PIPES and 3.7% formaldehyde (final concentration).

4. Blocking buffer containing 10% of normal donkey serum, 0.01% of sodium azide, and 0.1% of Triton X-100 in PBS.
5. Monoclonal anti-HIV-1 p24 antibody (AG3.0) that only binds to mature HIV-1 capsids (NIH HIV reagent program Cat#-ARP-4121).
6. Alexa Fluor 647 AffiniPure Donkey anti-mouse IgG.
7. Phosphate-buffered saline (PBS).
8. 96-well plate black walls, glass bottom. This is optional for a high sample number throughput.

3 Methods

Note that all the procedures involving cells are to be carried out at 37 °C and 5% CO_2 unless specified otherwise.

3.1 Dual-Labeled HIV and SIVmac VLP Production

Day −1: Seed 2 × 10^6 HEK-293T cells into a 100-mm diameter culture dish.

Day 0: Transfection of cells:

In a 1.5-mL microtube, mix:

1. 3 μg of pHIV-iGFP-Δenv.
2. 3 μg of pHIV-CARU-Δenv.
3. 4 μg of pCMV-VSV-g.
4. 1 mL of serum-free Opti-MEM or DMEM.
5. 40 μL of PEI.
6. Mix well and incubate for a minimum of 15 min (*see* **Note 1**).
7. Apply the mix gently on a 100-mm plate of HEK-293T cells while shaking the plate side to side.

In another 1.5-mL microtube, mix:

1. 6 μg of pSIV3.
2. 4 μg of pCMV-VSV-g.
3. 1 mL of serum-free Opti-MEM or DMEM.
4. 40 μL of PEI.
5. Mix well and incubate for a minimum of 15 min (*see* **Note 1**).
6. Apply the mix gently on another 100-mm plate of HEK-293T cells while shaking the plate side to side.

Day 1 (14–16 h Post-Transfection): Change the media with pre-warmed DMEM.

Day 2 (36–48 h Post-Transfection):

1. Harvest the supernatants from the two different viral preparations with a syringe and filter through a 0.45-μm PVDF membrane to remove cellular debris.
2. Aliquot into cryovials and store them at −80 °C.

3.2 Virus on Glass (VOG) Assay of HIV Fluorescent Samples

1. Insert glass coverslips into a 24-well plate (*see* **Note 2**).
2. Coat the coverslip with fibronectin, at 2 μg/cm^2 diluted in PBS, for 20 min and wash three times with PBS. Do not let the surface dry out.
3. Add 15 μL of viral preparation and 100 μL of PBS to the glass and spin the glass facing up for 1 h at 16 °C at 1600 *g* (*see* **Note 3**).
4. Remove the media and wash the glass with PBS three times.
5. Incubate the top of the glass with the fixing solution for 5 min.
6. Wash three times with PBS.
7. Permeabilize the virus with the blocking solution containing Triton X-100 for 5–10 min before staining them with anti-HIV-1 p24 (AG3.0) antibody (1:200 dilution into blocking solution for 1 h).
8. Wash the sample three times with PBS and incubate them for 1 h with anti-mouse AF647 antibody (1:1000 dilution into blocking solution).
9. Wash three more times with PBS and mount the coverslip with a semi-dry mounting media of choice (*see* **Note 4**).
10. Image using the microscope of choice using 2 μm total Z-stack with a 0.2–0.3 μm distance and analyze the data (Fig. 2) (*see* **Notes 5** and **6**).
11. Count the colocalization percentage of HIV-iGFP and HIV-CA-mRuby3 to mature HIV AG3.0 antibody-positive spots. Only samples showing >90% of triple signals per virus should be used for the IVT experiments (Fig. 2).

3.3 Preparing MDM for Live Cell Imaging

Day −10: Isolation of monocytes and differentiation into MDM.

Preparing the MDM using EasySep™ Direct Human Monocyte Isolation Kit

1. Collect blood using blood heparin-containing EDTA collection tubes and mix by inverting lavender cap.
2. Add 4 mL of whole blood into 14-mL polystyrene tubes and incubate 5 min at room temperature with 200 μL of

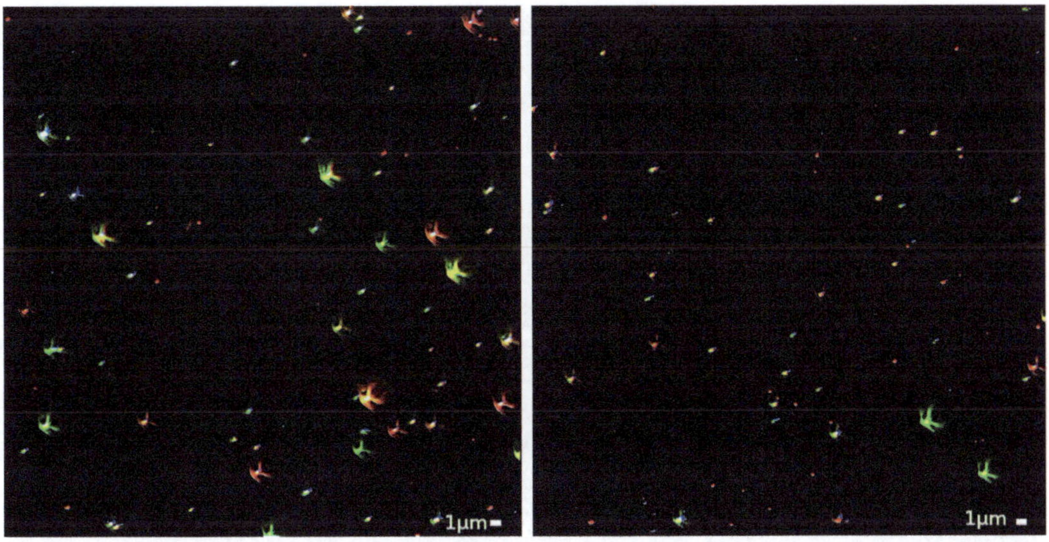

Fig. 2 Virus on glass imaging. Acquired images of viruses on glass were deconvolved and max intensity Z-projected. It is important to avoid virus clustering, represented by a higher number of higher intensity blobs (left) and where mostly individual particles are immobilized in the glass surface (right). Red: HIV-CA-mRuby3; green: HIV-iGFP; blue: AG3.0-AF647. Scale bar 1 μm

Isolation Cocktail and 200 μL of RapidSpheres (previously vortexed for 30 s). Pipette up and down for mixing.

3. Top the tube with 8 mL of PBS and mix gently by pipetting up and down two to three times and place the tube without lid in EasyEights™ magnet for 5 min at room temperature.
4. Carefully pipette the enriched cell suspension into a new tube (*see* **Note 7**).
5. Add 200 μL of RapidSpheres in the new tube containing the enriched suspension, mix by pipetting two to three times, and incubate for 5 min at room temperature.
6. Transfer the tube to the magnet for a second separation, 5 min at room temperature.
7. Carefully pipette only the clear fraction of the enriched cell suspension into a new tube (*see* **Note 8**).
8. Place the new tube in the magnet for 5 min at room temperature for a third separation.
9. Carefully pipette only the clear fraction of the enriched cell suspension into a 15-mL conical tube (*see* **Note 8**).
10. Centrifuge the tube for 5 min at 500 × *g* and discard the supernatant.
11. Resuspend into 2 mL of RPMI media containing M-CSF, transfer into a new 5-mL polystyrene tube, and incubate 5 min in the magnet at room temperature to remove the maximum of beads.

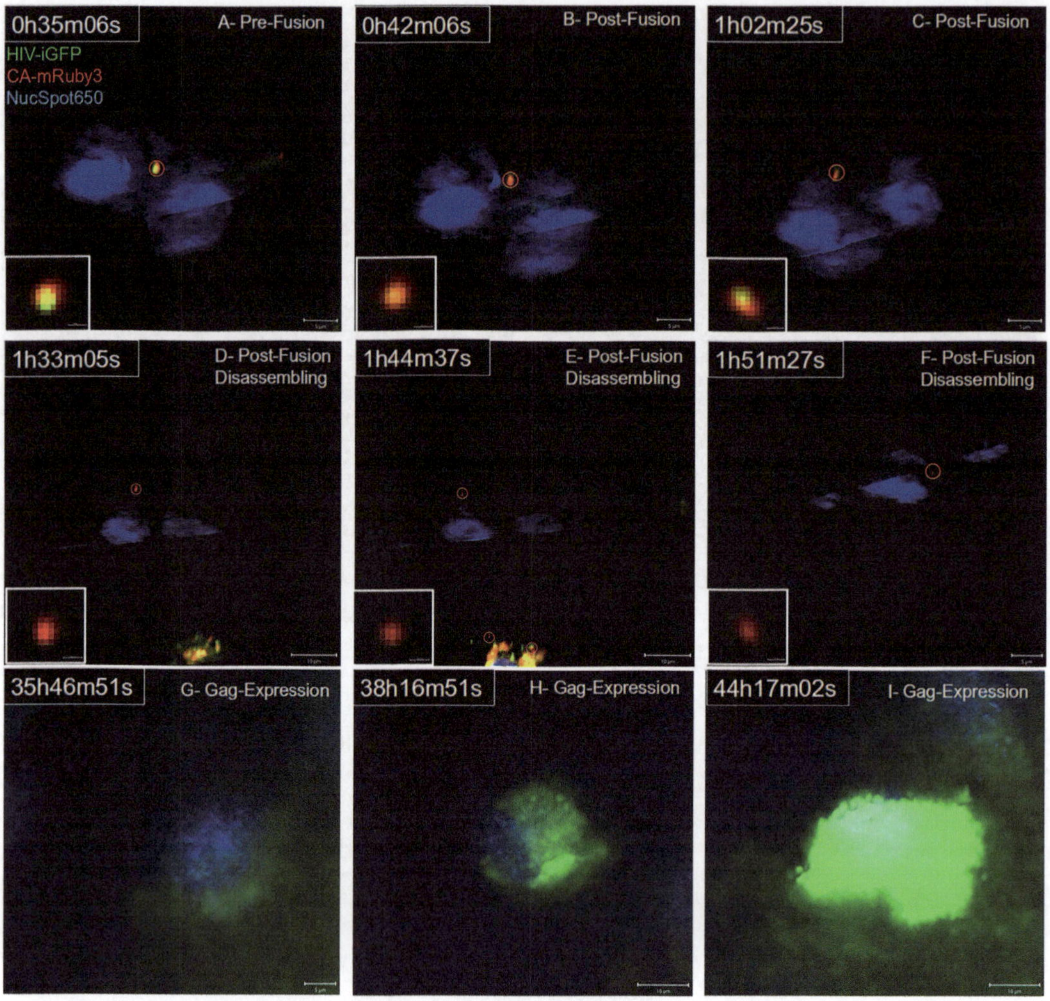

Fig. 3 IVT live-cell imaging. MDM cells were challenged with HIV-iGFP/HIV-CAmRuby3 dual-labeled viruses. The viruses contain fluid-phase GFP (green), reporting fusion and capsid integrity loss in two independent and consecutive intensity reductions of GFP signal, and include HIV CA fused to mRuby3 (red) quantifying the CA content of the viral core over a period of 48 h. Nuclei are detected via NucSpot650 (blue). The virion, seen in Panel (**a**), is the single particle infecting the *cell.* Fusion is prevented by adding NH4Cl prior to imaging and induced by fresh media washout after the first acquisition of all fields of view is completed. In this example, the highlighted labeled virion (marked by a red circle in the top and middle panels and the insets) has high GFP intensity prior to fusion and presented two independent GFP signal intensity reduction events over ~27 min (**a**–**d**), representing fusion and initiation of capsid disassembly, respectively. Following fusion, there is a gradual loss of CA-mRuby3 over 51 min (**c**–**f**) until the signal falls under the limit of detection (**f**) while at the cytoplasm. Production of virions is then observed by Gag-iGFP (shown from **g** to **i**) or GAG-mRuby3 fluorescence (not shown), which are expressed from the integration site. This particular example fits with a model of progressive capsid uncoating that is initiated at the cytoplasm

12. Plate the isolated macrophages directly into deltaT at 400,000 cells, with 1 mL of RPMI without phenol supplemented with 10% FBS and 1 U/mL of penicillin, 1 μg/mL streptomycin, 3.38 mg/mL glutamine, and 50 ng/mL of M-CSF (*see* **Note 9**).

Day 0: Infection and live-cell imaging

1. Prepare a master mix with 12 mL of phenol red-free RPMI culture media +6 μL of polybrene (stock at 10 mg/mL) + 12 μL of acid ascorbic (stock at 250 mM).
2. Prepare tube #1 with volume of virus wanted, 20 μL of SIVmac-VLPs, 1 μL of Nucspot60, 1 μL of verapamil, and master mix up to 500 μL (*see* **Note 10**).
3. Prepare tube #2 with 2.5 μL of NH4Cl and 10 μL of a well characterized fluorescent viral prep 987.5 μL of master mix (total 1 mL) (*see* **Note 11**).
4. Prepare tube #3 with 5 μL of NH4Cl and 1995 μL of master mix (total 2 mL).
5. Keep 500 μL of the media from the deltaT dish and incubate 1 h with solution #1.
6. Remove all the media and add 750 μL of tube #2 containing the fluorescently labeled viruses and NH4Cl on the cells and incubate for 1 h. The NH4Cl will prevent endosome acidification, which results in the blocking of viral fusion by VSV-G pseudotypes.
7. Remove all the media and wash with 1 mL of tube #3 containing the NH4Cl. This step will remove floating viruses while still blocking viral fusion. Optional: Repeat the previous step for further wash.
8. Set up the microscope's acquisition software to acquire multiple visit points for 48 h. We perform a 6-min time lapse for the first 7 h, followed by 15 min of time lapse from 7 to 48 h (*see* **Notes 12**, **13**, and **14**).
9. Rinse the cells using 9 mL of master mix (without NH_4Cl to allow synchronized fusion) with a peristaltic pump and flow 10 mL/min (*see* **Note 15**) at the moment the first time loop finishes the last visit point.
10. Identify the infected cells and analyze the data tracking viruses over time within an infected cell (Fig. 3) with the software of choice.

4 Notes

1. We find that longer periods of incubation do not affect transfection efficiency.
2. A 96-well plate glass bottom #1.5HS might be preferred for a higher number of samples.
3. The volume of incubation is dependent on the viral concentration. It is useful to do several dilutions to obtain identifiable independent viral particles.
4. If using 96-well glass bottom plates, add 50 μL of PBS and cover with ELISA/PCR 96-well plate adhesive lid.
5. We preferably image from bottom to top while setting the Z-coordinate of the microscope at the level of the coverslip surface. Thus, the Perfect Focus System will always image 2 μm starting from the surface where the viruses can attach.
6. Virus clustering is possible resulting in brighter spots due to multiple particles being present in the diffraction-limited areas (Fig. 2, Left). Avoid imaging clusters of viruses and acquire the images at dilutions where the particles are spaced (Fig. 2, Right) to avoid misrepresentation of the data.
7. Collect the entire clear fraction from top to bottom. For optimal recovery, also collect up to 10% (400 μL) of the starting sample volume of RBCs.
8. Collect the entire clear fraction from the opposite edge of the magnet. It is crucial not to disturb the red blood cells clustered with unwanted cells that are captured by the magnet.
9. Change half of the media every other day with culture media described at the previous point, the cells under differentiation attach to the bottom of the well within the first 2 days.
10. The amount of SIVmac-VLPs should be calculated with a previous VLP particle titration. Use the minimal amount of VLPs that result in maximal infection by HIV-iGFP+HIV-CA-mRuby3 viral preps.
11. For the first test on the viral preparation, it should be attempted to follow this protocol with different amounts of viruses to empirically establish the volume of a viral prep to have a low amount of viral particles per cell (optimal is 1 virus per 3 cells in average).
12. Phototoxicity is different for different cell types and different cell lines. We optimized this protocol to utilize the maximum of light that will keep 95% of the cells alive after 48 h in our specific microscopy system. We find that phototoxicity leading to apoptosis/cell death occurs under conditions of minimal

photobleaching (*see* **Note 13**), and it is the limiting factor to the light amount. More details are in ref. 15.

13. Photobleaching should be measured and avoided. Often, we use trapped viruses in the coverslips not associated with cells to measure the effects of photobleaching in each experiment. While higher amounts of light allow to detect dimmer signals, it might lead to the deterioration of the fluorescent protein signals. Conversely, using lower light amounts will conserve the signal but will result in a lower limit of detection. The experiments should be optimized to visualize all particles with minimal photobleaching. More details are in refs. 7, 16.
14. The number of fields of view depends on the speed of the microscope for acquisition and XYZ movement. In the microscope identified above with triggered acquisition, we can fit 54 fields of view (visit points) in a 9-by-6 area pattern, in the 7-min time lapse.
15. We find that higher flow rates for wash results in less debris after the wash allowing fusion.

Acknowledgement

This work was supported by NIH-NIAID grants K22-AI140963 to J.I.M and R01AI150998 to T.J.H. and subcontracts P50AI150464 to J.I.M (Center Development Award) and T.J.H.

References

1. Andrews SS, Dinh T, Arkin AP (2009) Stochastic models of biological processes. In: Meyers RA (ed) Encyclopedia of complexity and systems science. Springer, New York, pp 8730–8749
2. Noble D (2021) The role of stochasticity in biological communication processes. Prog Biophys Mol Biol 162:122–128. https://doi.org/10.1016/j.pbiomolbio.2020.09.008
3. Srivastava R, You L, Summers J, Yin J (2002) Stochastic vs. deterministic modeling of intracellular viral kinetics. J Theor Biol 218:309–321. https://doi.org/10.1006/jtbi.2002.3078
4. Khalili S, Armaou A (2008) Sensitivity analysis of HIV infection response to treatment via stochastic modeling. Chem Eng Sci 63:1330–1341. https://doi.org/10.1016/j.ces.2007.07.072
5. Campbell EM, Hope TJ (2008) Live cell imaging of the HIV-1 life cycle. Trends Microbiol 16:580–587. https://doi.org/10.1016/j.tim.2008.09.006
6. McDonald D, Vodicka MA, Lucero G et al (2002) Visualization of the intracellular behavior of HIV in living cells. J Cell Biol 159:441–452. https://doi.org/10.1083/jcb.200203150
7. Mamede JI, Cianci GC, Anderson MR, Hope TJ (2017) Early cytoplasmic uncoating is associated with infectivity of HIV-1. Proc Natl Acad Sci USA 114:E7169–E7178. https://doi.org/10.1073/pnas.1706245114
8. Francis AC, Marin M, Shi J et al (2016) Time-resolved imaging of single HIV-1 uncoating in vitro and in living cells. PLoS Pathog 12:e1005709. https://doi.org/10.1371/journal.ppat.1005709
9. Francis AC, Melikyan GB (2018) Single HIV-1 imaging reveals progression of infection through CA-dependent steps of docking at the nuclear pore, uncoating, and nuclear transport. Cell Host Microbe 23:536–548.e6.

https://doi.org/10.1016/j.chom.2018.03.009

10. Blanco-Rodriguez G, Gazi A, Monel B et al (2020) Remodeling of the core leads HIV-1 preintegration complex into the nucleus of human lymphocytes. J Virol 94:e00135-20. https://doi.org/10.1128/jvi.00135-20
11. Burdick RC, Li C, Munshi M et al (2020) HIV-1 uncoats in the nucleus near sites of integration. Proc Natl Acad Sci 117:5486–5493. https://doi.org/10.1073/pnas.1920631117
12. Boireau S, Maiuri P, Basyuk E et al (2007) The transcriptional cycle of HIV-1 in real-time and live cells. J Cell Biol 179:291–304. https://doi.org/10.1083/jcb.200706018
13. Xu H, Franks T, Gibson G et al (2013) Evidence for biphasic uncoating during HIV-1 infection from a novel imaging assay. Retrovirology 10:70. https://doi.org/10.1186/1742-4690-10-70
14. Hübner W, Chen P, Del Portillo A et al (2007) Sequence of human immunodeficiency virus type 1 (HIV-1) Gag localization and oligomerization monitored with live confocal imaging of a replication-competent, fluorescently tagged HIV-1. J Virol 81:12596–12607. https://doi.org/10.1128/JVI.01088-07
15. Tinevez J-Y, Dragavon J, Baba-Aissa L et al (2012) Chapter fifteen – A quantitative method for measuring phototoxicity of a live cell imaging microscope. In: Conn PM (ed) Methods in enzymology. Academic Press, pp 291–309
16. Diaspro A, Chirico G, Usai C et al (2006) Photobleaching. In: Pawley JB (ed) Handbook of biological confocal microscopy. Springer US, Boston, pp 690–702

Chapter 2

Imaging HIV-1 Nuclear Import, Uncoating, and Proviral Transcription

Ryan C. Burdick, Alice Duchon, Wei-Shau Hu, and Vinay K. Pathak

Abstract

Live-cell imaging has become a powerful tool for dissecting the behavior of viral complexes during HIV-1 infection with high temporal and spatial resolution. Very few HIV-1 particles in a viral population are infectious and successfully complete replication (~1/50). Single-particle live-cell imaging enables the study of these rare infectious viral particles, which cannot be accomplished in biochemical assays that measure the average property of the entire viral population, most of which are not infectious. The timing and location of many events in the early stage of the HIV-1 life cycle, including nuclear import, uncoating, and integration, have only recently been elucidated. Live-cell imaging also provides a valuable approach to study interactions of viral and host factors in distinct cellular compartments and at specific stages of viral replication. Successful live-cell imaging experiments require careful consideration of the fluorescent labeling method used and avoid or minimize its potential impact on normal viral replication and produce misleading results. Ideally, it is beneficial to utilize multiple virus labeling strategies and compare the results to ensure that the virion labeling did not adversely influence the viral replication step that is under investigation. Another potential benefit of using different labeling strategies is that they can provide information about the state of the viral complexes. Here, we describe our methods that utilize multiple fluorescent protein labeling approaches to visualize and quantify important events in the HIV-1 life cycle, including docking HIV-1 particles with the nuclear envelope (NE) and their nuclear import, uncoating, and proviral transcription.

Key words HIV-1, Live-cell imaging, Nuclear pore complex, Capsid

1 Introduction

Live-cell fluorescence microscopy has become a powerful tool to study the spatiotemporal movements of virus particles and has provided insight into several key events in the early stage of HIV-1 infection. Pioneering studies nearly two decades ago that utilized live-cell imaging of HIV-1 labeled with an enhanced green fluorescent protein (eGFP)-tagged Vpr, an accessory protein that is packaged into virions, revealed the potential of tracking single-virus particles in infected cells through space and time [1]. These studies concluded that viral complexes migrate through the cytoplasm

Vinayaka R. Prasad and Ganjam V. Kalpana (eds.), *HIV Protocols*, Methods in Molecular Biology, vol. 2807,
https://doi.org/10.1007/978-1-0716-3862-0_2,

using dynein and the microtubule network toward the nucleus [1]. Live-cell imaging methods have explored Gag translation [2], virus assembly [3, 4], viral genome packaging [2, 5], cell-to-cell virus infection [6], and other critical events in HIV-1 replication. A multitude of different HIV-1 labeling strategies, including labeling viral proteins with fluorescent proteins [1, 7–9] or fluorescent dyes [10, 11], labeling viral nucleic acid [8, 12, 13], labeling host proteins that interact with viral proteins [14], or host proteins that are incorporated into virions [15], have provided useful tools to enable tracking HIV-1 particles throughout the viral life cycle. In this chapter, we will detail protocols that we have developed or successfully implemented to visualize various steps in the early stage of HIV-1 replication. We will describe protocols for labeling HIV-1 with host restriction factor APOBEC3F fused to yellow fluorescent protein (A3F-YFP; [15]), direct labeling of HIV-1 capsid by fusion with GFP [8], labeling viral cores with content marker GFP (cmGFP) [16], and visualizing proviral transcription by specific recognition of RNA stem loops in the HIV-1 RNA with mCherry-tagged bacterial protein Bgl (Bgl-mCherry).

After fusion of the viral and host membranes, HIV-1 must enter and traverse the cytoplasm of an infected cell, reverse transcribe its genomic RNA into double-stranded DNA, enter the nucleus through a nuclear pore, undergo intranuclear transport to the site of integration, and integrate its DNA into the host genome. However, many critical steps in the early stage of HIV-1 infection, including docking of HIV-1 particles at nuclear pore complexes, mechanism of nuclear import, and uncoating, are not well understood and have only recently started to be elucidated using live-cell microscopy methods [8, 16, 17]. More than 10 years ago, we observed that viral preintegration complexes could be fluorescently labeled by virion incorporation of host restriction factor A3F fused to YFP [15]. These fixed-cell studies showed that A3F-YFP-labeled viral particles could be detected in the nuclei of infected cells and that reverse transcription is not required for the nuclear import of viral complexes. Subsequently, we utilized A3F-YFP labeling of virions and live-cell microscopy to analyze the dynamics of HIV-1 nuclear envelope docking, nuclear import, and intranuclear movements [17]. These studies established that HIV-1 complexes exhibit an unusually long nuclear envelope (NE) residence time prior to nuclear import (1.5 ± 1.6 h) compared to cellular cargos, which are imported in milliseconds, and that capsid mutations that influence capsid stability influence viral core docking to the NE and nuclear import. The results also showed that CA mutants that are defective in binding to host protein cyclophilin A are imported into the nucleus with faster kinetics (0.7 ± 1.0 h). After nuclear import, viral complexes initially exhibit fast movements for a short time (~10 min) as the viral complexes move away from the NE and toward the nucleus interior (short, fast phase) followed by a long, slow phase in which the viral complexes exhibit restricted movements for several hours. During this slow phase, the viral complexes remain near their nuclear point of entry and ~1.4 μm from the NE, suggesting that the viral complexes reach the site of integration shortly after nuclear import. HIV-1 capsid disassembly (uncoating) must occur so that viral DNA can access the host genome and integrate to form a provirus [18]. Until recently [8] and for several decades [19], HIV-1 uncoating was thought to occur before reverse transcription in the cytoplasm, in part because it was thought that viral cores were too big to fit through nuclear pores and that the capsid must disassemble to give reverse transcriptase access to substrate nucleotide triphosphates.

To determine the timing and location of HIV-1 uncoating, we developed a method to directly label HIV-1 capsid protein (CA) with GFP. We observed that HIV-1 cores enter the nucleus intact or nearly intact and that their nuclear import is dependent upon binding to cleavage and polyadenylation specificity factor 6 (CPSF6). We also developed a strategy to visualize proviral transcription, which allows us to identify rare infectious virus particles and determine the nuclear location of integration [8]. HIV-1 transcription sites were visualized by specific recognition of Bgl RNA stem loops (BglSL) in the HIV-1 vector RNA with Bgl-mCherry. After productive infection of HeLa cells expressing Bgl-mCherry with virus containing BglSL, the nascent BglSL-containing viral RNA was detected after binding to multiple Bgl-mCherry proteins. We observed that the nuclear HIV-1 cores uncoat near (<1.5 μm) the genomic site of integration and shortly before (<1.5 h) integration. Because we could not determine whether the nuclear GFP-CA-labeled viral cores lost integrity by rupture or via a small loss of CA before capsid disassembly, we also utilized content marker GFP (cmGFP) that is incorporated into capsids and is released upon loss of capsid integrity [16]. We observed that loss of cmGFP occurred only ~1–3 min prior to loss of GFP-CA, suggesting that intact HIV-1 capsids enter the nucleus and lose their integrity minutes before capsid disassembly. Overall, these studies indicate that the HIV-1 capsid remains intact through nuclear import and that uncoating occurs in the nucleus shortly before integration and near the integration site [8, 16]. Here, we describe our methods to visualize and quantify key events in the early stage of the HIV-1 life cycle, including docking of HIV-1 cores with the NE and their nuclear import, uncoating of nuclear viral cores, and proviral transcription.

2 Materials

2.1 Plasmids

1. pHDV-EGFP [20].
2. pHGFP-BglSL [8].
3. pHGFP-GFPCA-BglSL [8].
4. pHGFP-iGFP-BglSL [16].
5. pHGFP [8].
6. pHGFP-GFPCA [8].
7. NL4-3 Gag-iGFP ΔEnv [4].
8. pA3F-YFP [15].
9. pHCMV-G (VSV-G; [21]).

2.2 Cell Culture Reagents

1. 293T cells (ATCC Cat No. CRL-3216).
2. HeLa cells (ATCC Cat No. CCL-2).
3. HeLa:mRuby-LaminB [16].
4. HeLa:Bgl-mCherry [8].
5. Dulbecco's modified Eagle's medium (DMEM) with or without phenol red supplemented with 10% fetal bovine serum, 1% penicillin/streptomycin, and, for DMEM without phenol red, L-glutamine. The DMEM containing phenol red and supplemented with FBS and antibiotics is referred to as complete media (CM), whereas DMEM lacking phenol red and supplemented with FBS and antibiotics is referred to as clear CM.
6. Trypsin–EDTA (0.25%), phenol red.
7. 100-mm × 20-mm polystyrene culture plates.
8. Doxycycline hydrochloride.
9. Aphidicolin.
10. BioTracker 650 Red Nuclear Dye (Sigma-Aldrich, Cat. No. SCT119).

2.3 Virus Production and Collection

1. Polyethylenimine, Linear, MW 25000 (PEI; Polysciences, Inc. Cat. No. 23966-1).
2. 10-mL syringes with Luer-Lok tip.
3. 0.45 μM PVDF-sterilized syringe filters.
4. Dulbecco's phosphate-buffered saline (dPBS).
5. Sucrose.
6. 14 × 89-mm Ultra-Clear Centrifuge Tubes.
7. Ultracentrifuge and swinging bucket rotor.
8. Cryotube vials.

2.4 Quantitation of Virus and Virus Infection

1. ibiTreated 8-Well μ-Slides (IBIDI Cat No. 80825-90).
2. ST-8R Centrifuge with Swinging Bucket SW 41 Ti (Thermo Scientific).
3. Polybrene.

2.5 Image Collection

1. Nikon Eclipse Ti-E microscope equipped with a Yokogawa CSU-W1 spinning disk unit and a Plan-Apochromat 100× N.A. 1.49 oil objective (Nikon), using 488-nm (GFP), 514-nm (YFP), 561-nm (mRuby), and 594-nm (mCherry) lasers for illumination, a 565-nm splitter, and two ORCA-fusion BT cameras (Hamamatsu).
2. Stage Top Incubator with Lens Heater (Tokai Hit).
3. NIS-Elements software (Nikon).

2.6 Single-Virion Analysis Via MATLAB and Localize

1. MATLAB (MathWorks; MATLAB scripts are available upon request).
2. Localize [22].
3. Excel (Microsoft) and Prism 9 (GraphPad).

3 Methods

The following methods describe (1) the preparation of HIV-1 virions labeled with A3F-YFP, GFP-CA, or cmGFP; (2) virus infection and live-cell imaging of HIV-1 core docking at the NE and nuclear import; and (3) virus infection and live-cell imaging of HIV-1 uncoating, HIV-1 transcription sites, and GFP reporter expression.

3.1 Preparation of HIV-1 Virions Labeled with A3F-YFP, GFP-CA, or cmGFP

The steps involved in the preparation of HIV-1 labeled with A3F-YFP, GFP-CA, or cmGFP are described in Subheadings 3.1.1, 3.1.2, and 3.1.3. These include (1) virus production by PEI transfection of 293T cells, (2) virus collection and concentration, and (3) quantitation of virus using single-virion analysis.

3.1.1 PEI Transfection

Plate low passage (<1 month in culture) 293T cells at 3.5×10^6 cells per 100-mm dish in 10 mL CM. Twenty hours later, transfect cells with a combined total of 14 μg of DNA. To generate HIV-1 particles labeled with A3F-YFP, transfect plasmids encoding Vif-deficient HIV-1 (pHDV-EGFP;10 μg), VSV-G (pHCMV-G; 2 μg), and A3F-YFP (pA3F-YFP; 2 μg) using a 1:4 weight ratio of DNA to PEI (56 μg PEI) (Fig. 1a, b; *see* **Note 1**). HDV-EGFP is an HIV-1-derived vector, which is deficient for envelope glycoprotein gene (*env*) and the accessory genes *vif*, *vpr*, *vpu*, and *nef*, and, when supplemented with an appropriate envelope protein, can complete a single cycle of replication.

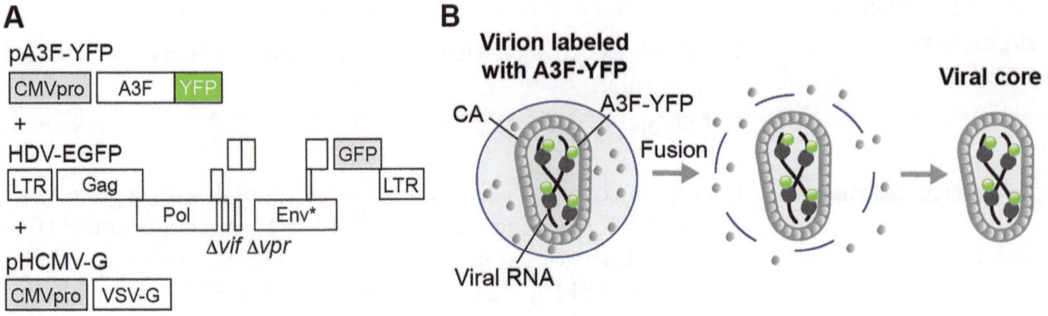

Fig. 1 Vectors used to generate virions labeled with A3F-YFP. (**a**) Plasmids used for transfection to generate HIV-1 virions labeled with A3F-YFP. The GFP reporter in the *nef* open reading frame of HDV-EGFP is expressed in virus producer cells but is not packaged into virions. Asterisk indicates mutation in the *env* gene that introduces a premature stop codon. (**b**) Schematics of an A3F-YFP-labeled virion and post-fusion viral core

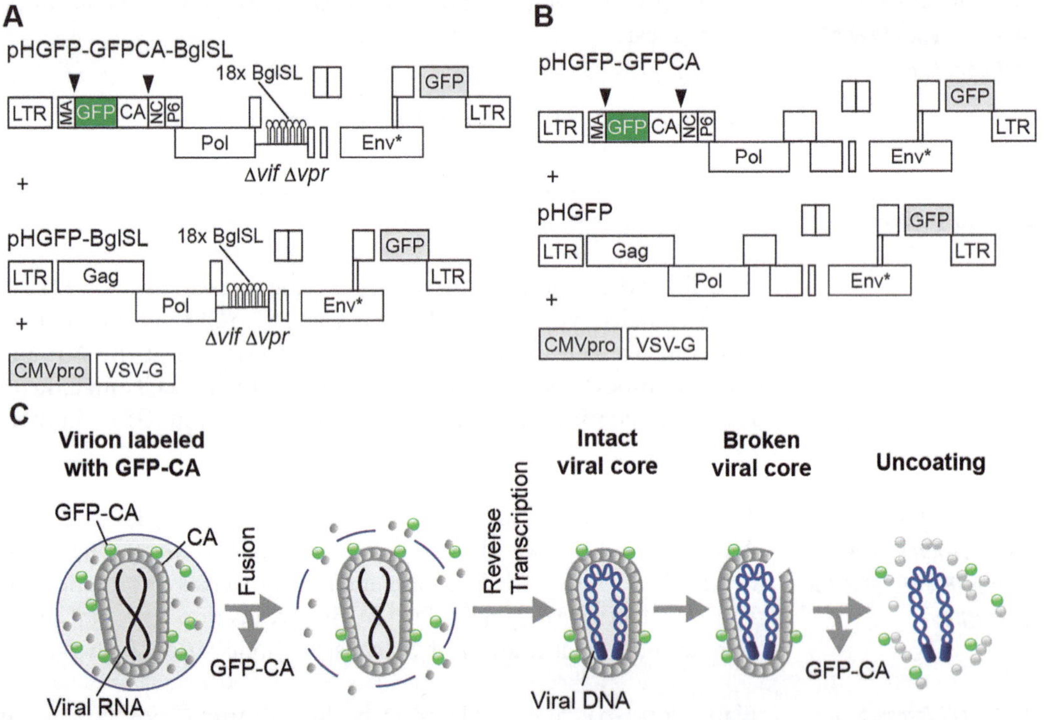

Fig. 2 HIV-1 vectors used to generate virions labeled with GFP-CA. (**a**) Plasmids used for transfection to generate virions with genomic RNAs containing BglSL and viral cores that are labeled with GFP-CA. (**b**) Plasmids used for transfection to generate virions with genomic RNAs lacking BglSL and are labeled with GFP-CA. For (**a**) and (**b**), HIV-1 plasmids expressing GFP-CA (pHGFP-GFPACA-BglSL or pHGFP-GFPCA) and WT Gag (pHGFP-BglSL or pHGFP) were transfected at a 1:10 plasmid ratio. Black triangles indicate protease cleavage sites; proteolytic cleavage of Gag results in the expression of GFP-CA upon virus maturation. Asterisk indicates mutation in the *env* gene that introduces a premature stop codon. GFP reporter in the *nef* open reading frame is expressed in virus producer cells but is not packaged into virions. (**c**) Schematics of a GFP-CA-labeled virion and post-fusion viral core; GFP-CA loss indicates capsid disassembly

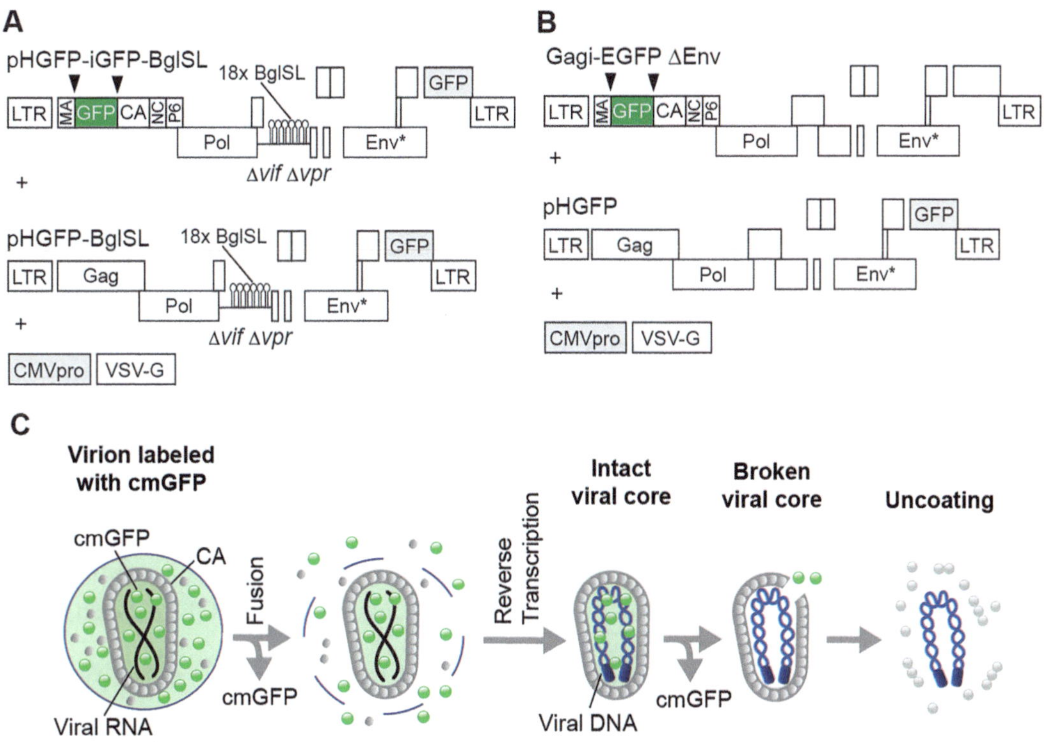

Fig. 3 HIV-1 vectors used to generate virions labeled with cmGFP. (**a**) Plasmids used for transfection to generate virions with genomic RNAs containing BglSL and viral cores that are labeled with cmGFP. (**b**) Plasmids used for transfection to generate virions with genomic RNAs lacking BglSL and are labeled with cmGFP. For (**a**) and (**b**), HIV-1 plasmids expressing cmGFP (pHGFP-iGFP-BglSL or Gagi-EGFP ΔEnv) and WT Gag (pHGFP-BglSL and pHGFP) were transfected at a 1:2 plasmid ratio. Black triangles indicate protease cleavage sites; proteolytic cleavage of Gag results in the expression of free GFP and CA upon virus maturation. Asterisk indicates mutation in the *env* gene that introduces a premature stop codon. GFP reporter in the *nef* open reading frame is expressed in virus producer cells but is not packaged into virions. (**c**) Schematics of a cmGFP-labeled virion and post-fusion viral core; cmGFP loss indicates loss of viral core integrity

Alternatively, use A3F-E289K-YFP, a mutant that is not sensitive to Vif-mediated degradation [23], to label HIV-1 in the presence or absence of the *vif* gene in HIV-1.

To generate HIV-1 particles containing genomes that encode BglSL and are labeled with GFP-CA, transfect plasmids encoding HIV-1 (pHGFP-BglSL; 9.1 μg, pHGFP-GFPCA-BglSL; 0.9 μg) and VSV-G (pHCMV-G; 2 μg) (Fig. 2a). Because the incorporation of GFP-CA into mature capsids upon virion maturation can inhibit virus infectivity, produce viruses by transfection of HIV-1 plasmids expressing GFP-CA and WT gag at a 1:10 plasmid ratio. The infectivity of these GFP-CA-labeled viruses is reduced less than twofold compared to unlabeled viruses, and the GFP-CA-labeled mature capsids are efficiently detected by live-cell microscopy [8]. To generate HIV-1 particles containing genomes that lack BglSL and are labeled with GFP-CA, transfect pHGFP-GFPCA and pHGFP at a 1:10 plasmid ratio (Fig. 2b). Schematics of a GFP-CA-labeled virion and post-fusion core are shown in Fig. 2c.

To generate HIV-1 particles containing genomes that encode BglSL and are labeled with cmGFP, transfect plasmids encoding HIV-1 (pHGFP-BglSL; 6.66 μg, pHGFP-iGFP-BglSL; 3.33 μg) and VSV-G (pHCMV-G; 2 μg) (Fig. 3a). Alternatively, use pHmNG-iGFP-BglSL, an HIV-1 vector similar to pHGFP-iGFP-BglSL that expresses mNeonGreen (mNG) in the *nef* open reading frame, to avoid deletion of sequence between the *gfp* gene located between MA and CA and the *gfp* gene in the *nef* open reading frame. Deletion occurs as a result of reverse transcriptase undergoing an intramolecular template switch between homologous *gfp* sequences during reverse transcription. Viruses are produced by transfection of HIV-1 plasmids expressing cmGFP and WT gag at a 1:2 plasmid ratio since the incorporation of cmGFP into mature capsids does not inhibit virus infectivity. To generate HIV-1 particles containing genomes that lack BglSL and are labeled with cmGFP, transfect NL4-3 Gag-iGFP ΔEnv and pHGFP at a 1: 2 plasmid ratio (Fig. 3b). Schematics of a cmGFP-labeled virion and post-fusion core are shown in Fig. 3c.

Gently mix DNA and PEI in 5-mL serum-free DMEM and incubate at room temperature for 20 min. Replace the supernatant with the DNA-PEI solution and incubate the cells at 37 °C for 3 h. Replace the DNA-PEI solution with 10 mL CM.

3.1.2 Virus Collection and Concentration

1. Collect the virus-containing supernatant ~20–24 h after transfection using 10-mL syringes containing a Luer-Lok tip and filter through 0.45-μm PVDF-sterilized syringe filters directly into 14 × 89-mm Ultra-Clear Centrifuge Tubes containing a 1-mL 20% sucrose cushion (in 1X dPBS). To minimize the disturbance of the sucrose cushion, carefully add the virus-containing supernatant dropwise to the wall of a tilted centrifuge tube.
2. Balance the centrifuge tubes and load into the swinging bucket rotor according to the manufacturer's instructions. Centrifuge the viruses at 25,000 rpm (~100,000 × *g*) for 1.5 h at 4 °C.
3. Carefully decant the supernatant and resuspend the virus pellet in 0.5 mL clear CM, which should result in a ~20-fold concentration of the original virus preparation. Aliquot the virus into cryotube vials to minimize the number of freeze–thaws and store at −80 °C.

3.1.3 Quantitation of Virus

One important consideration for live-cell imaging of virus particles is to infect the cells with an appropriate amount of virus particles so that individual particles can be faithfully tracked through the time-lapse images (*see* **Note 2**). We often quantify the number of fluorescent protein-labeled virus particles for each virus preparation using single-virion analysis [13] (*see also* Chapter 6).

1. Dilute a small volume of concentrated virus (~1–5 μL) in dPBS (final volume = 200 μL), briefly vortex, and add to a well in an ibiTreated 8-Well chamber slide.
2. Secure slides to a swinging bucket insert and centrifuge at 1200 × *g* for 30 min at room temperature.
3. Acquire confocal microscopy images of the virus particles using a Nikon Eclipse Ti-E microscope equipped with a CSU-W1 spinning disk unit and a Plan-Apochromat 100× N.A.1.49 oil objective using a 488-nm laser for excitation for GFP or 514-nm laser for excitation for YFP. Maintain the focal plane containing the fluorescent protein-labeled virus particles using the Nikon Perfect Focus System and acquire several fields of view at random.
4. Detect and quantify the number of diffraction-limited fluorescent particles in each image using Localize software. A MATLAB script can be used to automate image processing and summarize the particle counts from multiple images. The average number of fluorescent particles per field of view is directly proportional to the amount of fluorescent protein-labeled virus in each preparation [24]. Importantly, use this number to infect cells with a normalized number of fluorescent protein-labeled virus particles when multiple virus preparations are used for infections (e.g., a panel of capsid mutants). The number of virus particles used for infection will need to be determined empirically depending on the experimental goal and setup (*see* **Note 3**).

3.2 Virus Infection and Live-Cell Imaging of HIV-1 Core Docking at the NE and Nuclear Import

The steps involved in virus infection of HeLa:mRuby-LaminB cells and live-cell imaging of HIV-1 core docking at the NE and nuclear import are described in Subheadings 3.2.1 and 3.2.2. These include (1) virus infection of HeLa:mRuby-LaminB cells and (2) live-cell imaging of HIV-1 core docking at the NE and nuclear import.

3.2.1 Virus Infection of HeLa:mRuby-LaminB Cells

1. Plate HeLa:mRuby-LaminB at 3.0 × 10^4 cells per well in an ibiTreated 8-Well chamber slide in 200 μL clear CM. Although we have previously used HeLa cells that express POM121-mCherry, a nuclear pore protein, to identify the NE [17], we often use HeLa cells that express mRuby-LaminB because the signals are stronger and less prone to photobleaching. Alternatively, use HeLa cells that are stained with BioTracker 650 Red Nuclear Dye to visualize nuclear import. Another important consideration is the density of cells because crowded cells are often positioned at different focal planes. If this is occurring,

seed cells at a lower initial density so that all the cells within the field of view are adequately separated and are likely to be at the same focal plane.

2. Mix approximately 1–5 μL of the concentrated A3F-YFP-, GFP-CA-, or cmGFP-labeled virus (equivalent to ~1000 fluorescent particles per field of view) with 2 μg polybrene and CM to 200 μL final volume, briefly vortex, and add to each well after removing the original CM.
3. Secure slides to a swinging bucket insert and centrifuge at 1200 × *g* for 1 h at 15 °C, a temperature that permits attachment of virus particles to cells but not their endocytosis [25].
4. Immediately after the centrifugation, replace the media with clear CM + 10 μg/mL aphidicolin and transfer the slide to a prewarmed Stage Top Incubator (time 0). Aphidicolin is added to prevent cell division, which will interfere with the visualization of HIV-1 complexes in the infected cells. It is important to prewarm the microscope for >2 h prior to imaging so that the temperature of the incubation chamber is stabilized before the addition of the slide. Infection of cells with this amount of virus and using these conditions typically yields ~2 nuclear viral particles per cell, although adjustments to the amount of virus used for infections may be necessary depending upon the experimental goal.

3.2.2 Live-Cell Imaging of HIV-1 Core Docking at the NE and Nuclear Import

1. Acquire confocal z-stacks of the infected cells every 10 min for 10 h starting ~10 min after infection (time 0 defined by transfer of slide to 37 °C incubator) (Fig. 4a, b; *see* **Note 4**). Acquire confocal z-stacks of the cells (9 slices with 0.4-μm step interval centered near the equatorial plane of the cells) and multiple fields of view (~30) at every time point, resulting in the imaging of >300 cells per time point. Acquire GFP and mRuby images using 488- and 561-nm lasers for excitation, respectively.
2. Manually identify the virus particles inside the nuclei in the time-lapse images and record the frame in which the virus particle first arrived at the NE and entered the nucleus for each virus particle. The time when the virus particle first arrived at the NE represents the time at which the virus particle remained in the cytoplasm until stable docking with the NE (relative to the time of infection). Determine the NE residence time prior to import by subtracting the time when the virus particle is first observed in the nucleus from the time when the virus particle is first observed at the NE. We currently analyze ~50 nuclear import events per experimental condition, which often requires the acquisition of multiple movies. Representative live-cell microscopy images of a GFP-CA-labeled viral core entering the nucleus of a HeLa:mRuby-LaminB cell are shown in Fig. 4c.

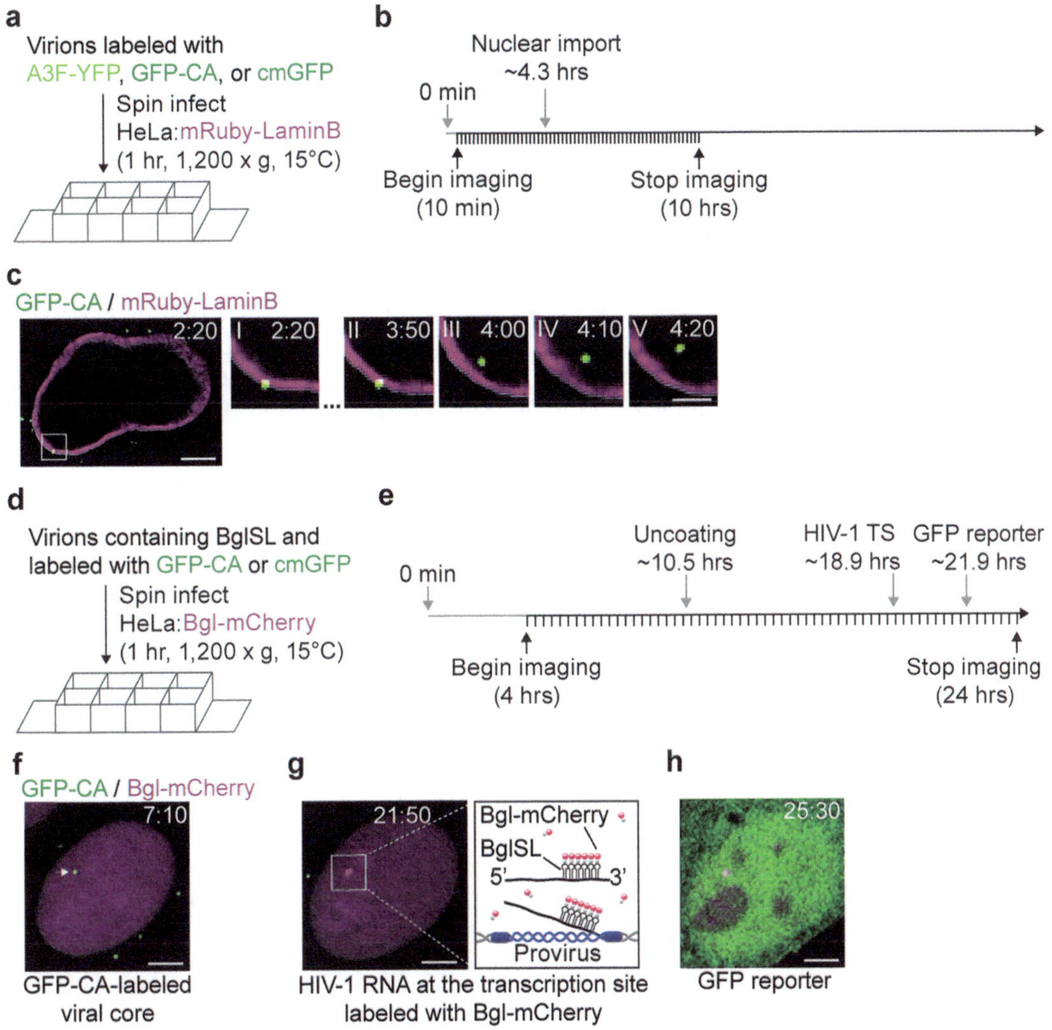

Fig. 4 Schematics of live-cell fluorescence imaging experiments to detect HIV-1 nuclear import, uncoating, proviral transcription, and GFP reporter expression. (**a**) HeLa:mRuby-LaminB cells are spin-infected with virions labeled with A3F-YFP, GFP-CA, or cmGFP. (**b**) Confocal z-stacks of the cells are acquired every 10 min for 10 h starting 10 min postinfection. The average time of nuclear import for A3F-YFP-labeled viral complexes is shown [17]. (**c**) Representative live-cell microscopy images of a GFP-CA-labeled viral core entering the nucleus of a HeLa:mRuby-LaminB cell. GFP-CA-labeled viral core, docked at NE 2:20 hpi (I), was last observed at the NE 3:50 hpi (II) and entered the nucleus (III–V). Scale bar, 5 μm. (**d**) HeLa:Bgl-mCherry cells are spin-infected with virions containing genomic RNAs that encode BglSL and the viral cores that are labeled with GFP-CA or cmGFP. (**e**) Confocal z-stacks of the cells are acquired every 20 min for 20 h starting at 4 hpi. The average time of uncoating of nuclear HIV-1 cores, HIV-1 transcription site appearance, and GFP reporter detection is shown [8]. (**f–h**) Representative live-cell microscopy images of a HeLa:Bgl-mCherry cell infected with virions labeled with GFP-CA. (**f**) Example of viral core labeled with GFP-CA inside the nucleus at 7:10 hpi. (**g**) Left, HIV-1 transcription site appeared at 21:50 hpi near the site of the GFP-CA disappearance, which occurred at 7:30 hpi. Right, schematic showing HIV-1 RNA containing BglSL at the transcriptionally active provirus that is bound to Bgl-mCherry. Bgl-mCherry protein that is not bound to HIV-1 RNA is diffusely distributed in the nucleus and is used to define the nuclear boundary. (**h**) GFP reporter expression detected at 25:30 hpi. Scale bar, 5 μm

3.3 Virus Infection and Live-Cell Imaging of Uncoating, HIV-1 Transcription Sites, and GFP Reporter Detection

The steps involved in virus infection of HeLa:Bgl-mCherry cells and live-cell imaging of uncoating, HIV-1 transcription sites, and GFP reporter detection are described in Subheadings 3.3.1 and 3.3.2. These include (1) virus infection of HeLa:Bgl-mCherry cells and (2) live-cell imaging of uncoating, HIV-1 transcription sites, and GFP reporter detection.

3.3.1 Virus Infection of HeLa:Bgl-mCherry Cells

1. Plate HeLa:Bgl-mCherry cells at 3.0×10^4 cells per well in an ibiTreated 8-Well chamber slide in 200 μL clear CM + 1 μg/mL doxycycline hydrochloride (Dox). HeLa:Bgl-mCherry cells express Bgl-mCherry protein under the control of a doxycycline-inducible promoter; Bgl-mCherry specifically binds to Bgl RNA stem loops inserted in the viral genome.
2. Mix approximately 1–5 μL of the concentrated GFP-CA- or cmGFP-labeled virus with 2 μg polybrene and CM to 200 μL final volume, briefly vortex, and add to each well after removing the original CM.
3. Centrifuge slides as described in Subheading 3.2.1, **step 3** and replace the media with clear CM + 10 μg/mL aphidicolin + 1 μg/mL Dox. To visualize the uncoating of rare infectious viral particles, it is imperative to use an amount of virus that results in <5% of the cells becoming infected and express the GFP reporter (measured at 48 h postinfection by flow cytometry). The low infectivity results in an average of <1 GFP-CA-labeled nuclear particle per cell.

3.3.2 Live-Cell Imaging of Uncoating, HIV-1 Transcription Sites, and GFP Reporter Detection

1. Acquire confocal z-stacks (9 slices with 0.4-μm step interval centered near the equatorial plane of the cells) of the infected cells every 20 min for 24 h starting ~4 h after infection (Fig. 4d, e). Acquire multiple fields of view (~50) at every time point, resulting in the imaging of >500 cells per time point. Acquire GFP and mCherry images sequentially using 488- and 594-nm lasers for excitation, respectively.
2. Identify productively infected cells expressing the GFP reporter in the time-lapse images (average time of GFP reporter detection is ~22 h postinfection). Next, identify the HIV-1 transcription site, a single bright focus in the mCherry channel that appears ~3 h before GFP reporter detection, in the same cells. We find that most, if not all, infected cells expressing the GFP reporter will also contain a detectable HIV-1 transcription site, indicating a high efficiency of HIV-1 transcription site detection. Next, inspect the earlier time points to determine whether a GFP-CA-labeled or cmGFP-labeled viral core disappeared in the same cell. We discovered that HIV-1 transcription sites are often associated with the disappearance of a nearby GFP-CA-labeled or cmGFP-labeled viral core ~8.4 h prior to the

appearance of the HIV-1 transcription site. Record the time when the GFP-CA (or cmGFP) signal disappeared, when the HIV-1 transcription site is first detected, and when the GFP reporter expression is first detected. A similar experimental setup is used to analyze the kinetics of uncoating (e.g., by using viral genomes that do not contain BglSL) or the kinetics of HIV-1 transcription site detection (by using viruses that are not labeled with GFP-CA or cmGFP). HeLa:mRuby-LaminB cells can also be used to visualize HIV-1 uncoating. Representative live-cell microscopy image of a GFP-CA-labeled viral core in the nucleus of a HeLa:Bgl-mCherry cell, HIV-1 transcription site, and GFP reporter expression is shown in Fig. 4f–h.

4 Notes

1. Effect of fluorescent labeling of virions on virus behavior being studied. It is important to determine whether the virus label may be altering the specific behavior of the virus particle under investigation. For example, it is known that A3F incorporation into virus particles can inhibit reverse transcription [26, 27]. It was unknown at the time of this early study whether reverse transcription was required for the nuclear import of subviral complexes. Therefore, we performed an experiment in which we measured the nuclear import efficiency and kinetics when reverse transcription was inhibited by nevirapine treatment of cells and showed that inhibition of reverse transcription did not affect the kinetics of nuclear import [15]. We later discovered that blocking reverse transcription with the reverse transcriptase inhibitor nevirapine significantly decreased HIV-1 uncoating efficiency [8], indicating that reverse transcription is required for efficient uncoating of nuclear HIV-1 cores. Therefore, while A3F-YFP labeling may be an appropriate label for capturing HIV-1 nuclear import, it may not be suitable to study HIV-1 uncoating.
2. Another important consideration is the amount of virus used for infections. The amount of virus used for infections highly depends on the experimental objective. For example, we often use higher amounts of virus for visualizing nuclear import than we do for visualizing uncoating and proviral transcription, largely because we are not specifically tracking the rare infectious particles in these experiments.
3. Detection sensitivity of the microscope. It is important to determine the detection sensitivity of the microscope for proper interpretation of imaging assays. Generally, virus particles are labeled with a relatively low number of fluorescent

proteins and high laser powers are required for the efficient detection of fluorescent protein-labeled virus particles. However, laser powers must be minimized to reduce phototoxicity and photobleaching of fluorophores during live-cell movies. Thus, detection efficiency must often be sacrificed. The appropriate detection sensitivity needed will depend on specific experimental goals and assay conditions. We often find that imaging and quantifying virus particles on a slide using different combinations of laser powers and camera exposure times is a useful exercise to determine the detection sensitivity of the microscope.

4. The determination of the frame rates and time window for imaging specific events in the virus life cycle must be considered. Ideally, one would acquire the infected cell using high frame rates for an extended period of time to capture many events in the early stage of HIV-1 infection at high temporal resolution. However, due to the photobleaching of the fluorophores and phototoxicity associated with repeated laser exposure, the optimal time frame for imaging key steps in the viral life cycle and optimal frame rates must be carefully considered. For example, we previously utilized a fixed-cell imaging assay in which we quantified the number of nuclear A3F-YFP-labeled virus particles at several time points after infection and concluded that nuclear import of HIV-1 occurred <12 h after infection [15]. Therefore, we designed our live-cell imaging experiments such that time-lapse images were acquired of the infected cells every 3 min for the first 10 h of infection and observed the average time of nuclear import is 4.3 hpi. The average time of nuclear import for other cell types may vary, so it will be important to first measure nuclear import at multiple time points using fixed-cell assays and then design a live-cell experiment that captures images during the appropriate time window and at the appropriate frame rate. Ultimately, live-cell imaging is a delicate balance between maximizing detection sensitivity while minimizing photobleaching of fluorophores and phototoxicity. Typically, higher frame rates will be required to track fast-moving objects (e.g., RNAs), whereas lower frame rates will be sufficient to track immobile objects (e.g., HIV-1 cores docked at nuclear pore complex) or slow-moving objects (e.g., nuclear HIV-1 cores).

Acknowledgements

The content of this publication does not necessarily reflect the views or policies of the Department of Health and Human Services, nor does mention of trade names, commercial products, or organizations imply endorsement by the U.S. Government. This work was supported in part by the Intramural Research Program of the NIH, National Cancer Institute, Center for Cancer Research, and funding from the Office of AIDS Research to V.K.P. and W.S.H.

References

1. McDonald D, Vodicka MA, Lucero G, Svitkina TM, Borisy GG, Emerman M, Hope TJ (2002) Visualization of the intracellular behavior of HIV in living cells. J Cell Biol 159(3): 441–452. https://doi.org/10.1083/jcb.200203150
2. Chen J, Liu Y, Wu B, Nikolaitchik OA, Mohan PR, Chen J, Pathak VK, Hu WS (2020) Visualizing the translation and packaging of HIV-1 full-length RNA. Proc Natl Acad Sci USA 117(11):6145–6155. https://doi.org/10.1073/pnas.1917590117
3. Jouvenet N, Bieniasz PD, Simon SM (2008) Imaging the biogenesis of individual HIV-1 virions in live cells. Nature 454(7201): 236–240. https://doi.org/10.1038/nature06998
4. Hubner W, Chen P, Del Portillo A, Liu Y, Gordon RE, Chen BK (2007) Sequence of human immunodeficiency virus type 1 (HIV-1) Gag localization and oligomerization monitored with live confocal imaging of a replication-competent, fluorescently tagged HIV-1. J Virol 81(22):12596–12607. https://doi.org/10.1128/JVI.01088-07
5. Chen J, Rahman SA, Nikolaitchik OA, Grunwald D, Sardo L, Burdick RC, Plisov S, Liang E, Tai S, Pathak VK, Hu WS (2016) HIV-1 RNA genome dimerizes on the plasma membrane in the presence of Gag protein. Proc Natl Acad Sci USA 113(2):E201–E208. https://doi.org/10.1073/pnas.1518572113
6. Hubner W, McNerney GP, Chen P, Dale BM, Gordon RE, Chuang FY, Li XD, Asmuth DM, Huser T, Chen BK (2009) Quantitative 3D video microscopy of HIV transfer across T cell virological synapses. Science 323(5922): 1743–1747. https://doi.org/10.1126/science.1167525
7. Albanese A, Arosio D, Terreni M, Cereseto A (2008) HIV-1 pre-integration complexes selectively target decondensed chromatin in the nuclear periphery. PLoS One 3(6):e2413. https://doi.org/10.1371/journal.pone.0002413
8. Burdick RC, Li C, Munshi M, Rawson JMO, Nagashima K, Hu WS, Pathak VK (2020) HIV-1 uncoats in the nucleus near sites of integration. Proc Natl Acad Sci USA 117(10): 5486–5493. https://doi.org/10.1073/pnas.1920631117
9. Francis AC, Melikyan GB (2018) Single HIV-1 imaging reveals progression of infection through CA-dependent steps of docking at the nuclear pore, uncoating, and nuclear transport. Cell Host Microbe 23(4):536–548.e536. https://doi.org/10.1016/j.chom.2018.03.009
10. Schifferdecker S, Zila V, Muller TG, Sakin V, Anders-Osswein M, Laketa V, Krausslich HG, Muller B (2022) Direct capsid labeling of infectious HIV-1 by genetic code expansion allows detection of largely complete nuclear capsids and suggests nuclear entry of HIV-1 complexes via common routes. mBio 13(5):e0195922. https://doi.org/10.1128/mbio.01959-22
11. Arhel N, Genovesio A, Kim KA, Miko S, Perret E, Olivo-Marin JC, Shorte S, Charneau P (2006) Quantitative four-dimensional tracking of cytoplasmic and nuclear HIV-1 complexes. Nat Methods 3(10):817–824. https://doi.org/10.1038/nmeth928
12. Muller TG, Zila V, Peters K, Schifferdecker S, Stanic M, Lucic B, Laketa V, Lusic M, Muller B, Krausslich HG (2021) HIV-1 uncoating by release of viral cDNA from capsid-like structures in the nucleus of infected cells. Elife 10:e64776. https://doi.org/10.7554/eLife.64776
13. Chen J, Nikolaitchik O, Singh J, Wright A, Bencsics CE, Coffin JM, Ni N, Lockett S, Pathak VK, Hu WS (2009) High efficiency of HIV-1 genomic RNA packaging and heterozygote formation revealed by single virion analysis. Proc Natl Acad Sci USA 106(32): 13535–13540. https://doi.org/10.1073/pnas.0906822106
14. Francis AC, Marin M, Shi J, Aiken C, Melikyan GB (2016) Time-resolved imaging of single HIV-1 uncoating in vitro and in living cells. PLoS Pathog 12(6):e1005709. https://doi.org/10.1371/journal.ppat.1005709
15. Burdick RC, Hu WS, Pathak VK (2013) Nuclear import of APOBEC3F-labeled HIV-1 preintegration complexes. Proc Natl

Acad Sci USA 110(49):E4780–E4789. https://doi.org/10.1073/pnas.1315996110

16. Li C, Burdick RC, Nagashima K, Hu WS, Pathak VK (2021) HIV-1 cores retain their integrity until minutes before uncoating in the nucleus. Proc Natl Acad Sci USA 118(10):e2019467118. https://doi.org/10.1073/pnas.2019467118
17. Burdick RC, Delviks-Frankenberry KA, Chen J, Janaka SK, Sastri J, Hu WS, Pathak VK (2017) Dynamics and regulation of nuclear import and nuclear movements of HIV-1 complexes. PLoS Pathog 13(8):e1006570. https://doi.org/10.1371/journal.ppat.1006570
18. Muller TG, Zila V, Muller B, Krausslich HG (2022) Nuclear capsid uncoating and reverse transcription of HIV-1. Annu Rev Virol 9(1):261–284. https://doi.org/10.1146/annurev-virology-020922-110929
19. De Clercq E, Yamamoto N, Pauwels R, Baba M, Schols D, Nakashima H, Balzarini J, Debyser Z, Murrer BA, Schwartz D et al (1992) Potent and selective inhibition of human immunodeficiency virus (HIV)-1 and HIV-2 replication by a class of bicyclams interacting with a viral uncoating event. Proc Natl Acad Sci USA 89(12):5286–5290. https://doi.org/10.1073/pnas.89.12.5286
20. Unutmaz D, KewalRamani VN, Marmon S, Littman DR (1999) Cytokine signals are sufficient for HIV-1 infection of resting human T lymphocytes. J Exp Med 189(11):1735–1746. https://doi.org/10.1084/jem.189.11.1735
21. Yee JK, Miyanohara A, LaPorte P, Bouic K, Burns JC, Friedmann T (1994) A general method for the generation of high-titer, pantropic retroviral vectors: highly efficient infection of primary hepatocytes. Proc Natl Acad Sci USA 91(20):9564–9568. https://doi.org/10.1073/pnas.91.20.9564
22. Zenklusen D, Larson DR, Singer RH (2008) Single-RNA counting reveals alternative modes of gene expression in yeast. Nat Struct Mol Biol 15(12):1263–1271. https://doi.org/10.1038/nsmb.1514
23. Smith JL, Pathak VK (2010) Identification of specific determinants of human APOBEC3F, APOBEC3C, and APOBEC3DE and African green monkey APOBEC3F that interact with HIV-1 Vif. J Virol 84(24):12599–12608. https://doi.org/10.1128/JVI.01437-10
24. Burdick R, Smith JL, Chaipan C, Friew Y, Chen J, Venkatachari NJ, Delviks-Frankenberry KA, Hu WS, Pathak VK (2010) P body-associated protein Mov10 inhibits HIV-1 replication at multiple stages. J Virol 84(19):10241–10253. https://doi.org/10.1128/JVI.00585-10
25. O'Doherty U, Swiggard WJ, Malim MH (2000) Human immunodeficiency virus type 1 spinoculation enhances infection through virus binding. J Virol 74(21):10074–10080. https://doi.org/10.1128/jvi.74.21.10074-10080.2000
26. Holmes RK, Koning FA, Bishop KN, Malim MH (2007) APOBEC3F can inhibit the accumulation of HIV-1 reverse transcription products in the absence of hypermutation – comparisons with APOBEC3G. J Biol Chem 282(4):2587–2595. https://doi.org/10.1074/jbc.M607298200
27. Mbisa JL, Bu W, Pathak VK (2010) APOBEC3F and APOBEC3G inhibit HIV-1 DNA integration by different mechanisms. J Virol 84(10):5250–5259. https://doi.org/10.1128/Jvi.02358-09

Chapter 3

RNA-FISH for HIV Transcription/Localization Analysis

Pavitra Ramdas and Ajit Chande

Abstract

RNA fluorescence in situ hybridization (FISH) serves as a method for visualizing specific RNA molecules within cells. Its primary utility lies in the observation of messenger RNA (mRNA) molecules associated with particular genes of significance. This technique can also be applied to examine viral transcription and the localization of said transcripts within infected cells. In this context, we provide a comprehensive protocol for the detection, localization, and quantification of HIV-1 transcripts in mammalian cell lines. This encompasses the preparation of required reagents, cellular treatments, visualization, and the subsequent analysis of the data acquired. These parameters play a pivotal role in enhancing our comprehension of the molecular processes during infection, particularly at the crucial transcription phase of the viral life cycle.

Key words RNA-FISH, HIV-1 transcription, Transcript localization, Quantification of viral transcripts

1 Introduction

The inception of in situ hybridization (ISH) techniques can be traced back to 1969 when Pardue and Gall pioneered their development to ascertain the chromosomal positioning of specific DNA sequences [1]. Subsequently, this technique has undergone a gradual evolution aimed at enhancing both precision and the simplicity of detection. Its applicability extends to the identification of both DNA and RNA. The ISH method, when targeting RNA, operates on the principle of nucleic acid hybridization, wherein a probe possessing a complementary sequence promptly anneals to form either DNA: RNA hybrids (with DNA probes) or RNA: RNA hybrids (when employing RNA probes) [2, 3].

Initially, radioactive probes found utility in ISH, with the sites of hybridization being detected through autoradiography. However, these probes proved expensive, demanded extended exposure durations, and raised safety concerns for users [4]. A significant advancement occurred in 1977 with the introduction of probes containing fluorophores for target detection [5]. This

Vinayaka R. Prasad and Ganjam V. Kalpana (eds.), *HIV Protocols*, Methods in Molecular Biology, vol. 2807,
https://doi.org/10.1007/978-1-0716-3862-0_3,

breakthrough marked a considerable stride forward, opening up opportunities for applying this technique across diverse fields.

The human immunodeficiency virus (HIV-1) capitalizes on the host's cellular machinery to accomplish transcription of the integrated proviral genetic material, constituting a pivotal phase within the viral life cycle [6]. Scrutinizing the gene expression of HIV-1 provides insights into the viral replication within the cells. Comparable to host mRNA, transcribed HIV-1 mRNA also undergoes posttranscriptional modifications, including 5′ capping and 3′ polyadenylation. Ensuing these alterations, the processed mRNA is transported to the cytoplasm for translation and subsequent assembly [7]. RNA-FISH makes for an elegant tool for the examination of the localization of viral mRNA, which can furnish valuable information regarding the succeeding stages of the viral life cycle [8].

A significant advancement in this technique is the ability to visualize RNA molecules at the single-cell level [9]. Furthermore, developments in fluorescence-based detection have enabled the targeting of multiple nucleic acids and the monitoring of RNA within distinct cell compartments [10]. In this context, we elucidate the process of RNA-FISH using fluorescent probes hybridizing with target HIV-1 RNA sequences and subsequently visualization using a fluorescence microscope. This protocol provides a comprehensive outline of the sequential steps to be undertaken, encompassing infection of target cells, preparation of fluorescent probes, fixation and permeabilization of infected cells, hybridization of probes with viral RNA, and microscopy-based visualization and analysis.

2 Materials

All solutions must be prepared using nuclease-free water and reagents of cell culture grade. While stock solutions can be prepared a few days in advance, the final buffers must be freshly prepared right before the experiment.

2.1 Cell Culture

1. Coverslips: Acid-treated 18 mm × 18 mm coverslips (square), 1 mm thickness.
2. Poly-L-Lysine Working Solution: Add one part of 0.1% poly-L-lysine and nine parts of nuclease-free water. Store the solution at 4 °C. Used for coating coverslips.
3. 70% Ethanol Solution: In a nuclease-free, 50-mL centrifuge tube, take 35 mL of 100% molecular biology-grade ethanol and 15 mL of nuclease-free water to make a 50 mL solution of 70% ethanol for sterilizing the coverslips.

4. Culture Dishes: Two 100-mm cell culture dishes for seeding cells for virus production. One 6-well plate for seeding target cells on coverslips for infection.
5. Cell Culture Media: Dulbecco's modified Eagle's medium (DMEM) or Roswell Park Memorial Institute (RPMI-1640) supplemented with fresh L-GlutaMAX and fetal bovine serum (10%).

2.2 Calcium Phosphate Transfection

1. 1× Tris–EDTA (TE) Buffer: 100 mM Tris–HCI and 10 mM EDTA diluted from stocks (1 M Tris–HCI and 0.5 M EDTA, pH 8.0) in nuclease-free water. Adjust the pH of the solution to 8.0.
2. 2.5 M Calcium Chloride Solution: Weigh 5.8 g calcium chloride and add it to 80 mL nuclease-free water. Mix it thoroughly and make up the volume to 100 mL.
3. 2× HEPES-Phosphate Buffer (HPB): To prepare 100 mL of 2× HPB solution, add 5.6 mL of 5 M sodium chloride (0.28 M final), 10 mL of 0.5 M solution; pH 7.0 (0.05 M final), and 1 mL of 0.15 M disodium hydrogen phosphate (Na_2HPO_4) (0.0015 M final). Adjust the pH of this solution to 7.1 and make up the volume to 100 mL with nuclease-free water (*see* **Note 1**).

All the solutions are to be sterile-filtered.

2.3 SYBR-Green-I-Based Product-Enhanced Reverse Transcriptase (SGPERT) Assay

1. 2× Lysis Buffer: 0.25% Triton X-100, 50 mM potassium chloride, 100 mM Tris–HCI (pH 7.4), and 40% glycerol. Add 2 μL RiboLock RNase inhibitor (40 U/μL) for 100 μL of 2× lysis buffer (*see* **Note 2**).
2. 10× Core Buffer: 50 mM ammonium sulfate, 200 mM potassium chloride, and 200 mM Tris–HCl (pH 8.3).
3. 2× Reaction Buffer: 1× core buffer, 10 mM magnesium chloride, 0.2 mg/mL bovine serum albumin (BSA), 1/10,000× SYBR Green-I, 400 μM deoxynucleotide triphosphates (dNTPs), 1 μM forward primer for MS2 phage (5′-TCCTGC TCAACTTCCTGTCGAG-3′), and 1 μM reverse primer for MS2 phage (5′- CACAGGTCAAACCTCCTAGGAATG -3′), and 7 pmoles/mL MS2 phage RNA.

2.4 Cell Fixation

1. 1× Phosphate Buffered Saline (PBS): Dilute 10× PBS to make a solution of 1× PBS (nine parts 10× PBS and one part nuclease-free water). pH 7.4.
2. Fixation Solution: From a stock of 37% paraformaldehyde (PFA), take 5.4 mL and add to it 44.6 mL of 1× PBS to make a 4% PFA solution.

2.5 Cell Permeabilization

1. Permeabilization Solution: From 10% Triton X-100, dilute to prepare 0.2–0.5% Triton X-100 in 1× PBS to permeabilize the cells. This solution is supplemented with 2 mM RiboLock RNase Inhibitor and 2 mM EGTA (*see* **Note 3**).

2.6 Fluorescence In Situ Hybridization

1. Twenty nucleotide long oligonucleotides ($n = 13$) complementary to the HIV-1 RNA sequence (corresponding to NL4–3) designed (*see* **Note 4**).
2. The oligonucleotides are labeled with Alexa 488 using a Ulysis Alexa 488 Nucleic Acid Labelling Kit. Alternatively, labeled probes can be procured (*see* **Note 5**).
3. 2× Saline Sodium Citrate (SSC) Buffer: Prepare a stock of 20× SSC buffer by adding 17.53 g of sodium chloride and 7.74 g of sodium citrate to 80 mL nuclease-free water. Adjust the pH to 7.0 using 12 N HCl. Make up the volume to 100 mL with water. Dilute the 20× SSC buffer to 2× SSC buffer with water.
4. Formamide Working Solution: Prepare a 50% formamide by mixing formamide with nuclease-free water for hybridization.
5. 10% Dextran Sulfate (Weight/Volume): Add 10 g of dextran sulfate to 80 mL nuclease-free water. Mix thoroughly and make up the volume to 100 mL (*see* **Note 6**).
6. Hybridization Solution: 50% formamide, 2× SSC buffer, 10% w/v dextran sulfate.
7. Wash Solution: 10% formamide prepared in 2× SSC. This solution can be stored at 4 °C.
8. Hoechst Solution: A 10 mg/mL stock of Hoechst 33342 solution is prepared. This stock solution is diluted 1:10,000 for staining the nuclei of cells.

2.7 Mounting of Coverslips on Slides

1. Microscopic Glass Slides: 25 mm × 75 mm slides cleaned with 70% ethanol using lint-free, antistatic tissue and dried for an hour before mounting the coverslips.
2. Mounting Media: Home-made mounting media can be prepared by adding nine parts glycerol to one part 1× PBS, or alternatively, commercial mounting media can be used to mount the coverslips onto slides for imaging.

2.8 Imaging and Analysis

1. Wide-field fluorescence microscope with 63× magnification and 355 and 488 nm lasers/light source to visualize Hoechst and hybridized probes, respectively.
2. ImageJ or Fiji software for analysis of obtained images from fluorescence microscopy.

3 Methods

3.1 Cells

Culture all cell lines in their respective growth media (DMEM or RPMI) supplemented with 10% heat-inactivated fetal bovine serum (FBS, US Origin) and 2 mM L-GlutaMAX. The cells are to be grown in a humidified incubator at 37 °C with 5% CO_2 (*see* **Note 7**).

3.2 Calcium Phosphate Transfection

This approach offers an economical and effective means of introducing foreign DNA plasmids into cells. In essence, the DNA combines with calcium phosphate, forming precipitates that are then internalized by cells.

1. On the day preceding the transfection, seed 2.2×10^6 cells in a 10-cm plate to achieve a 60% confluency during transfection.
2. For the transfection process, take a 1.5-mL (preferably low-bind) microcentrifuge tube (Tube 1) and combine plasmid DNA with 450 μL of 1× TE buffer.
3. Thoroughly mix this concoction and add 50 μL of 2.5 M calcium chloride. Vortex the mixture and give it a brief spin to ensure any solution adhering to the tube wall settles down.
4. In a separate 1.5-mL microcentrifuge tube, add 500 μL of 2× HPB (Tube 2).
5. Then, in a drop-wise manner, add the mixture from Tube 1 to the solution in Tube 2.
6. Let this mixture incubate at room temperature for 20 min, and then, add it drop by drop to the cell culture dish, gently mixing it in.

3.3 Virus Production

For single-cycle infectivity experiments, HEK293T cells serve as the hosts for producing HIV-1 virions [11] through calcium phosphate transfection. All the experiments on virus production and infection have to be carried out strictly following biosafety guidelines.

1. Transfect 1 μg of an HIV-1 envelope glycoprotein encoding plasmid along with 7 μg of NL4-3 Luciferase Env(-) Vpr(-) plasmid (HIV-1 variant deficient in Envelope, Vpr, and with Nef replaced by luciferase) (ARP2128 NIBSC).
2. In parallel, produce a mock supernatant by transfecting cells with 8 μg of empty vector, such as pBluescript.
3. At 12 h post-transfection, gently remove the medium from the transfected cells, wash them with 1× PBS, and replace the medium with a fresh medium (*see* **Note 8**).
4. Then, 48 h after transfection, collect the supernatant containing the virions in a 15-mL microcentrifuge tube.

5. Centrifuge it at 300 × *g* for 5 min to remove any cellular debris. Additionally, filter the viral supernatant using a 0.22-μm syringe filter.

3.4 SGPERT Assay

To verify the production of functional viral particles, perform an SGPERT assay to quantify the reverse transcriptase units within the produced HIV-1 virions. This assay operates based on the principle that, upon lysing the produced virions, reverse transcriptase (RT) is released, which can then convert the MS2 phage RNA into DNA. The resulting DNA is subsequently subjected to PCR using primers designed to bind to the MS2 DNA [12, 13]. The quantification is performed using quantitative PCR (qPCR), thus providing an indirect estimation of the RT units present in the viral supernatant (*see* **Note 9**).

1. To perform the assay, take 5 μL of filtered virus and mix it with 5 μL of 2× lysis buffer.
2. Incubate this mixture for 10 min at room temperature and then dilute it with 90 μL 1× core buffer.
3. Take 10 μL from this mixture and add it to 10 μL 2× reaction buffer and briefly spin the tube to mix the contents.
4. Subject the mixture to a one-step real-time qPCR with the following program: 42 °C—20 min (for RT), 95 °C—2 min, 40× (95 °C—5 s, 60 °C—5 s, 72 °C—15 s, 80 °C—7 s). Melting curve: 65–95 °C, increment 0.5 °C for 5 s (*see* **Note 10**).

3.5 Infection of Target Cells

Initially, determine the infectivity units of the virus produced and based on the obtained infectivity units, and introduce a sufficient quantity of virus to the wells seeded for RNA-FISH, ensuring a minimum of 40% infection (*see* **Notes 11** and **12**). For this:

1. Seed the target cells in a 96-well plate with 16 wells designated for each condition, ensuring four replicates of, at least, four dilutions per condition. After leaving the plate overnight, add the neat virus to the first four wells, and then proceed to dilute the virus further using a dilution factor of 5, resulting in dilutions of 5, 25, and 125 times (*see* **Note 13**).
2. After 2 days of infection, lyse the cells in the 96-well plate to quantify luciferase activity as a measure of infectivity (as per the manufacturer's protocol). Calculate the infectivity units by the following formula:

$$\text{Infectivity units}\left(\frac{\text{I.U.}}{\text{mu RT}}\right)=\frac{\text{Average luciferase value}\times 1{,}000{,}000}{\text{RT value (From SGPERT)}\times\text{volume of virus added}}$$

3. For infection of cells to perform RNA-FISH, first place coverslips in all the wells of a 6-well plate. Perform this step and the following steps in an aseptic environment.
4. To sterilize the coverslips, add 1 mL 70% ethanol to the wells and swirl gently for a couple of minutes.
5. Carefully discard the ethanol with a 1-mL micropipette, add 1 mL nuclease-free water, and wash off any residual alcohol.
6. Discard the water and let the wells dry of water. Open the lid of the 6-well plate and place it under UV treatment for 10 min (*see* **Notes 14** and **15**).
7. Following this, treat the coverslips with poly-L-lysine by adding 2 mL of the poly-L-lysine working solution (materials) and incubate for 20 min at room temperature.
8. Remove the poly-L-lysine solution and wash the coverslips with 2 mL 1× PBS. Repeat this step twice.
9. Once the PBS dries, seed the target cells for infection and subsequent RNA-FISH, allocating three wells for the mock condition and three wells for the infection condition.

3.6 Preparation of Fluorescent Hybridization Probes

Various types of probes, such as double-stranded DNA and single-stranded RNA, can be employed for FISH. However, using oligonucleotides offers a distinct advantage over the other options due to their compact size, enabling precise control over binding energy and specificity. In particular, DNA nucleotides exhibit superior stability compared to RNA [14].

To incorporate fluorophores into oligonucleotides that are complementary to HIV-1 mRNA, use the Ulysis Alexa 488 Nucleic Acid Labeling Kit or a similar kit, following the manufacturer's protocol. The oligonucleotides used here are listed in Table 1. Alternatively, one can opt to procure commercially available fluorescent probes.

3.7 RNA-FISH

Twenty hours after infection, the HIV-1 transcript levels peak in the cytoplasm for the subsequent stage of translation. Consequently, if tracking transcript export is a focal point of your experiment, we suggest fixing the infected cells and conducting FISH at this specific time point. However, it is essential to note that the experiment's timing can be adjusted according to the goals, allowing for exploration at various time points following infection.

1. For Adherent Cells: Discard the medium carefully and gently rinse the cells with 1× PBS. For Suspension Cells: Centrifuge the cells at 300 × *g* for 5 min and wash with 1× PBS for 300 × *g* for 5 min.
2. Cell Fixation: Fix the cells with 1 mL of 4% PFA solution by incubating at room temperature for 20 min.

Table 1
List of oligonucleotides used for HIV-1 RNA-FISH

S. no.	Sequence of oligonucleotide
1.	TCCACAGATCAAGGATATCTTGTC
2.	CTGACTAAAAGGGTCTGAGG
3.	CTTCAGCAAGCCGAGTCCT
4.	TCCACAGATCAAGGATATCTTGTC
5.	ACCTGAAGCTCTCTTCTGG
6.	TACCATTTGCCCCTGGAGGTT
7.	TAGAGTGGTGGTTGCTTCC
8.	CTCTCGCACCCATCTCTCTC
9.	CACCTGCCATCTGTTTTCCA
10.	CTGTCCTCTGTCAGTTTCCTAA
11.	TGATTACTATGGACCACACAACT
12.	CCTGTGTAATGACTGAGGTGT
13.	CCATACTATATGTTTTAGTTTATAT
14.	CTGTGCGTTACAATTTCTGGGTC
15.	CTATGTCGACACCCAATTCT

3. Discard the PFA and rinse the cells with 1× PBS thrice.
4. Cell Permeabilization: Following fixation, permeabilize the cells with the permeabilizing solution by incubating them on ice for 10 min.
5. Wash the cells with 1× PBS thrice for 5 min each.
6. Wash the cells with 2× SSC buffer for 10 min.
7. Hybridization: Add 75 ng of labeled DNA probes per 1 mL of the hybridization buffer and incubate it at 37 °C for 10 min.
8. Follow this, by incubating the mixture at 90 °C for 10 min and chill on ice immediately.
9. Add this mixture to the permeabilized cells such that the coverslip is completely covered by the hybridization solution containing the probes.
10. Incubate this in a dark, humid chamber at 37 °C overnight (~15–16 h).
11. The next day, discard the hybridization solution and wash the cells with wash solution (materials) for 10 min at 37 °C. Repeat this step two more times.
12. Wash the cells with 1× SSC buffer thrice for 5 min at 37 °C.

13. Stain the cells with Hoechst by immersing the coverslip with the working Hoechst solution (materials) for 10 min at room temperature.
14. Discard the Hoechst solution and wash the cells with 4× SSC solution for 10 min at room temperature.
15. Mount the coverslips on a glass slide using mounting media and let the slide dry overnight (*see* **Note 16**).

3.8 Data Visualization and Analysis

Images can be captured on any wide-field fluorescence microscope with the 63× or 100× objective. We use a Zeiss Apotome.2 fluorescence microscope. Images are captured using a 63× oil-immersion objective and a CCD camera. Thirty z-stacks with a step size of 0.3 μm were obtained.

For analyzing the localization and counting the hybridization spots in each cell compartment, Image J or Fiji software is used [15]. In the case of our experiment, we only wanted to analyze the transcripts within and outside the nucleus (Fig. 1). Therefore, we stained the cells with Hoechst and analyzed them by plotting a ratio of the puncta in the nucleus to the total number of puncta (Fig. 2). The "Analyze particles" manual is available on the website of Image J https://imagej.nih.gov/ij/docs/menus/analyze.html, which gives a step-by-step procedure to score the RNA-FISH spots in the image (*see* **Note 17**).

For troubleshooting the experiment, *see* **Notes 18–25**.

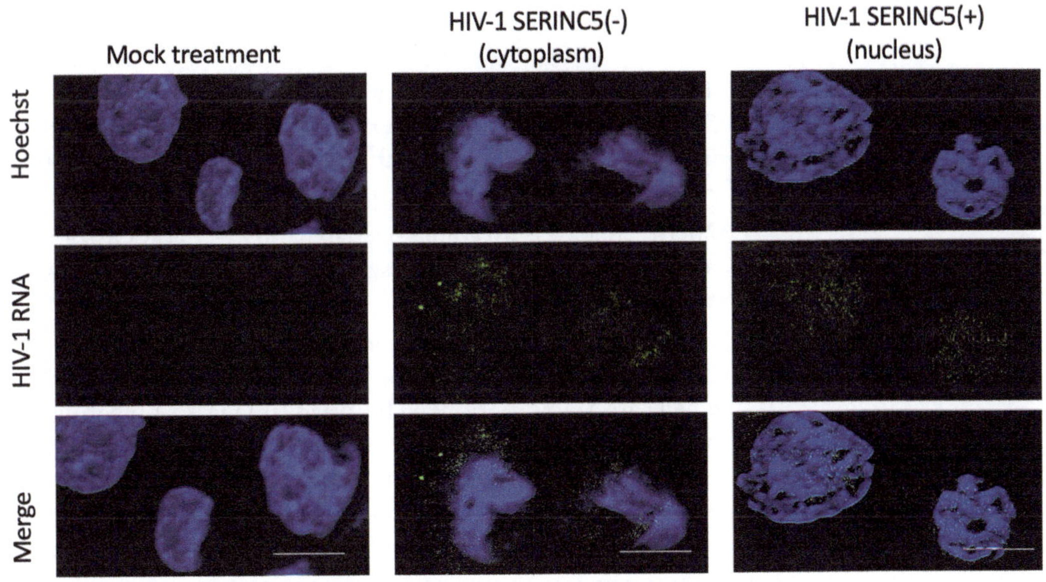

Fig. 1 Hybridization of probes with HIV-1 RNA seen as green spots/puncta by RNA-FISH in cells stained with Hoechst (nuclei) in mock-treated (left panel) and infected conditions (middle and right panel). The incorporation of a host restriction factor, SERINC5, in the virions results in nuclear retention of HIV-1 transcripts in these cells and hence shows different localization of the viral RNA (right panel). The scale bar represents 10 μm

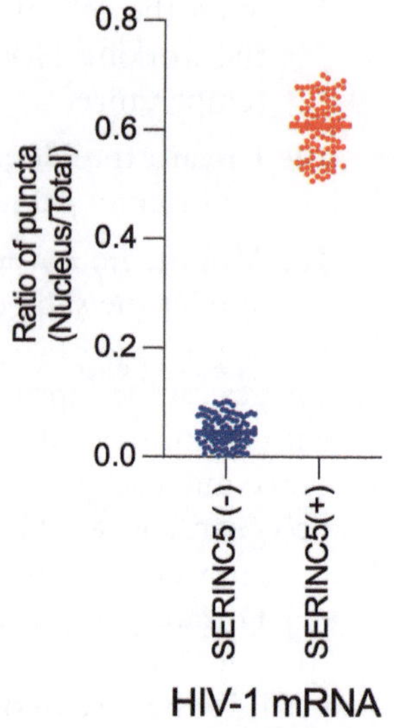

Fig. 2 To analyze the localization of HIV-1 RNA by FISH, Fiji 2 software was used to count the foci of viral RNA (as seen in Fig. 1), which is represented as a ratio of nuclear puncta to total puncta in infected cells (*n* = 100)

4 Notes

1. When preparing the 2× HPB solution, ensure precise pH adjustment, as its sensitivity can impact transfection efficiency.
2. RNase inhibitors are stable in lysis buffer for several weeks because this solution does not freeze. Add them while preparing and storing aliquots.
3. The Triton X-100 concentration range may require titration for different cell types. For example, THP-1 macrophage cell lines need 0.4% Triton X-100 for efficient permeabilization.
4. Optimal DNA oligonucleotide probe size should range from 20 to 50 bases, avoiding overlap with the target region. A minimum distance of two bases should be maintained between adjacent probes.
5. Utilizing labeled DNA oligonucleotides as opposed to purchasing labeled probes is a cost-effective alternative, both of which are viable options. The disadvantage to preparing the probes in-house, as mentioned above, is that they are time-consuming and the concentration of labeled oligos may be lesser than what can be obtained commercially.
6. When preparing the dextran sulfate solution, gently agitate it after mixing in water at room temperature until complete dissolution.

7. Ensure both virus-producing and target cells are mycoplasma-free, as mycoplasma can negatively impact viral infectivity.
8. While replacing medium post-transfection for virus production, adhere to all protocols essential for handling infectious viral particles, including discarding tips in bleach and preventing aerosol formation.
9. Guard against bacteriophage contamination in 2× reaction buffer by performing the procedures in a laminar cabinet for a sterile environment. Employ a negative control (No RT) and standards of varying concentrations (positive control-commercial RT enzyme) for each SGPERT assay.
10. Adjust the extension temperature and time according to the DNA polymerase used, modifying the PCR program accordingly.
11. For all the procedures involving cells, wear gloves and appropriate personal protective equipment (PPE) while handling.
12. Follow the discard procedures for each of the chemicals used as per the product safety sheet.
13. Check the infectivity of different dilutions of the virus to account for the saturated infectivity of the produced viral particles.
14. During UV treatment of coverslips before poly-L-lysine application, ensure the plate is devoid of media or cells. Likewise, no media bottles should be present in the hood during this process.
15. UV treatment is an additional step to ensure complete sterility and can be omitted if necessary.
16. Mount coverslips onto glass slides using adequate mounting media and proper technique, preventing the formation of air bubbles between the coverslip and slide.
17. Supplementary to the mentioned analysis technique, installable plugins are accessible for ImageJ/Fiji to count hybridization spots.
18. Before initiating the main experiment, use the following as negative controls for a mock hybridization:
    (a) Hybridization reaction lacking probes.
    (b) Hybridization reaction with sense strand probes (complementary to the antisense probes used in the experiment), mimicking sequences similar to the target mRNA instead of being complementary to them.
19. Address excessive background signal by incorporating a blocking agent like yeast tRNA (1 mg/mL) during hybridization. This is particularly useful under saturation infection conditions or when targeting abundantly expressed host transcripts.

20. In cases of lower concentrations of labeled probes, consider placing a substantial drop of hybridization solution only onto the coverslip, relying on surface tension to keep it in place.
21. Address high background resulting from nonspecific interaction by:
    (a) Increasing wash times after hybridization.
    (b) Titrating probe concentration in the hybridization solution through a pilot experiment.
    (c) Evaluating and potentially reducing hybridization time.
22. Enhance hybridization stringency by reducing formamide concentration or increasing washing temperature.
23. Adjust probe concentrations based on target transcript abundance (lower concentration for higher transcript abundance).
24. Throughout the procedure, keep the coverslips from drying out.
25. The entire experiment, including virus production, can be downscaled. For instance, use round 18-mm coverslips in 12-well plates or smaller round 12-mm coverslips in 24-well plates.

References

1. Pardue ML, Gall JG (1969) Molecular hybridization of radioactive DNA to the DNA of cytological preparations. Proc Natl Acad Sci 64(2):600–604. https://doi.org/10.1073/pnas.64.2.600
2. Jensen E (2014) Technical review: In situ hybridization. Anat Rec 297(8):1349–1353. https://doi.org/10.1002/ar.22944
3. Milman G, Langridge R, Chamberlin MJ (1967) The structure of a DNA-RNA hybrid. Proc Natl Acad Sci 57(6):1804–1810. https://doi.org/10.1073/pnas.57.6.1804
4. Huber D, Voithenberg LV, Kaigala GV (2018) Fluorescence in situ hybridization (FISH): history, limitations and what to expect from micro-scale FISH? Micro Nano Eng 1:15–24. https://doi.org/10.1016/j.mne.2018.10.006
5. John B, Randolph ASW (1997) Stability, specificity and fluorescence brightness of multiply-labeled fluorescent DNA probes. Nucleic Acids Res 25(14):2923–2929. https://doi.org/10.1093/nar/25.14.2923
6. Ramdas P et al (2020) From entry to egress: strategic exploitation of the cellular processes by HIV-1. Front Microbiol 11:559792. https://doi.org/10.3389/fmicb.2020.559792
7. Chen L, Keppler OT, Scholz C (2018) Post-translational modification-based regulation of HIV replication. Front Microbiol 9:2131. https://doi.org/10.3389/fmicb.2018.02131
8. Ramdas P, Chande A (2023) SERINC5 mediates a postintegration block to HIV-1 gene expression in macrophages. mBio 14(2):e00166-00123. https://doi.org/10.1128/mbio.00166-23
9. Levsky JM et al (2002) Single-cell gene expression profiling. Science 297(5582):836–840. https://doi.org/10.1126/science.1072241
10. Raj A et al (2008) Imaging individual mRNA molecules using multiple singly labeled probes. Nat Methods 5(10):877–879. https://doi.org/10.1038/nmeth.1253
11. Rosa A et al (2015) HIV-1 Nef promotes infection by excluding SERINC5 from virion incorporation. Nature 526(7572):212–217. https://doi.org/10.1038/nature15399
12. Pizzato M et al (2009) A one-step SYBR Green I-based product-enhanced reverse transcriptase

assay for the quantitation of retroviruses in cell culture supernatants. J Virol Methods 156(1–2):1–7. https://doi.org/10.1016/j.jviromet.2008.10.012
13. Vermeire J et al (2012) Quantification of reverse transcriptase activity by real-time PCR as a fast and accurate method for titration of HIV, lenti- and retroviral vectors. PLoS One 7(12):e50859. https://doi.org/10.1371/journal.pone.0050859
14. Tripathi V et al (2015) RNA fluorescence in situ hybridization in cultured mammalian cells. Methods Mol Biol 1206:123–136. https://doi.org/10.1007/978-1-4939-1369-5_11
15. Schindelin J et al (2012) Fiji: an open-source platform for biological-image analysis. Nat Methods 9(7):676–682. https://doi.org/10.1038/nmeth.2019

Chapter 4

Single-Cell Single-Molecule RNA-FISH Combined with Immunofluorescence and High-Speed and High-Resolution Scanning Analysis to Visualize the Reactivation of Latent HIV-1

Rajiv Pathak, Carolina Eliscovich, Robert H. Singer, and Ganjam V. Kalpana

Abstract

Latent HIV-1 reservoirs are a major obstacle to the eradication of HIV-1. Several cure strategies have been proposed to eliminate latent reservoirs. One of the key strategies involves the reactivation of latent HIV-1 from cells using latency-reversing agents. However, currently it is unclear whether any of the latency-reversing agents are able to completely reactivate HIV-1 provirus transcription in all latent cells. An understanding of the reactivation of HIV-1 provirus at single-cell single-molecule level is necessary to fully comprehend the reactivation of HIV-1 in the reservoirs. Furthermore, since reactivable viruses in the pool of latent reservoirs are rare, combining single-cell imaging techniques with the ability to visualize a large number of reactivated single cells that express both viral RNA and proteins in a pool of uninfected and non-reactivated cells will provide unprecedented information about cell-to-cell variability in reactivation. Here, we describe the single-cell single-molecule RNA-FISH (smRNA-FISH) method to visualize HIV-1 gag RNA combined with the immunofluorescence (IF) method to detect Gag protein to characterize the reactivated cells. This method allows the visualization of subcellular localization of RNA and proteins before and after reactivation and facilitates absolute quantitation of the number of transcripts per cell using FISH-quant. In addition, we describe a high-speed and high-resolution scanning (HSHRS) fluorescence microscopy imaging method to visualize rare and reactivated cells in a pool of non-reactivated cells with high efficiency.

Key words smRNA-FISH, Immunofluorescence, smRNA-FISH+IF, Single-cell single-molecule RNA- FISH, Single-cell imaging, High-speed high-resolution scanning, HIV-1 latency, Reactivation, Gag RNA

1 Introduction

HIV-1 reservoirs are a major obstacle to eliminate AIDS from the face of the earth. HIV-1 persists in cells present in various anatomical locations in the form of integrated and transcriptionally silent

Vinayaka R. Prasad and Ganjam V. Kalpana (eds.), *HIV Protocols*, Methods in Molecular Biology, vol. 2807, https://doi.org/10.1007/978-1-0716-3862-0_4,

proviral DNA, which can be activated spontaneously to generate new progeny of HIV-1. The cells harboring these long-lived, transcriptionally silent proviruses constitute latent reservoirs [1–3]. The presence of these latent reservoirs can be observed when patients on continuous suppressive therapy occasionally exhibit "blips" in plasma viremia, suggesting the reactivation of latent reservoirs by unknown stimuli and by the rebound of viremia when therapy is discontinued in aviremic patients [4]. Many strategies have been proposed to eliminate these latent reservoirs [5]. One such strategy, the "shock and kill" therapy, aims to eliminate HIV-1 reservoirs by activating latent provirus using "latency-reversing agents" (LRAs) and simultaneously targeting the reactivated cells for killing [2, 6]. There are several obstacles to achieving this strategy including variability in the reactivation of provirus by using LRAs, different mechanisms of latency in different latently infected cells requiring different types of LRAs, and differential ability to reactivate due to the site of integration of provirus. One study has reported that only a tiny fraction of the HIV reservoir (~$1/60^{th}$ of replication-competent proviruses) produces infectious progeny virions [7]. The reason for this low and variable level of activation has been attributed to cellular/viral heterogeneity including differences in epigenetic regulation, sites of integration, viral mutations, cellular factors, and/or transcriptional/posttranscriptional blocks [2]. Similarly, cell culture and ex vivo models of latency showed reactivation of a small fraction of proviruses [8]. These studies highlight the importance of understanding the reactivation kinetics and mechanisms of latency at the single-cell single-molecule level.

To understand the mechanism of latency establishment, reactivation, and the effects of various LRAs, we have applied smRNA-FISH methodology to visualize latent cells. We have adapted two different microscopy instruments, one that allows the detection of a single molecule of RNA and the other that allows high-throughput scanning of the entire slide. Finally, to quantitatively analyze the reactivation at single-cell single-molecule level, we have utilized a modular tool, the FISH-quant. Combinations of these methodologies and tools together provide images and valuable information about latency reactivation. Here, we describe these methodologies, microscopy analysis, and application of the analysis tools in three different sections. First, we describe the smRNA-FISH method using probes specific for HIV-1 *gag* RNA that detects the expression of unspliced HIV-1 RNA. In this method, we have combined smRNA-FISH with immunofluorescence using an HIV α-p24 antibody to simultaneously visualize the HIV-1 RNA and Gag proteins within the same reactivated cells. Here, we would like to note that any antibody of interest against viral or cellular proteins can be used in combination with RNA-FISH. Here, we have described a method of using α-p24 antibody for IF in combination with gag RNA probes for RNA-FISH. Images are acquired using a microscope that allows the detection of diffraction-limited single molecules of HIV-1 gag RNA. Second, once the images are acquired, we describe the analysis of these images using FISH-quant tool that allows quantitation of the number of mature HIV-1 gag transcripts per cell and the nascent transcripts per transcription site [9]. Third, we describe the use of high-speed high-resolution scanning (HSHRS) fluorescent microscope to scan the entire slide to acquire images of a million cells, allowing sampling of a large pool of cells to identify rare, reactivated cells. These three methods and analyses are described here as related to studies on HIV-1 latency and reactivation. We would like to indicate that these methods are highly applicable to any viral or cellular systems as long as the probes and antibodies are prepared according to the specifications of the system of choice.

2 Materials

2.1 HIV-1 Latent T Cells Used in the Study

The latent T-cell line models ACH2, J1.1, and J-Lat were obtained from the NIH AIDS Reagent program.

2.2 Cell Culture Media

Cell Culture Medium: RPMI-1640 supplemented with 10% fetal bovine serum, 2 mM L-glutamine, 10 mM HEPES, 100 U/mL penicillin, and 100 μg/mL streptomycin.

2.3 Latency Reversing Agents

The latently infected cells were induced by the addition of the following drugs, 30 nM panobinostat, 25 ng/mL bryostatin, or 50 ng/mL PMA. DMSO was used as no drug control.

2.4 Equipment, Plasticware, Slides, and Cover Glass

1. Incubators: Standard tissue culture incubators (37 °C with 5% carbon dioxide) and regular incubator for smRNA-FISH hybridization (37 °C).
2. Nunc™ Lab-Tek™ sterile chambered slide with the cover either 2-chambered or 4-chambered.
3. Fisherbrand™ premium cover glasses (dimensions ($L \times W$): 50 × 22 mm; thickness range: 0.13–0.16 mm) that cover the entire slide.
4. Serological pipettes (1, 5, and 10 mL).
5. Syringe filter 0.2 μm.
6. 10-mL syringe.
7. Large (150 mm) Petri dish for creating a humid chamber for hybridization.

2.5 Reagents for Fixation and Buffers for RNA-FISH and Immunofluorescence

1. 1× PBSM (1× PBS + 5 mM $MgCl_2$).
   To Make: Add 250 μL of 1 M $MgCl_2$ in 50 mL of 1× PBS (PBS made from 10× PBS stock for FISH).
2. Fixing Solution: 4% paraformaldehyde in 1× PBSM. To Make: Add 1 mL of paraformaldehyde (20%) in 4 mL of 1× PBSM. The tube should be protected from light.
3. Quenching Solution: 0.1 M glycine/PBSM: To Make: Put 0.19 g glycine in 50 mL 1× PBSM.
4. Permeabilization (Triton) Solution: 0.1% Triton X-100 in 1× PBSM.
   To Make: Make 10% stock of Triton X-100. Put 50 μL of 10% stock in 5 mL 1× PBSM.
5. Pre-Hybridization Buffer (10% deionized formamide in 2× SSC): To Make: 40 mL DEPC-treated water, 5 mL 20× SSC, 5 mL formamide.
6. Hybridization Buffer (HB):
   10% deionized formamide.
   1 mg/mL competitor *E. coli* tRNA MRE 600.
   10% dextran sulfate.
   0.2 mg/mL UltraPure BSA.
   2× sodium chloride–sodium citrate (SSC) buffer.
   (Dilute 20× SSC 1:10 with double distilled water. Store at room temperature).
   2 mM ribonucleoside vanadyl complex.
   10 U/mL SUPERase·In.
   UltraPure DEPC-treated water.
7. smRNA-FISH Probe Mix: HIV-1 gag RNA probes (250 nM final concentration, Biosearch Technologies, Custom Stellaris Probes).
8. HIV-1 gag RNA Probe Set: Each of the 48 smRNA-FISH HIV-1 Gag RNA probes (5′ → 3′) was tagged with Quasar 570, and all the probes were mixed in an equimolar ratio and will anneal to different, nonoverlapping positions of the gag RNA of HIV-1 genome.
   (ttaatactgacgctctcgca, tcccatcgatctaattctcc, atttttctttcccc ctggc, ctgcttgcccatactatatg, aactgcgaatcgttctagct, atgtttctaac aggccagga, ccagtatttgtctacagcct, tgtctgaagggatggttgta, gagggttgctactgtattat, ctctatcctttgatgcacac, aagcttccttggtgtc tttt, tgctcttcctctatcttgtc, ctgcttgctgtgctttttc, ttttggctgacc tgattgct, gatgttctgcactatagggt, atggcctgatgtaccatttg, ctgaaagcc ttctcttctac, gctgaaaacatgggtatcac, ttaaatcttgtggggtggct, cccac tgtgtttagcatggt, tttgcatggctgcttgatgt, cttcctcattgatggtctct, atgcactctatcccattctg, tcatctggcctggtgcaata, tatgtcacttcccc ttggtt, gaagggtactagtagttcct, ctactgggataggtggatta, ctggtaggg ctatacattct, gtccttgtcttatgtccaga, catagtctctaaagggttcc, gctcgg ctcttagagtttta, caacaaggtttctgtcatcc, aatctgggttcgcatttgg,

gctggtcccaatgcttttaa, cctgacatgctgtcatcatt, aaaactcttgccttatg gcc, tacttggctcattgcttcag, ctgcatcattatggtagctg, ctttggttcctaaaa ttgcc, cccttctttgccacaattga, tccaacagcccttttttccta, tttggtgtccttcc tttcca, cctgtctctcagtacaatct, ggccagatcttccctaaaaa, aagaaaattccc tggccttc, acctgaagctctcttctggt, tacagttccttgtctatcgg, tgacctga gggaagttaaag).

Resuspend the probe sets in TE buffer to a final concentration of 25 μM and store it as aliquots protected from light at − 20 °C.

9. Primary Antibody: 1:500 dilution of goat α-p24 antibody.
10. Secondary Antibody: Donkey α-goat, Alexa Fluor 647 (1:1000 dilutions).
11. DAPI (1.0 μg/mL DAPI in 2× SSC).
12. ProLong Gold Antifade Mounting Media.
13. Parafilm.

2.6 Fluorophore Combinations for Multiplexing smRNA-FISH and IF

Combination 1 (Probe 1: Quasar 570 + IF: Alexa Fluor 647).

Combination 2 (Probe 1: Quasar 670 + IF: Alexa Fluor 488).

Combination 3 (Probe 1: Quasar 570 + IF: Alexa Fluor 488).

Combination 4 (Probe 1: Quasar 570 + Probe 2: Quasar 670 + IF: Alexa Fluor 488).

2.7 High-Speed and High-Resolution Scanning (HSHRS) Image Acquisition

1. PANNORAMIC 250 Flash III Slide Scanner.
2. PANNORAMIC scanner software for image acquisition.

2.8 Single-Molecule Image Acquisition

1. Wide-field epifluorescence microscope: Olympus BX-63 Microscope equipped with a Super Apochromatic 60×/1.35 NA Olympus Objective (UPLSAPO60XO), a SOLA light engine (Lumencor), and an ORCA-R2 Digital Interline CCD Camera (C10600-10B; Hamamatsu).
2. 95% ethanol for cleaning slides.
3. Immersion oil.
4. Metamorph software.

2.9 Image Analysis Software

1. CaseViewer 2.4 (64-bit version) software for visualization of scanned slides.
2. ImageJ and/or Fiji software.
3. FISH_QUANT.

3 Methods

3.1 Activation of Cells (ACH2)

1. Warm complete RPMI media (with 10 mM HEPES).
2. Thaw the latency-reversing agents or drugs of choice (e.g., panobinostat, bryostatin, and/or PMA).
3. Determine the overall cell count by employing trypan blue staining.
4. Determine the volume (in mL) of the existing cell suspension required to achieve the target live cell concentration (e.g., between 200,000 and 1 million cells /mL needed per well).
5. Pellet the cells and discard the supernatants.
6. Add the specified drug (e.g., 30 nM panobinostat, 25 ng/mL bryostatin, and/or 50 ng/mL PMA) to the cell culture medium and mix gently but thoroughly (in the tube). Resuspend the cell pellets with 1 mL of medium containing drug. (*see* **Notes 1** and **2**).
7. Place the respective samples into individual wells of a 12-well plate. Subsequently, incubate the plate in a cell culture incubator at 37 °C for the designated time intervals.
8. Collect the samples at specified time points post-induction and then proceed with RNA-FISH+IF assay.

3.2 RNA-FISH+IF Protocol

This protocol spans 2 days, with specific steps outlined for each day.

Day 1

1. Transfer the cells into 1.5-mL Eppendorf tubes.
2. Wash the wells with 1× PBSM after cells are collected, ensuring the collection of all the cells from the wells. Pool the wash with the samples in the same Eppendorf tubes.
3. Centrifuge them for 2 min in a microcentrifuge.
4. Wash cells with 1 mL PBSM 2× minimum.
5. Resuspend cells in 50 μL fixing solution (4% paraformaldehyde in 1× PBSM) and fix them at room temperature in darkness for 30 min.
6. Dispense cells onto the slide drop by drop making sure that droplets are evenly distributed within the same well. Gently tap the slides to ensure even cell distribution and allow them to air-dry for 5–10 min at room temperature (*see* **Note 3**).
7. Place 1 mL of 0.1 M glycine/PBSM into each well of the slide, or fill each well to its capacity, and let it incubate at room temperature for 10 min. This step is needed for quenching.
8. Wash with 1× PBSM (1 mL/well) 2× for 10 min each at room temperature.

9. Permeabilize cells for 10 min with 0.1% Triton X-100 in PBSM (1 mL/sample).
10. Wash 2× 10 min at room temperature with PBSM (1 mL/sample).
11. Add 1 mL of pre-hybridization solution, which consists of 10% deionized formamide in 2× SSC for ACH2 cells, and let it incubate at room temperature for 30 min (*see* **Note 4**).
12. While incubating, prepare the hybridization buffer, maintaining the cold storage of all its components in an ice bucket.
13. *Preparation of Hybridization Buffer (HB)*: Prepare the hybridization buffer, adjusting the volume based on the number of samples. Utilize double-distilled or ultrapure DEPC-treated water when preparing the necessary buffer solutions (*see* **Notes 5** and **6**; Table 1).
14. *Preparing the Hybridization Chamber*: Hybridization is carried out in a humid chamber set up using a Petri dish as follows: Place a large 150-mm Petri dish covered with aluminum foil on a clean surface. Inside each Petri dish, place a parafilm square. To maintain high humidity in the Petri dishes, place two 15-mL Falcon tube lids filled with 2 mL of ddH_2O and

Table 1
Composition of hybridization buffer used to carry out RNA-FISH in this study for detecting single-molecule HIV-1 gag RNA

Reagents for hybridization buffer (HB)	Required volume (for making 1 mL HB)	Final concentration/special notes
UltraPure DEPC-treated water	186 μL	
Competitor *E. coli* tRNA MRE 600 (10 mg/mL)	100 μL	1 mg/mL
20% dextran sulfate	500 μL	10%
UltraPure BSA (50 mg/mL)	4 μL	0.2 mg/mL
Sodium chloride–sodium citrate (SSC) buffer (20×)	100 μL	2×
Ribonucleoside vanadyl complex (200 mM)	10 μL	2 mM (*see* **Note 5**)
Deionized formamide	100 μL	10% (*see* **Note 6**)
SUPERase.In (20 U/mL)	0.5 μL	10 U/mL
RNA-FISH probe (25 μM) (HIV-1 gag RNA probe mixture)	10 μL	Use 1 μL probe/100 μL HB (250 nM probe)
Goat α-p24 antibody	2 μL	Use (1:500) dilution

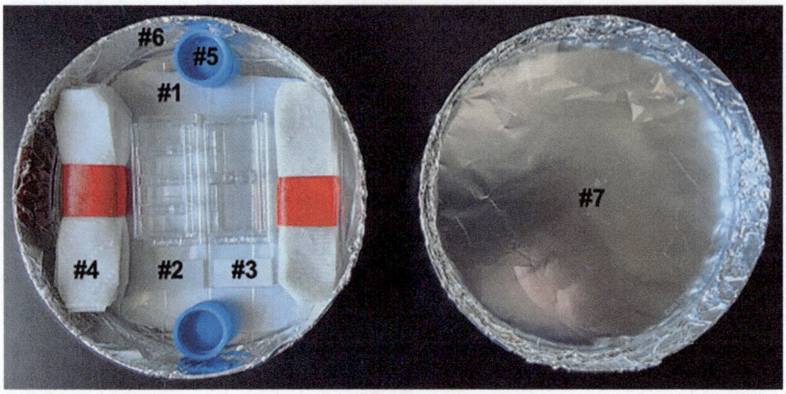

Fig. 1 Representative image illustrating components required for setting up the humid hybridization chamber for smRNA-FISH analysis. The components include a parafilm square (#1), chambered slides (#2 and #3), pieces of damp paper towel placed around the dish's perimeter (#4), and two 15-mL Falcon tube lids filled with ddH_2O (#5). The Petri plate bottom (#6) and the lid (#7) both are covered with aluminum foil. These elements collectively maintain optimal high humidity crucial for the hybridization process. Aluminum foil cover serves to shield the slides from light exposure

position damp paper towel pieces along the dish's edge (Fig. 1) (*see* **Note** 7). These prepared Petri dishes will serve as humid chambers for the RNA-FISH slides. Once the chamber is set up, proceed with the incubation process.

15. Combine 4 μL of the FISH probe with 400 μL of hybridization solution (HB) (*see* **Note 8**) and add this mixture along with the primary antibody (diluted at 1:500 for α-p24) to each sample.
16. Gently dispense combined hybridization solution into the respective wells, ensuring that the liquid rests at the center.
17. Seal the hybridization chamber with parafilm.
18. Incubate the cells overnight at 37 °C incubator.

Day 2

19. Prepare enough fresh pre-hybridization buffer containing 10% deionized formamide (for ACH2) in 2× SSC, for all following steps including for secondary antibodies and for all the washes.
20. Wash 3× quickly with pre-hybridization buffer (~500 μL/well).
21. Wash 2× 20 min at 37 °C with pre-hybridization buffer + α-goat, Alexa Fluor 647 secondary antibody (1:1000 dilutions); 400 μL/well.
22. Perform four quick rinses at room temperature with 2× SSC. During the fourth wash, incubate chambered slides with 2× SSC for 2 h at room temperature in the dark, covered with a lid.
23. Add 1 mL each of DAPI solution (1.0 μg/mL DAPI in 2× SSC) (*see* **Note 9**) for wells in 2 well-chambered slides and incubate samples for 2 min at room temperature.

24. Wash for 5 min at room temperature with 2× SSC.
25. Completely remove any residual 2× SSC solution within the chambered well slide.
26. Next, carefully detach the wells from the slide and thoroughly clean the slides clearing any adhesive residue, such as glue, from the well boundaries.
27. Place 1–2 drops of ProLong Gold Antifade Mounting Medium onto the slide and then position the cover slip over the cells. Ensure that the coverslip is properly aligned with the slide, with no edges extending beyond the slide.
28. If you are using multiple slides in your experiment, then process all the slides in a similar manner.
29. Allow the slide to air-dry overnight at room temperature in the dark, by placing them inside a box covered with aluminum foil. This is a curing step that is needed when using a mounting medium such as Prolong Gold.
30. Store the slides at 4 °C in darkness until they are imaged.
31. Day before imaging the slides, take them out of storage and keep them at room temperature in darkness for several hours till the slides get equilibrated to room temperature. After completing scanning and other microscopy tasks, return the slides to 4 °C storage. These slides are good for up to a year at 4 °C.

3.3 Single-Molecule Image Acquisition

1. Utilize an upright Olympus BX-63 Microscope with a Super Apochromatic 60×/1.35 NA Olympus Objective (UPLSAPO60XO), a SOLA light engine from Lumencor, and an ORCA-R2 Digital Interline CCD Camera (C10600-10B by Hamamatsu). Employ zero-pixel shift filter sets, including DAPI-5060C-Zero, Cy3-4040C-Zero (for Quasar 570 detection), and Cy5-4040C-Zero (for Quasar 670 detection).
2. Configure optical sectioning for cell acquisition with a 200 nm Z step size (~41 optical sections) to cover the full cell thickness.
3. Employ Metamorph software (Molecular Devices) to manage microscope automation and capture images.
4. For smRNA-FISH with Quasar 570 (CY3) or Quasar 670 (CY5), set the illumination to 100% for Cy3 and Cy5, and 12% for DAPI.
5. Exposure times and gain settings for CY5, CY3, and DAPI may vary based on the signal strength. As a guideline, employ exposure times of approximately 25 ms for CY5, 200 ms for CY3, and 10 ms for DAPI. Adjust the gain settings to 2 for CY5, 20 for CY3, and 2 for DAPI.

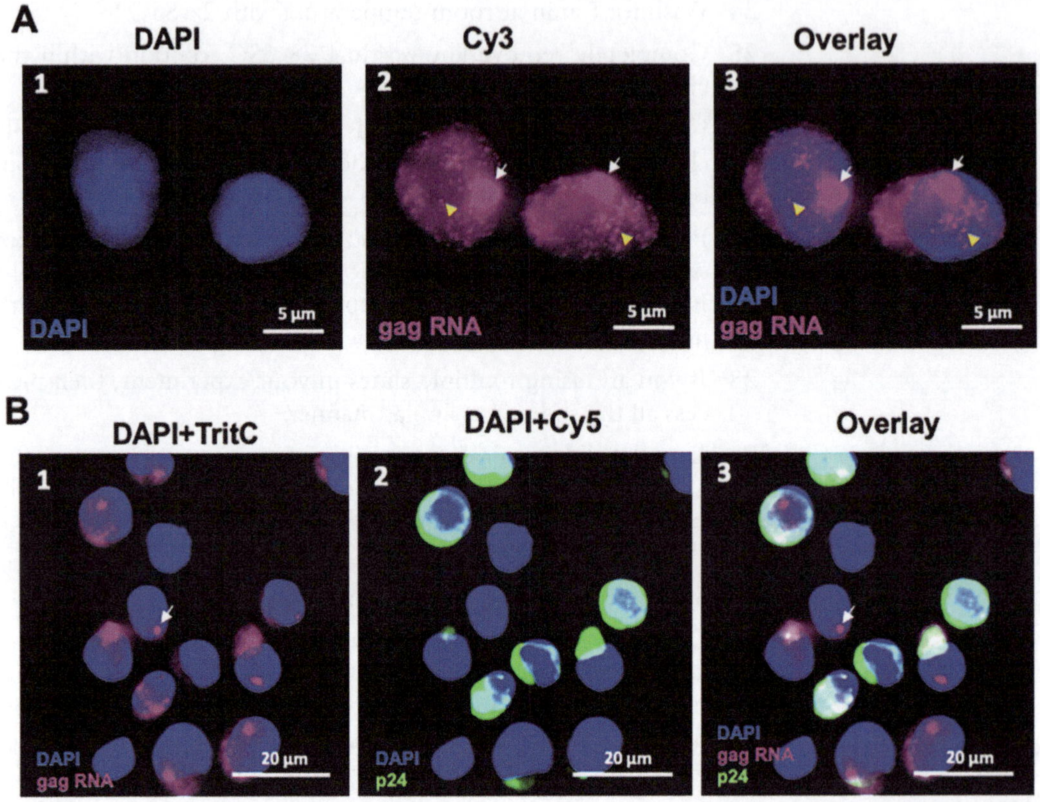

Fig. 2 Expression of HIV-1 RNA and Gag protein in reactivated latent T cells: Representative images of ACH2 cells after treatment with PMA. (**a**) Single-molecule image acquisition. ACH2 cells were treated with PMA and subjected to smRNA-FISH using gag probes and images were acquired utilizing an upright Olympus BX-63 Microscope. Blue color indicates DAPI staining of the nuclei. Magenta color (Cy3 channel) indicates HIV-1 gag RNA staining by the HIV-1 gag probe. White arrows point to transcription sites and yellow arrowheads point to diffraction-limited single molecules of RNA. (**b**) Simultaneous detection of gag RNA and Gag protein using RNA-FISH+IF method. ACH2 cells were treated with PMA and subjected to smRNA-FISH+IF using gag probes and α-p24 antibodies, and images were acquired utilizing HSHRS fluorescent microscope. Trit-C channel detects HIV-1 gag RNA (magenta), Cy5 channel detects Gag protein (green), and DAPI channel detects nuclear staining (blue). White arrows point to transcription sites

6. After acquiring single-molecule images, you can store the slides at 4 °C, where they remain viable for up to a year.
7. *See* Fig. 2a for an example of RNA-FISH and single-molecule image acquisition. *See* Fig. 2b for an example of RNA-FISH+IF using gag RNA probe and α-p24 antibody.

3.4 Workflow for Single-Molecule Analysis

Employ FISH-quant software available in MATLAB to analyze the single-molecule images to quantitate a number of RNA transcripts per reactivated cell [9]. Here, we describe a method of quantification of HIV-1 gag mRNA molecules in the nucleus and in the cytoplasm of reactivated latent ACH2 cells. In addition, the number of nascent transcripts within the transcription sites (TS) can also be determined using this method.

1. Open *MATLAB* software and then open the selected image for which analysis needs to be done (*see* **Note 10**).
2. Define the outline of the nucleus using DAPI staining and the cells using either the Cy3 or the Cy5 staining.
3. Transcription sites can also be outlined if you need to include them in your analysis.
4. Subtract the background signals using the 3D_LoG option (size 5; standard deviation 1) and then save the filtered image.
5. Adjust the program parameters to detect the expected number of mRNAs and then apply the settings to the entire image.
6. Apply a 3D Gaussian fitting to each spot to determine the coordinates of the mRNAs. Ensure that the selected parameters effectively identify the majority of mRNAs. If the parameters appear suitable, save both the detection settings and the identified spots.
7. If many images need to be analyzed, then first create the outlines for all your images. Then, open in Tools > Batch processing. Load your previously saved settings for all your outlines. Then, proceed to process data for all your images.
8. Utilize the "average of thresholded spots" feature to calculate the mean intensity of all mRNAs, which can then be employed for assessing the intensity of individual transcription sites.
9. To configure transcription site selection settings, go to Tools > TxSite quantification. Load the "average of thresholded spots" image, quantify TxSites for all locations, and save these settings. These saved settings can then be used in "Batch processing" mode for automated transcription site quantification across all images.
10. Save the outcomes by choosing Batch > Save > Summary of both nascent and mature mRNAs, which provides cell-specific spot counts. Also, choose Batch > Save > Summary of all thresholded spots to access spot position and intensity information. These are .txt files, making it convenient to transfer data to other analysis tools.
11. If smRNA-FISH was successful, the intensity distribution of cytoplasmic spots should follow a Gaussian pattern.

3.5 High-Speed and High-Resolution Scanning (HSHRS) of the Whole Slide

1. When images from a large number of cells need to be acquired, use the PANNORAMIC 250 Flash III Slide Scanner for fluorescence microscopy of the entire slide.
2. First, wipe and clean your slides with ethanol to remove dust particles and then load your slides into the cassette. You can put up to 25 slides in a single cassette and up to 10 slides in this machine (*see* **Note 11**).

3. Click on the Panoramic Scanner icon, and then check all the scanning parameters for fluorescent scanning of the slides using the "Manual mode" option.
4. Insert the cassette into the instrument, followed by inserting your first slide or the slide you want to use to test all the settings.
5. Once the cassette is inserted, click on the next green button to insert your slide.
6. In the scan options menu, adjust the settings as follows: Set image quality to excellent, choose a focus frequency of 5, make sure to uncheck the skip empty fields of view option, enable stitching, activate flat field correction, disable sharpening, and decide whether to change filters after each field of view or after the entire slide scan, based on your specific needs.
7. Load these scan settings and proceed with user-controlled scanning mode.
8. Then, select the 20× objective position (or change it to 40× if doing 40×) and select the Live option to further adjust the settings for the required exposure time (ms) for all your fluorescent dyes.
9. The three filters that we used for RNA-FISH+IF methods were DAPI, Cy5-Q, and TRITC-Dendra. To select the exposure time for each filter, choose between 2 and 2000 ms, depending on the sensitivity and intensity of the fluorescent signals in your RNA-FISH experiments. Please *see* Table 2, as one example of the exposure times that we standardized with our samples to detect HIV-1 gag RNA and Gag protein in reactivated latent cells (*see* Table 2).
10. Adjust your focus channel based on DAPI signals and select DAPI as a stitching channel. We recommend using autofocus to scan the entire slide, or you can choose a focus range for scanning, which will save the exposure duration.
11. Save the final settings because you will need them for scanning the entire slide (*see* **Note 12**).

Table 2
Exposure times for DAPI, Cy5-Q, and TRITC-Dendra used in this study to detect HIV-1 gag RNA and Gag protein in reactivated latent cells

Fluorescent dyes/filter	Exposure time (ms)	Detection for
DAPI	10–20	Host cell nucleus
Cy5-Q	60–80	HIV-1 gag protein
TRITC-Dendra	300–400	HIV-1 gag RNA molecules

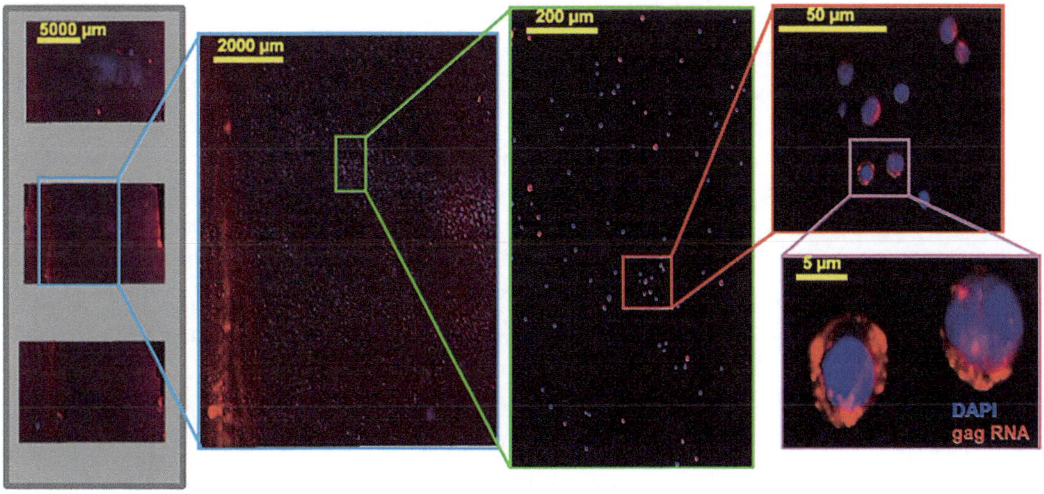

Fig. 3 Representative illustration displaying images scanned by the PANNORAMIC 250 Flash III Slide Scanner, showcasing fluorescence microscopy scanning of the entire slide. Because of the high-resolution scanning, the images from the entire slide are viewable and can be magnified for comprehensive analysis using CaseViewer 2.4 (64-bit version) software. In the depicted images, the blue color indicates DAPI-positive host cell nuclei, while the red color represents cells with positive HIV gag RNA

12. To scan all your slides, go to fluorescent scan mode and select automatic scanning mode.
13. Choose the barcode area and provide names for each of your slides.
14. Load the saved scan settings and then select the scanned area for all the slides that are being processed for the scan.
15. Proceed with automatic scanning mode (*see* **Note 13**).
16. Scanned slides can be visualized using CaseViewer 2.4 (64-bit version) (Fig. 3 and *see* **Note 14**).

4 Notes

1. Final concentration of PMA should be 50 ng/mL (in 1 mL of media-containing cells).
2. To prepare a 50 ng/mL PMA solution, begin by diluting 5 µL of a 1 mg/mL PMA stock solution in 495 µL of medium (a 1: 100 dilution). Then, take 5 µL from this diluted PMA solution and combine it with 1 mL of fresh medium. A corresponding dilution using DMSO was prepared as a negative control.
3. For adherent cells, grow and induce the cells in chambered wells and wash well before fixing. Since they are already attached to the surface, no need to use the drop method to make them adhere to the slide.

4. Incubation with a pre-hybridization buffer is a blocking step, and you can block it for longer time.
5. Prior to utilizing ribonucleoside vanadyl complex (200 mM), warm it in a heat block at 40 °C for 5 min until it turns dark black in color. Subsequently, allow it to cool to room temperature.
6. After opening the bottle, deionized formamide is good for up to 3 months when stored at 4 °C. After 3 months, discard the bottle and obtain a new one for use.
7. Lids of the 15-mL Falcon tubes are used because they are easily available and are sterile.
8. Mix the probe and antibody with the necessary amount of HB required for the experiment.
9. To prepare a 1:10,000 DAPI solution, combine 1 μL of DAPI (stock concentration: 10 mg/mL) with 10 mL of 2× SSC and thoroughly mix the solution using vortexing.
10. For smFISH analysis, we highly recommend utilizing FISH-quant, a freely available software package developed in the MATLAB programming language by MathWorks.
11. Ensure that the size of your slide is compatible with the machine, and for that, we recommend using the slide scale bar. We recommend using the slide with dimensions ($L \times W$): 75 × 25 mm.
12. First, you need to stop the live feed before loading or unloading the slides (to change the slide) and do the same for other slides as well.
13. Under the service tab, select "show images during scanning" at the beginning to verify proper focusing. Also, you might want to check the section that indicates the limited focus range to ensure it covers the entire range and focuses on different positions later in the process.
14. For setting the parameters for image acquisition on CaseViewer 2.4 (64-bit version), use mock control to set the background level of the fluorescence in the test images.

Acknowledgments

We thank Dr. Prasad for critically reading the manuscript. This work was supported by funding from NIH grants R01DA043169, 1R21AI152826, R01AI170206 to GVK, R01DK110063 to CE, and 5R35GM136296 to RHS. We thank Andrea Briceno, Hillary Guzik, and Vera DesMarais at the Analytical Imaging Facility (AIF) for facilitating the use of 3DHISTECH P250 FLASH III automated slide scanner, which was funded by NIH grant 1S10OD026852-01A1.

References

1. Bruner KM, Wang Z, Simonetti FR, Bender AM, Kwon KJ, Sengupta S et al (2019) A quantitative approach for measuring the reservoir of latent HIV-1 proviruses. Nature 566(7742): 120–125
2. Sengupta S, Siliciano RF (2018) Targeting the latent reservoir for HIV-1. Immunity 48(5): 872–895
3. Finzi D, Hermankova M, Pierson T, Carruth LM, Buck C, Chaisson RE et al (1997) Identification of a reservoir for HIV-1 in patients on highly active antiretroviral therapy. Science 278(5341):1295–1300
4. Van Lint C, Bouchat S, Marcello A (2013) HIV-1 transcription and latency: an update. Retrovirology 10:67
5. Maina EK, Adan AA, Mureithi H, Muriuki J, Lwembe RM (2021) A review of current strategies towards the elimination of latent HIV-1 and subsequent HIV-1 cure. Curr HIV Res 19(1): 14–26
6. Archin NM, Liberty AL, Kashuba AD, Choudhary SK, Kuruc JD, Crooks AM et al (2012) Administration of vorinostat disrupts HIV-1 latency in patients on antiretroviral therapy. Nature 487(7408):482–485
7. Ho YC, Shan L, Hosmane NN, Wang J, Laskey SB, Rosenbloom DI et al (2013) Replication-competent noninduced proviruses in the latent reservoir increase barrier to HIV-1 cure. Cell 155(3):540–551
8. Cillo AR, Sobolewski MD, Bosch RJ, Fyne E, Piatak M Jr, Coffin JM et al (2014) Quantification of HIV-1 latency reversal in resting CD4+ T cells from patients on suppressive antiretroviral therapy. Proc Natl Acad Sci USA 111(19): 7078–7083
9. Mueller F, Senecal A, Tantale K, Marie-Nelly H, Ly N, Collin O et al (2013) FISH-quant: automatic counting of transcripts in 3D FISH images. Nat Methods 10(4):277–278

Chapter 5

Visualizing HIV-1 Assembly at the T-Cell Plasma Membrane Using Single-Molecule Localization Microscopy

Rayane Dibsy, Kaushik Inamdar, Cyril Favard, and Delphine Muriaux

Abstract

The 20-year revolution in optical fluorescence microscopy, supported by the optimization of both spatial resolution and timely acquisition, allows the visualization of nanoscaled objects in cell biology. Currently, the use of a recent generation of super-resolution fluorescence microscope coupled with improved fluorescent probes gives the possibility to study the replicative cycle of viruses in living cells, at the single-virus particle or protein level. Here, we highlight the protocol for visualizing HIV-1 Gag assembly at the host T-cell plasma membrane using super-resolution light microscopy. Total internal reflection fluorescence microscopy (TIRF-M) coupled with single-molecule localization microscopy (SMLM) enables the detection and characterization of the assembly of viral proteins at the plasma membrane of infected host cells at the single protein level. Here, we describe the TIRF equipment, the T-cell culture for HIV-1, the sample preparation for single-molecule localization microscopies such as PALM and STORM, acquisition protocols, and Gag assembling cluster analysis.

Key words HIV-1, Viral Gag protein assembly, TIRF-M, Single-molecule light microscopy, PALM, STORM

1 Introduction

The human immunodeficiency virus type 1 (HIV-1) is an enveloped single-stranded (+)RNA virus belonging to the family *Lentiviridae* that mainly infects CD4+ T lymphocytes, as well as macrophage–lineage cells. The HIV-1 Gag structural protein is a 55 kDa polyprotein containing four domains: matrix (MA), capsid (CA), nucleocapsid (NC), and p6, in addition to two small spacer peptides, SP1 and SP2. Approximately 1500–2000 Gag polyproteins assemble to form a single immature capsid shell [1] that will mature upon budding through the action of the viral protease. This maturation process allows a structural rearrangement of the particle, with the spatiotemporal cleavage of the MA, CA, NC, and spacer peptides, that allows MA to stay bound to the lipid membrane and CA into a conical-shaped core containing the NC and the

Vinayaka R. Prasad and Ganjam V. Kalpana (eds.), *HIV Protocols*, Methods in Molecular Biology, vol. 2807,
https://doi.org/10.1007/978-1-0716-3862-0_5,

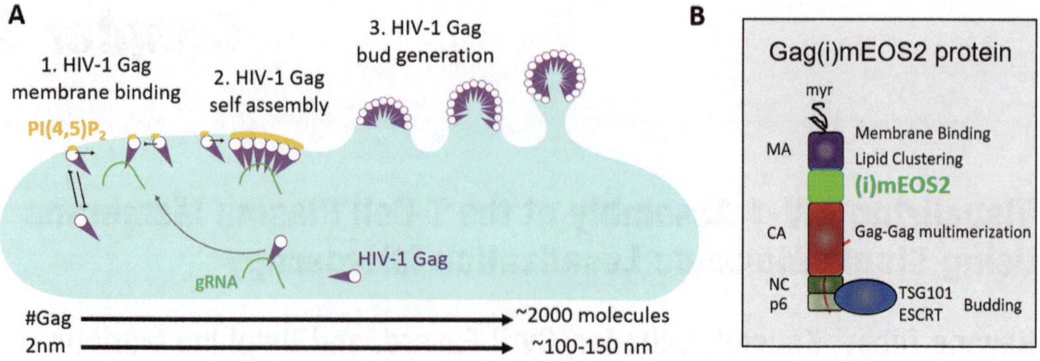

Fig. 1 (**a**) Scheme of HIV-1 Gag assembly process and particle budding at the cell plasma membrane. (**b**) Scheme of HIV-1 Gag(i)mEOS2 reporter protein. The retroviral Gag protein is composed of four main domains: matrix (MA) for membrane binding, capsid (CA) for Gag-Gag interactions, nucleocapsid (NC) for multimerization of Gag on the genomic RNA, and recruitment of Tsg101 and p6 domain for particle budding with the recruitment of the endosomal sorting complexes required for transport (ESCRT) machinery. The photoconvertible "green-to-red" fluorescent mEOS2 protein is located between the MA and CA domains of Gag

genomic RNA. Only HIV-1 Gag is required for immature particle assembly (Fig. 1). The NC-p6 domain of Gag is required for particle budding and release by recruiting the endosomal sorting complexes required for transport (ESCRT) machinery host cell factors [2]. Membrane targeting of Gag requires the N-terminal MA domain, thanks to the myristate and a stretch of highly basic residues (the highly basic region, HBR), which interacts with the acidic head groups of phospholipids at the inner leaflet of the cell plasma membrane (PM), especially including the phosphatidylinositol 4,5-bisphosphate (PI(4,5)P2) enriched in the cell plasma membrane [3–6]. CA contains residues that are critical for Gag–Gag interactions and NC is required for viral genomic RNA packaging, as well as nonspecific interactions with RNA [7], both being essential for Gag multimerization and particle assembly. Gag is the only viral protein required for particle assembly and release of immature viral particles in cells, although the production of infectious viruses requires other viral proteins, such as the protease PR, the Env proteins, and the genomic RNA. Thus, tagging just the viral Gag protein enables the visualization of HIV-1 assembly processes in cells [8].

The HIV-1 particle is 100–150 nm in diameter in size [9–11], and thus, the assembly site is below the resolution limit of conventional optical microscopy methods. Electron microscopy has long been a standard to study HIV-1 assembly and the particle structure [12]. However, despite its stellar contribution to our understanding of HIV-1 particle structures, electron microscopy does not allow for the study of the dynamics and the kinetics of nanoscale events taking place during virus assembly at the host cell plasma membrane in living cells.

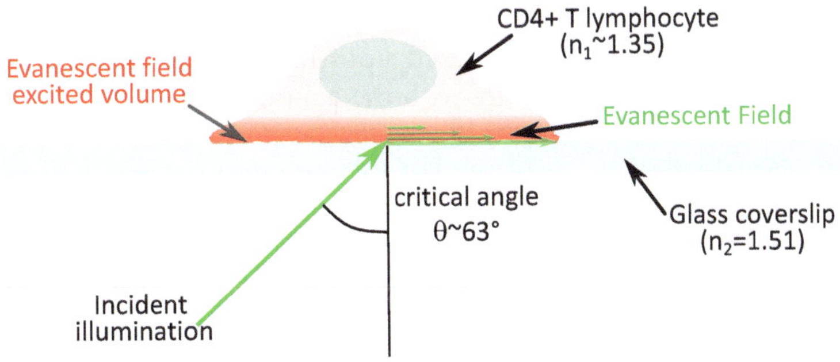

Fig. 2 Principle of total internal reflection fluorescence microscopy (TIRF-M) imaging. Total internal reflection occurs when the sample is illuminated at a critical angle ($\theta \geq 63°$) (green arrow) greater than arcsin (n_1/n_2), with n_1, the refractive index of the cell ($n_1 \sim 1.35$), and n_2, the refractive index of the glass coverslip ($n_2 \sim 1.515$). An evanescent field is then generated at the glass/cell interface (in green), which, due to its small axial penetration ($z \sim$ 100–200 nm), excites the fluorescence coming from a small volume of the cell (in dark red). This illumination angle can be reached in high numerical aperture objectives of inversed microscopes, to facilitate the use of TIRF-M for living cells

Recent fluorescence microscopy advances, coupled with genetically encoded fluorescence proteins, have offered new insights into the dynamics underlying HIV-1 assembly (for reviews, *see* [13–16]).

1.1 Visualization Using TIRF-M

Since HIV-1 assembly occurs at the cell plasma membrane, most of the improvement in deciphering this process was achieved using total internal reflection fluorescence microscopy (TIRF-M) (Fig. 2). Indeed, imaging in TIRF mode allows better contrast thanks to a low penetration depth (100–300 nm). To achieve total internal reflection microscopy, the sample has to be illuminated with an incidence angle (θ_c) greater than arcsin (n_1/n_2), with n_1, the refractive index of the cell ($n_1 \sim 1.35$), and n_2, the refractive index of the glass coverslip ($n_2 \sim 1.515$). This gives $\theta_c = 63°$. In an objective-based TIRF microscope (e.g., excitation is achieved through the objective), this illumination angle can only be reached using high numerical aperture objectives (NA $\geq$ 1.45). Different manufacturers have their own objective-based TIRF microscopes (Zeiss, Leica, Nikon, Olympus, and Abbelight) using dedicated objectives and dedicated optical setup for laser illumination.

Thanks to TIRF microscopy, the kinetics of HIV-1 Gag assembly and particle release could be visualized, for the first time, using fluorescent-tagged Gag proteins in a transfected adherent model cell line [8]. With live two-color TIRF microscopy, the timing of the association of Gag with the genomic RNA at the cell membrane [17, 18] could be observed for the first time and thereafter with the ESCRTs for particle budding [19, 20]. Using multicolor frame-by-frame alternating TIRF/wide-field imaging, Ivanchenko et al. visualized a possible pre-association of several cytosolic Gag molecules with the genomic RNA [21].

1.2 Visualization Using PALM and STORM

In the last two decades, far-field optical microscopy has overcome the old resolution barrier, e.g., the diffraction limit, enabling super-resolution visualization of previously invisible molecular details in biological systems. These microscopy techniques have been used to answer questions involving organization, interaction, stoichiometry, and dynamics of individual molecular building blocks and their integration into functional cellular machineries [22]. Super-resolution microscopy (SRM) techniques PALM [9] and STORM [10] are based on single-molecule detection and localization (often called single-molecule localization microscopies, SMLM) taking advantage of the fluorophore ability to blink or to be photoconverted [11]. Single-molecule localization microscopies PALM and STORM are based on emission blinking of the fluorophores, allowing time sequential identification of single emitters (for review, *see* [23]). Among all the possible processes leading to blinking, the most common are photoconversion (PALM) or dark state trapping (dSTORM). In both cases, this requires a high level of excitation energy (photon flux or illumination), either to pump the molecule into the dark state or to photobleach it after photoconversion. Therefore, PALM and/or STORM microscopes are equipped with high-power lasers (from 100 mW to 1 W, generally emitting at 561 nm and from 630 to 640 nm) and a photoconverting or photoswitching/photoactivation excitation wavelength (405 nm for photoconversion of mEos proteins or photoswitching/photoactivation of organic dyes such as Alexa Fluor 647 and Cy5B).

By detecting each single fluorophore and localizing precisely their center of mass, single-molecule positions from each of thousands of images can be obtained. These positions are then used to reconstruct the final image with single-molecule localization precision (for a review, *see* [23]). This localization precision is inversely proportional to the number of photons detected for each single molecule. Therefore, using very bright fluorophores, 10–20 nm localization precisions can be achieved. Since a decade, these super-resolution microscopies have been successfully applied to the study of virus replication cycle, such as HIV-1, at the single protein level (for reviews, *see* [14, 15]).

Using Gag-mEos2 and PALM microscopy, we and others could quantify the nanoscale morphology of the assembling HIV-1 virus-like particles in Cos7 [24] and Jurkat T cells [25]. The same results were found in A3.01 T cells using STORM microscopy [26]. Dual-color PALM/STORM microscopy also allowed HIV researchers to characterize and quantitatively image Gag/Env [27], or tetherin, a restriction factor, [28] and more recently IRSP53, a membrane curvature protein and an HIV-1 assembly cofactor [29], as well as the interplay between HIV-1 Gag and the cortical actin [30]. SMLM is nowadays the imaging method of choice to visualize at the nanometric scale of HIV-1 assembly and budding in cells.

2 Materials

2.1 Disposables

1. Cell culture dishes (T-75, 10-cm^2 culture dish, 12-well tissue culture-treated plates).
2. Glass-bottom 35-mm cell culture-treated plates.
3. 15- and 50-mL plastic conical centrifuge tubes.
4. 1.5- and 2-mL conical centrifuge plastic tubes.
5. Micropipettes (0.2–2, 1–20, 20–200, 200–1000 μL) and tips.
6. Syringes (10 or 20 mL without needle).
7. 0.45- and 0.22-μm disposable filters.
8. SW41Ti Ultracentrifugation Tube Ultra Clean Beckman.
9. 25-mm glass coverslip #1.5H.
10. Disposable cell counting slides and automated cell counter.
11. Concave coverslip.
12. 2-component Twinsil® Sealing Kit.

2.2 Equipment

1. Cell culture incubator with humidified atmosphere (37 °C, 5% CO_2).
2. Standard bench-top centrifuge.
3. Weighing machine (balance).
4. A fridge.
5. A sterile cell culture laminar flow hood biosafety level 2 and level 3.
6. An ultracentrifuge with SW41Ti rotor Beckman.
7. An inverted Nikon TIRF microscope with a 561-nm laser.
8. 100× 1.49 numerical aperture (NA) objective with oil immersion (1.45 NA is sufficient for TIRF-M).

2.3 Reagents

2.3.1 Cell Culture Media

1. Dulbecco's modified Eagle's medium (DMEM) supplemented with 10% fetal bovine serum (FBS), sodium pyruvate, and antibiotics (penicillin–streptomycin).
2. RPMI 1640 plus GlutaMAX supplemented with 10% fetal bovine serum (FBS) and sodium pyruvate and antibiotics (penicillin–streptomycin).

3. Fetal bovine serum (FBS).
4. Antibiotics (penicillin–streptomycin).

2.3.2 Buffers

1. Phosphate-buffered saline (PBS).
2. Lebovitz L15 medium without phenol red (transparent L15 medium) for cell culture.
3. HEPES (1 M).
4. Sodium pyruvate.
5. CaCl2 250 mM/HBS2X Transfection Kit.
6. Ultrapure sucrose.
7. TNE (Tris–HCl Ph 7.4, NaCl 100 mM, EDTA 1 mM).

2.4 Plasmids

Plasmid expressing Env-deleted HIV-1(i) GFP was used for STORM, named pHIV-1 Gag(i)GFPΔEnv (NL4.3 Strain), and is from NIH (Catalog Number: 12455, Lot Number: 140409) (as in [30]). Plasmid pCMV-VSV-G (#8454) is commercially available from Addgene. Plasmid pHIV-1Gag(i)mEOS2 (as in [25]) used for PALM can be provided by Addgene.

2.5 Others

1. HIV-1 p24 AlphaLISA kit (PerkinElmer).
2. AlphaLISA 96-well plate (PerkinElmer).
3. Poly-L-lysine.
4. Paraformaldehyde (PFA).
5. NH4Cl2 50 mM.
6. Triton X-100.
7. Bovine serum albumin (BSA).
8. Nanobody GFP booster coupled to Alexa Fluor 568.
9. Fiducial markers TetraSpeck microspheres 100-nm beads (Life Technologies), used as fiducial markers during image acquisition.
10. STORM buffer (Everspark by Idylle).
11. Smart kit for STORM from Abbelight.

3 Methods

3.1 Cell Culture

1. Harvest 1 million human Jurkat T lymphocytes (human T-cell leukemia cell line) (ATCC-CRL-2899TM) and culture them in T-75 culture flask in 10 mL of RPMI 1640 plus GlutaMAX supplemented with 10% fetal bovine serum (FBS), sodium pyruvate, and antibiotics (penicillin–streptomycin).
2. Harvest human embryonic kidney cells (293THEK cell) (HEK 293T-ATCC-CRL-1575TM) and culture them in T-75 culture

flask in 10-mL Dulbecco's modified Eagle's medium (DMEM) supplemented with 10% fetal bovine serum (FBS), sodium pyruvate, and antibiotics (penicillin–streptomycin).

3. Grow cells at 37 °C in a 5% CO_2 atmosphere.

3.2 Virus Production

1. Plate 2.5 million 293T human embryonic kidney (HEK) cells in 10 mL of growth media 1 day before transfection in a 10-cm culture dish (*see* **Note 1**).
2. At 50–70% confluence (*see* **Note 2**), transfect the adherent HEK293T cells with calcium phosphate precipitation method (*see* Inamdar et al. [29]) by mixing 8 μg total amount of plasmid for both pCMV-VSV-G and pHIV-1Gag(i)GFP/mEOS2ΔEnv (at a ratio of pCMV-VSV-G: pHIV-1Gag(i)-GFPΔEnv equivalent to 1:4) with equal volumes (500 μL) of $CaCl_2$ solution and HBS2X buffer (500 μL).
3. Incubate the 1-mL mix for 10 min at RT and add dropwise on the cells.
4. Transfer the cells to a biosafety level 3/2+ security laboratory (BSL3/2+).
5. Incubate the dish containing the transfected cells in a humidified incubator with 5% CO_2 at 37 °C for 12–14 h in a BSL3 laboratory.
6. At the end of 12–14 h, remove the medium and discard. Wash cells once with warm PBS and then add 10 mL of warm complete DMEM.
7. Store in a humidified incubator with 5% CO_2 at 37 °C overnight to 48 h post-transfection.
8. At 48 h post-transfection, collect the cell culture supernatant containing viral particles.
9. Filter this media with a 0.45-μm filter with the help of a 10-mL syringe.
10. Add 9 mL of this supernatant to a 10-mL ultracentrifuge tube.
11. Deposit 1 mL of 25% sucrose in TNE buffer (10 mM Tris–HCl [pH 7.4], 100 mM NaCl, 1 mM EDTA, filtered 0.22 μm) carefully at the bottom of the tube, forming the sucrose cushion.
12. Weigh the tubes carefully to balance them for ultracentrifugation.
13. Cool the ultracentrifuge to 4 °C.
14. Spin the samples at 100,000*g* for 90 min in a vacuum at 4 °C in a SW41Ti Beckman Coulter rotor.
15. Discard the supernatant and let the pellet air dry at room temperature.

16. Resuspend the dry pellet with RPMI media, without serum, at 4 °C overnight.
17. Vortex briefly the next day and recover the 100 μL solution of the virus into a 1.5-mL microcentrifuge tube.
18. This contains the virus and should be stored at −80 °C in a BSL3 laboratory (*see* **Note 3**).
19. Quantify the viral titer by alphaLISA-CAp24 immunoassay (PerkinElmer) according to the manufacturer. Recombinant HIV-1 p24 protein is used to construct a standard curve (*see* **Note 4**).

3.3 T-Cell Infection

1. Plate 1 million Jurkat T cells in 2 mL RPMI media in a 12-well culture plate.
2. Transfer into a BSL3 laboratory (or BSL2+ laboratory, depending on the country).
3. Infect with 500 ng/mL of HIV-1 CAp24.
4. Incubate for 2 h at 37 °C in the incubator at 5% CO_2.
5. Centrifuge the cells at 800*g* for 5 min.
6. Wash three times with PBS to remove the unattached virus, recentrifuge, and resuspend the infected cells in 2 mL of warm RPMI media.
7. Incubate the infected Jurkat T cells for 48 h at 37 °C in the incubator at 5% CO_2.
8. 48 h postinfection, discard the supernatant, wash the T cells with PBS as above, and resuspend the cell pellet in transparent L15 medium supplemented with 20 mM HEPES (*see* **Note 5**).

3.4 Preparation of Cover Slips

1. In a 6-well plate, wash 25-mm coverslip #1.5H with ethanol 100%.
2. Wash the glass coverslip with PBS (proceed with an excess of PBS) three times.
3. Cover the coverslip with 1 mL of poly-L-lysine for 30 min at 37 °C. Of course, at this stage other types of coating can be tried depending on the desired biological condition.
4. Wash the coverslip with PBS three times.
5. Store in PBS at 4 °C.

3.5 Sample Preparation for PALM/STORM/TIRF Microscopy

See Fig. 3 for workflow.

1. Deposit 250, 000–300, 000 infected T cells in L15 buffer on the poly-L-lysine glass coverslip (in a BSL3 or BSL2+ laboratory).
2. Incubate for 30 min at 37 °C in an incubator with 5% CO_2 in the BSL3 or BSL2+.

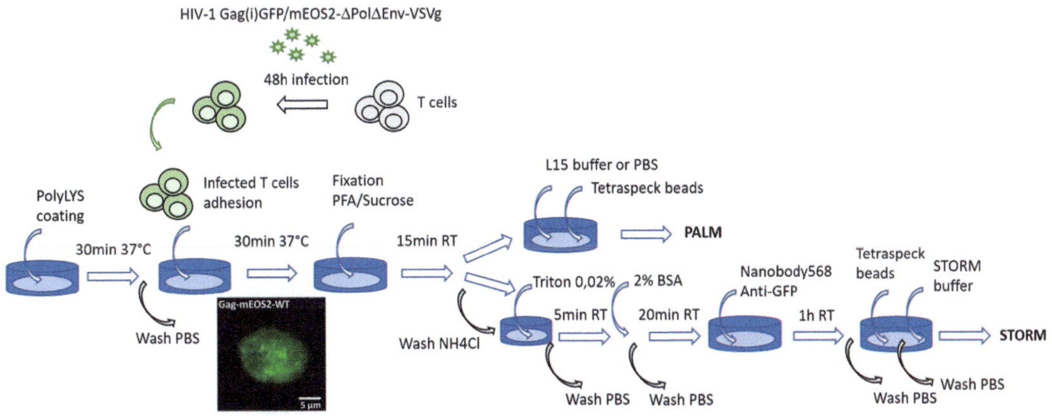

Fig. 3 PALM-STORM experimental workflow. Schematic representation of the experimental workflow to image HIV-1 Gag assembly platforms at the cell surface of infected Jurkat T cells using PALM or STORM procedures

3. Fix the infected T cells with a solution of 4% PFA with 4% sucrose in PBS for 20 min at room temperature.
4. Discard PFA–sucrose buffer.
5. Wash with PBS two times and leave in PBS.
6. Exit the BSL3 laboratory.
7. Quenched T-cell autofluorescence with 50 mM NH_4Cl in PBS for 5 min.
8. Wash with PBS.

For PALM, continue as follows:

1. Add diluted TetraSpeck microspheres 100 nm (named as beads afterward) in L15 medium (or PBS) as fiducial markers used for drift correction during image acquisition and incubate for 5 min at room temperature. The dilution of these beads depends on the microsphere's original concentration. The goal is to get around one bead per 2–3 cells (*see* **Note 6**).
2. Remove the unattached beads and store live cells in L15 medium supplemented with 20 mM HEPES at 37 °C.
3. If fixed cells, treat the cells with 4% PFA in PBS for 15 min, wash the cells twice with PBS, and store the cells in PBS at room temperature.
4. Go to image acquisition.

For STORM, continue as follows (*see* Fig. 4):

5. Permeabilize the cell membrane with 0.02% Triton in PBS for 5 min.
6. Block with 2% BSA in PBS for 20 min.
7. Wash PBS.

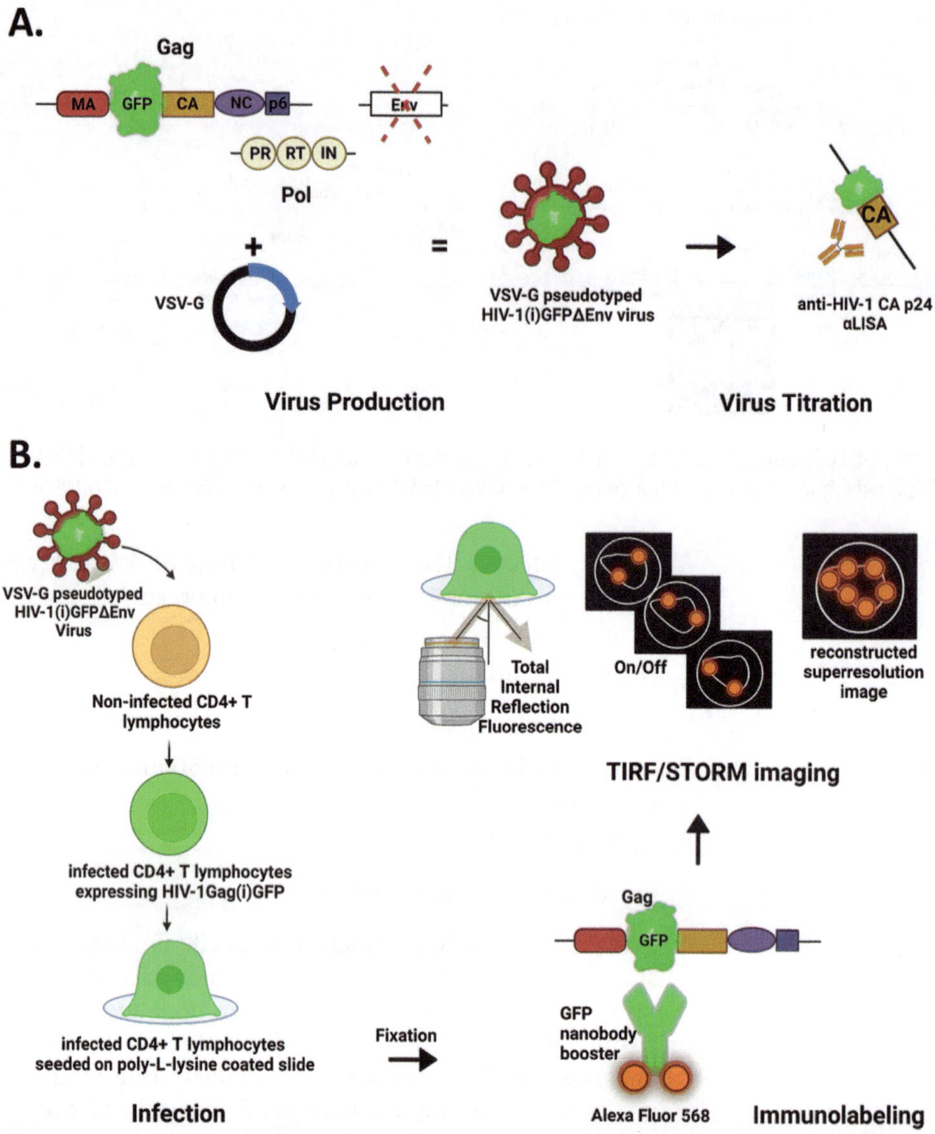

Fig. 4 TIRF/STORM protocol to visualize HIV-1 assembly clusters at the plasma membrane of HIV-1-infected CD4+ T lymphocytes. (**a**) HIV-1 internally tagged with GFP depleted of its envelope (Env) pseudotyped with VSV-G produces VSV-G-HIV-1(i)GFPΔEnv virus. The virus is titrated using anti-capsid p24 AlphaLISA kit. (**b**) Infection of CD4+ T lymphocytes with purified VSV-G-HIV-1(i)GFPΔEnv virus. Infected cells are seeded on poly-L-lysine-coated slide and then fixed. Immunolabeling of the virus using nanobody GFP conjugated with Alexa Fluor 568 suitable for STORM microscopy. The sample is subjected to TIRF/STORM imaging enabled by stochastic blinking of Alexa Fluor 568 using a 100× oil immersion 1.49 NA objective. Reconstruction is performed on 30,000 frames, resulting in a super-resolved image of HIV-1 Gag assembly clusters at the T-cell plasma membrane (*see* Fig. 5d for an example)

8. Incubate the cells on the coverslip with 1:500 nanobody–GFP booster–Alexa 568 in L15 medium for 1 h at room temperature (in the dark).
9. Wash three times with PBS.
10. Add the 100 nm beads (as in 3.5-PALM.1) at the appropriate dilution and incubate for 5 min at room temperature.
11. Remove the beads and wash once with L15 buffer or PBS.
12. Mount the coverslip with the adequate STORM buffer (*see* **Note 7**) according to the manufacturer (either Everspark buffer by Idylle, France, or Smart kit for STORM by Abbelight, France) (*see* **Note 8**).
13. In the case of Everspark mounting, use a concave coverslip and seal the sample with Twinsil, according to the manufacturer (the sample can be stored for weeks at 4 °C) (*see* **Note 9**).

3.6 Image Acquisitions (See Fig. 5)

1. To acquire PALM/STORM/TIRF images, use an inverted microscope with an oil immersion objective with NA > 1.45 (ideally NA = 1.49) (*see* **Note 10**).
2. Set the illumination in a TIRF mode by progressively increasing the angle to reach the critical angle. An increased contrast in the cell is a qualitative signature of correctly reached angle.
3. Acquire a minimum of 20, 000–30, 000 frames for each cell, using illumination irradiance between 1 and 10 kW/cm^2 on the sample using a 561 nm laser, with 10- to 30-ms exposure time on an EMCCD camera when using Alexa Fluor 568 as the fluorescent dye, to obtain the best compromise between resolution, precision, and photon budget (*see* [31] for details) (*see* **Note 11**).
4. Acquire images where the 100-nm beads are present in the field of view since these fiducial markers are needed to correct for drift occurring during acquisition (*see* **Note 12**).

3.7 STORM Post-Processing, Image Reconstruction, and Cluster Analysis at a Glance

It is of high importance to note here that there is no absolute protocol for image reconstruction and cluster analysis in SMLM. In addition to the commercially available software released by different companies (Nikon, Zeiss, and Abbelight), many different software packages exist (*see* https://srm.epfl.ch/SoftwareList for a detailed list) that differ in how they perform depending on the image acquisition modality and the object to be imaged. A benchmark of these different software packages has been released thanks to a challenge performed on synthetic dataset (*see* [32]). For our purpose (low/high-density 2D SMLM acquisitions), we found ThunderSTORM [33], a Fiji plugin (https://imagej.net/software/fiji/downloads), to be a good compromise between efficiency and simplicity of use.

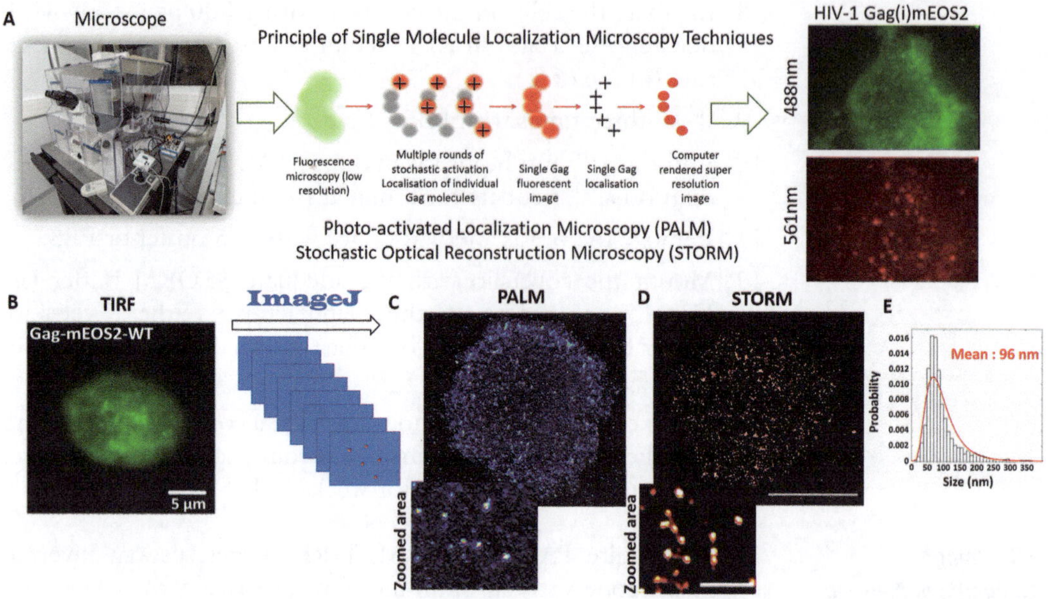

Fig. 5 Principle of TIRF-PALM-STORM, an example of raw images and reconstructed images of HIV-1 Gag assembly platforms. (**a**) TIRF-M with PALM/STORM acquisitions, the principle of PALM/STORM (adapted from S. Manley and collaborators). Raw images after photoconversion of "green-to-red" fluorescent Gag(i)mEOS2 at the cell plasma membrane. TIRF images were taken with 488-nm excitation (in green) and 561-nm (in red) of a cell expressing HIV-1 Gag-mEos2. (**b**) HIV-1(i)GFP-infected Jurkat T cells imaged by TIRF-M (low resolution) at 488 nm excitation. (**c**) PALM total localizations of HIV-1 Gag-mEos2 in Jurkat T cells as acquired by TIRF/PALM microscopy (high-resolution microscopy) and after image processing. The image displays ~2.5 million Gag total localizations (superimposed), 120 viral assembly platforms of 100 nm of diameter average, and 524 in the range of 50 nm. (**d**) STORM total localization of HIV-1(i)GFP-infected Jurkat T cells as acquired by TIRF/STORM microscopy (high-resolution microscopy). (**e**) Example of HIV-1 Gag assembly cluster size distribution using DBSCAN (probability of occurrence in the function of the assembly cluster size in nm)

After image reconstruction, image analysis has to be performed to identify and quantify the size of Gag clusters. To achieve this task, again different methods (*see* [34]) and software are available (*see* https://srm.epfl.ch/SoftwareList again). We use the DBSCAN method included in the SR-Tesseler software [35] that provides a graphical user interface. However, this method is also available in the LAMA software [36].

1. Use ThunderSTORM to reconstruct and post-process STORM acquisition data. Tutorials of the plugin can be found at https://github.com/zitmen/thunderstorm/wiki/Tutorials.
2. Eliminate the background noise within 50 nm radius.
3. Remove duplicate localization and merge together the repeating ones within 20 nm, which is the precision localization we usually obtain in our experiments.
4. Correct the drift using the TetraSpeck beads fiducial marker.

5. Save the coordinate files.
6. Open the coordinate files with SR-Tesseler.
7. In SR-Tesseler software, use DBSCAN to analyze the size distribution of the virus particles present in the STORM images.

3.8 Concluding Remarks

PALM-STORM and double-color STORM can be done more easily nowadays on fixed infected cells, as we recently reported [14, 15, 25, 29, 30], for the study of HIV-1 assembly and host cell factors in CD4 T cells. Fixing cells and observing events at the cell surface using TIRF-M makes it more practical to study when using single-molecule light microscopy (i.e., increasing signal-to-noise ratio). On the other hand, while live cell STORM imaging is technically difficult to achieve, thanks to the use of fluorescent chimeric proteins, live cell PALM is possible and could be used to follow the motions of single Gag molecules. This has been originally done on HIV-1 by S. Manley [37]. More recently, live PALM has been used to monitor the time course of assembly at the single-molecule level, e.g., Gag after Gag. This "tour de force" has been realized in a previous effort [25], visualizing and analyzing hundreds of thousands of single Gag trajectories occurring at the cell surface of HIV-1-infected host T lymphocytes. Recently, Interferometric PhotoActivation Localization Microscopy (iPALM), which localizes single fluorescent molecules with 20 nm lateral and 10 nm axial resolution, revealed HIV-1 Gag lattice dynamics in single-virus particles [38].

4 Notes

1. For optimal virus production, use early passaged 293T HEK cells (maximum at third passage after thawing).
2. When cells are transfected at 50% of confluence, it is recommended to supplement the media with 25 mM HEPES after cell washing and media change. This helps healthier and faster cell growth while cells start to produce viruses.
3. Freeze the aliquots (10–20 μL of virus per microtube) at −80 °C, right after collecting the virus, and store them at −80 °C. Make sure to titrate the virus stock after one freezing cycle and not right after collecting. This will avoid any loss in virus infectivity, such as VSV-G envelope, by the freeze–thaw process, and make sure that the concentration used for infection is correct.
4. Avoid refreezing–thawing virus aliquots. If this is the case, it is recommended to titrate viruses again before infection.

5. Filter your L15 medium prior to use. This will avoid some artifacts that could be detected during imaging.
6. To make sure that beads are homogeneously spread over the glass coverslip, and to avoid getting too many beads around a single cell, incubate beads while shaking with a very low speed.
7. Strictly avoid bubbles in the STORM buffer-mounted slide, to maintain buffer efficiency for longer time.
8. If you use Abbelight or other homemade STORM buffer, blinking is optimal up to the first 6 h of acquisition. If needed, do not hesitate to renew the buffer.
9. If you use Everspark STORM buffer, imaging the mounted sample after a minimum of 4 h of mounting gives optimal results, as compared to immediate imaging.
10. Always make sure that your sample is at room temperature before starting imaging.
11. T cells are generally autofluorescent, and make sure to always get your TIRF image showing virus buds before starting your STORM acquisition. In the case of imaging an autofluorescent noninfected cells, blink will appear but will be detectable only for a very few frames (generally less than 1 min).
12. TetraSpeck beads can aggregate together, and this can appear as >100 nm spot of high fluorescence intensity. In this case, avoid taking acquisition next to those types of beads that could interfere with the detection of single blinks during post-processing and increase the risk of biasing the posttreatment analysis.

Acknowledgments

We thank ANRS, SIDACTION, and CNRS for grant support. The following reagent was obtained through the NIH HIV Reagent Program, NIAID, NIH: Human Immunodeficiency Virus Type 1 (HIV-1) Non-Infectious Molecular Clone NL4-3 Gag-iGFPΔEnv, HRP-12455, contributed by Dr. Benjamin K. Chen. We also thank the BSL3 Microscopy facility at CEMIPAI CNRS Montpellier France (www.cemipai.fr).

References

1. Briggs JAG, Simon MN, Gross I et al (2004) The stoichiometry of Gag protein in HIV-1. Nat Struct Mol Biol 11:672–675. https://doi.org/10.1038/nsmb785
2. von Schwedler UK, Stuchell M, Müller B et al (2003) The protein network of HIV budding. Cell 114:701–713
3. Olety B, Ono A (2014) Roles played by acidic lipids in HIV-1 Gag membrane binding. Virus Res 193:108–115. https://doi.org/10.1016/j.virusres.2014.06.015
4. Ono A, Ablan SD, Lockett SJ et al (2004) Phosphatidylinositol (4,5) bisphosphate regulates HIV-1 Gag targeting to the plasma

membrane. Proc Natl Acad Sci USA 101:14889. https://doi.org/10.1073/pnas.0405596101

5. Yandrapalli N, Lubart Q, Tanwar HS et al (2016) Self assembly of HIV-1 Gag protein on lipid membranes generates PI(4,5)P2/cholesterol nanoclusters. Sci Rep 6:39332. https://doi.org/10.1038/srep39332
6. Saad JS, Miller J, Tai J et al (2006) Structural basis for targeting HIV-1 Gag proteins to the plasma membrane for virus assembly. Proc Natl Acad Sci USA 103:11364–11369. https://doi.org/10.1073/pnas.0602818103
7. Comas-Garcia M, Davis SR, Rein A (2016) On the selective packaging of genomic RNA by HIV-1. Viruses 8:246. https://doi.org/10.3390/v8090246
8. Jouvenet N, Bieniasz PD, Simon SM (2008) Imaging the biogenesis of individual HIV-1 virions in live cells. Nature 454:236–240
9. Betzig E, Patterson GH, Sougrat R et al (2006) Imaging intracellular fluorescent proteins at nanometer resolution. Science 313:1642. https://doi.org/10.1126/science.1127344
10. Rust MJ, Bates M, Zhuang X (2006) Sub-diffraction-limit imaging by stochastic optical reconstruction microscopy (STORM). Nat Methods 3:793–795. https://doi.org/10.1038/nmeth929
11. Ambrose WP, Moerner WE (1991) Fluorescence spectroscopy and spectral diffusion of single impurity molecules in a crystal. Nature 349:225
12. Briggs JAG, Kräusslich H-G (2011) The molecular architecture of HIV. J Mol Biol 410:491–500. https://doi.org/10.1016/j.jmb.2011.04.021
13. Chojnacki J, Müller B (2013) Investigation of HIV-1 assembly and release using modern fluorescence imaging techniques. Traffic 14:15–24. https://doi.org/10.1111/tra.12006
14. Inamdar K, Floderer C, Favard C, Muriaux D (2019) Monitoring HIV-1 assembly in living cells: insights from dynamic and single molecule microscopy. Viruses 11:72. https://doi.org/10.3390/v11010072
15. Arone C, Dibsy R, Inamdar K et al (2021) Illuminating the nanoscopic world of viruses by fluorescence super-resolution microscopy. Virologie 25:47–60. https://doi.org/10.1684/vir.2021.0908
16. Saffarian S (2021) Application of advanced light microscopy to the study of HIV and its interactions with the host. Viruses 13:223. https://doi.org/10.3390/v13020223
17. Jouvenet N, Simon SM, Bieniasz PD (2009) Imaging the interaction of HIV-1 genomes and Gag during assembly of individual viral particles. Proc Natl Acad Sci USA 106:19114–19119. https://doi.org/10.1073/pnas.0907364106
18. Kemler I, Meehan A, Poeschla EM (2010) Live-cell coimaging of the genomic RNAs and Gag proteins of two lentiviruses. J Virol 84:6352–6366. https://doi.org/10.1128/JVI.00363-10
19. Bleck M, Itano MS, Johnson DS et al (2014) Temporal and spatial organization of ESCRT protein recruitment during HIV-1 budding. Proc Natl Acad Sci USA 111:12211–12216. https://doi.org/10.1073/pnas.1321655111
20. Baumgärtel V, Ivanchenko S, Dupont A et al (2011) Live-cell visualization of dynamics of HIV budding site interactions with an ESCRT component. Nat Cell Biol 13:469–474. https://doi.org/10.1038/ncb2215
21. Ivanchenko S, Godinez WJ, Lampe M et al (2009) Dynamics of HIV-1 assembly and release. PLoS Pathog 5:e1000652. https://doi.org/10.1371/journal.ppat.1000652
22. Sahl SJ, Hell SW, Jakobs S (2017) Fluorescence nanoscopy in cell biology. Nat Rev Mol Cell Biol 18:685
23. Lelek M, Gyparaki MT, Beliu G et al (2021) Single-molecule localization microscopy. Nat Rev Methods Primers 1:39. https://doi.org/10.1038/s43586-021-00038-x
24. Gunzenhäuser J, Olivier N, Pengo T, Manley S (2012) Quantitative super-resolution imaging reveals protein stoichiometry and nanoscale morphology of assembling HIV-Gag virions. Nano Lett 12:4705–4710. https://doi.org/10.1021/nl3021076
25. Floderer C, Masson J-B, Boilley E et al (2018) Single molecule localisation microscopy reveals how HIV-1 Gag proteins sense membrane virus assembly sites in living host CD4 T cells. Sci Rep 8:16283. https://doi.org/10.1038/s41598-018-34536-y
26. Malkusch S, Muranyi W, Müller B et al (2013) Single-molecule coordinate-based analysis of the morphology of HIV-1 assembly sites with near-molecular spatial resolution. Histochem Cell Biol 139:173–179. https://doi.org/10.1007/s00418-012-1014-4
27. Muranyi W, Malkusch S, Müller B et al (2013) Super-resolution microscopy reveals specific recruitment of HIV-1 envelope proteins to viral assembly sites dependent on the envelope C-terminal tail. PLoS Pathog 9:e1003198. https://doi.org/10.1371/journal.ppat.1003198
28. Lehmann M, Rocha S, Mangeat B et al (2011) Quantitative multicolor super-resolution

microscopy reveals tetherin HIV-1 interaction. PLoS Pathog 7:e1002456. https://doi.org/10.1371/journal.ppat.1002456

29. Inamdar K, Tsai F-C, Dibsy R et al (2021) Full assembly of HIV-1 particles requires assistance of the membrane curvature factor IRSp53. Elife 10:e67321. https://doi.org/10.7554/eLife.67321
30. Dibsy R, Bremaud E, Mak J et al (2023) HIV-1 diverts cortical actin for particle assembly and release. Nat Commun 14:6945
31. Diekmann R, Kahnwald M, Schoenit A et al (2020) Optimizing imaging speed and excitation intensity for single-molecule localization microscopy. Nat Methods 17:909–912. https://doi.org/10.1038/s41592-020-0918-5
32. Sage D, Pham T-A, Babcock H et al (2019) Super-resolution fight club: assessment of 2D and 3D single-molecule localization microscopy software. Nat Methods 16:387–395. https://doi.org/10.1038/s41592-019-0364-4
33. Ovesný M, Křížek P, Borkovec J et al (2014) ThunderSTORM: a comprehensive ImageJ plug-in for PALM and STORM data analysis and super-resolution imaging. Bioinformatics 30:2389–2390. https://doi.org/10.1093/bioinformatics/btu202
34. Khater IM, Nabi IR, Hamarneh G (2020) A review of super-resolution single-molecule localization microscopy cluster analysis and quantification methods. Patterns 1:100038. https://doi.org/10.1016/j.patter.2020.100038
35. Levet F, Hosy E, Kechkar A et al (2015) SR-Tesseler: a method to segment and quantify localization-based super-resolution microscopy data. Nat Methods 12:1065–1071. https://doi.org/10.1038/nmeth.3579
36. Malkusch S, Heilemann M (2016) Extracting quantitative information from single-molecule super-resolution imaging data with LAMA—LocAlization Microscopy Analyzer. Sci Rep 6: 34486. https://doi.org/10.1038/srep34486
37. Manley S, Gillette JM, Patterson GH et al (2008) High-density mapping of single-molecule trajectories with photoactivated localization microscopy. Nat Methods 5:155–157. https://doi.org/10.1038/nmeth.1176
38. Saha I, Saffarian S (2020) Dynamics of the HIV Gag lattice detected by localization correlation analysis and time-lapse iPALM. Biophys J 119: 581–592. https://doi.org/10.1016/j.bpj.2020.06.023

Chapter 6

Single-Virion Analysis: A Method to Visualize HIV-1 Particle Content Using Fluorescence Microscopy

Alice Duchon, Ryan C. Burdick, Vinay K. Pathak, and Wei-Shau Hu

Abstract

HIV-1 virions incorporate viral RNA, cellular RNAs, and proteins during the assembly process. Some of these components, such as the viral RNA genome and viral proteins, are essential for viral replication, whereas others, such as host innate immune proteins, can inhibit virus replication. Therefore, analyzing the virion content is an integral part of studying HIV-1 replication. Traditionally, virion contents have been examined using biochemical assays, which can provide information on the presence or absence of the molecule of interest but not its distribution in the virion population. Here, we describe a method, single-virion analysis, that directly examines the presence of molecules of interest in individual viral particles using fluorescence microscopy. Thus, this method can detect both the presence and the distribution of molecules of interest in the virion population. Single-virion analysis was first developed to study HIV-1 RNA genome packaging. In this assay, HIV-1 unspliced RNA is labeled with a fluorescently tagged RNA-binding protein (protein A) and some of the Gag proteins are labeled with a different fluorescent protein (protein B). Using fluorescence microscopy, HIV-1 particles can be identified by the fluorescent protein B signal and the presence of unspliced HIV-1 RNA can be identified by the fluorescent protein A signal. Therefore, the proportions of particles that contain unspliced RNA can be determined by the fraction of Gag particles that also have a colocalized RNA signal. By tagging the molecule of interest with fluorescent proteins, single-virion analysis can be easily adapted to study the incorporation of other viral or host cell molecules into particles. Indeed, this method has been adapted to examine the proportion of HIV-1 particles that contain APOBEC3 proteins and the fraction of particles that contain a modified Gag protein. Therefore, single-virion analysis is a flexible method to study the nucleic acid and protein content of HIV-1 particles.

Key words Virus assembly, Gag, RNA genome packaging, HIV particle imaging, Fluorescence microscopy

1 Introduction

HIV-1 must generate infectious particles to infect new host cells. During the virus assembly process, HIV-1 incorporates viral and host components into virions, which include two copies of unspliced HIV-1 RNA as the viral genome. HIV-1 genomic RNA packaging and virus assembly are intricate and well-regulated processes relying on both host and viral components. The main viral

Vinayaka R. Prasad and Ganjam V. Kalpana (eds.), *HIV Protocols*, Methods in Molecular Biology, vol. 2807,
https://doi.org/10.1007/978-1-0716-3862-0_6,

driver of particle assembly and RNA genome packaging is the Gag polyprotein, which is composed of matrix (MA), capsid (CA), nucleocapsid (NC), two spacer peptides (SP1 and SP2) that flank NC, and p6 domains [1]. MA contains basic patches and a myristate moiety at the N-terminus that targets and anchors Gag to the plasma membrane. During assembly, CA and SP1 play multiple roles in oligomerization and have been implicated in facilitating RNA packaging. NC directly binds to HIV-1 unspliced RNA (hereafter referred to as HIV-1 RNA) through interactions with the packaging signal in the 5′ untranslated region; p6 enables virus particle release from the cell. In addition to serving as the translation template to generate Gag/Gag-Pol polyproteins, which contain viral structural and enzymatic proteins, HIV-1 RNA is packaged as a dimer into virus particles and functions as the virion genome [1].

Currently, the process by which HIV-1 RNA is selected from a vast excess of cellular RNA by Gag is not well understood. Most studies have used biochemical assays to measure the amount of viral RNA in particles and compared them to viral RNA in wild-type viruses. However, it was difficult to use biochemical assays to determine the proportion of wild-type particles that contain viral RNAs. To address this issue and to better understand HIV-1 RNA packaging, we developed single-virion analysis [2]. The schematic of single-virion analysis is shown in Fig. 1. Briefly, viral particles are visualized by fusing cerulean fluorescent protein (CeFP) to the C-terminus of HIV-1 Gag (Gag-CeFP) (Fig. 1a). When Gag-CeFP is expressed at a 1:1 ratio with untagged Gag, efficient assembly of morphologically normal virions is retained [3, 4]. To examine HIV-1 genomic RNA packaging, 18 copies of an RNA stem loop specifically recognized by BglG (BSL) are inserted in *pol*, downstream of the Gag open reading frame [2]. This ensures that only full-length unspliced HIV-1 RNA contains BSL. Expression of a fragment of the BglG protein fused to a yellow fluorescent protein containing a nuclear localization signal (NLS) (Bgl-YFP) enables tagging of the BSL in the unspliced HIV-1 RNA. Therefore, when plasmids expressing Gag and Gag-CeFP are cotransfected into cells with a plasmid expressing Bgl-YFP, the resulting virus particles are labeled with CeFP, and viral RNA-containing particles are labeled with both the CeFP and YFP. Using this assay, we have shown that HIV-1 packages its RNA genome very efficiently, as >90% of CeFP virus particles were found to contain fluorescently tagged HIV-1 RNA [2]. Co-localization and fluorescence intensity of packaged fluorophores are determined using two custom MATLAB programs [2]. These experiments required the examination of RNA packaged into thousands of individual virus particles and could not have been carried out with standard biochemical measurements.

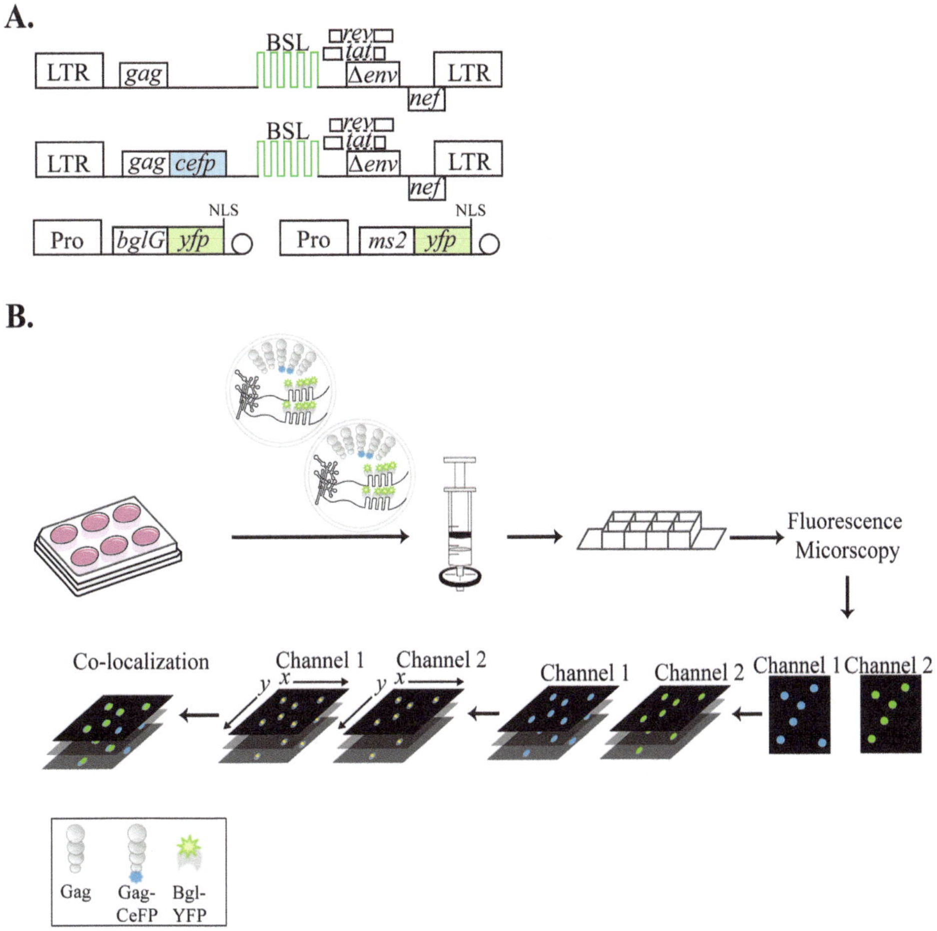

Fig. 1 Schematic of single-virion analysis. (**a**) HIV-1 constructs expressing Gag and Gag-CeFP contained 18 copies of BSL in the *pol* gene. BglG-YFP and MS2-YFP fusion proteins were expressed from separate plasmids and contained nuclear localization signals (NLS) at the C-terminus. Pro, promoter; the open circle indicates polyadenylation signal. CeFP and YFP are depicted with blue and green coloring, respectively. (**b**) Outline of the experimental procedure. Following the transfection of 293T cells in 6-well plates, media containing virus particles were collected, filtered, and plated onto chamber slides. Images of virus particles were captured via fluorescent microscopy and analyzed by a custom MATLAB program. Briefly, each image is split into individual channels. This process was repeated for each image and combined to create stacks of individual channels. These were used to determine spatial coordinates for each identified particle (shown in yellow) via the Localize application and subsequently utilized for co-localization and particle analysis purposes. Ledged indicates labeled viral components

Single-virion analysis can be easily modified to address a variety of experimental questions. By varying the *cis*- and *trans*-acting elements, single-virion analysis was used to define 5′ untranslated region sequences and Gag properties important for genome packaging [5, 6]. Additionally, we have modified the system to detect two HIV-1 RNAs derived from different constructs. Using this approach, we have shown that the complementarity of the dimerization initiation signals dictates the efficiency of RNA copackaging; therefore, RNA dimerization occurs prior to genome packaging [7, 8]. Furthermore, most HIV-1 particles contain two copies of RNAs. In addition to RNA packaging, single-virion analysis can also be modified to detect the incorporation of proteins. For example, by labeling host protein apolipoprotein B mRNA-editing enzyme catalytic subunits (APOBEC) 3G and 3F (A3G and A3F, respectively), we have shown that these two host innate immune proteins can be copackaged into particles [9]. As these examples illustrate, additional variations can be employed to further expand the system and gain insights into virion content.

2 Materials

2.1 Plasmids

1. Gag-BSL [2].
2. Gag-CeFP-BSL [2].
3. Bgl-YFP [10].
4. MS2-YFP [2].

2.2 Cell Culture and Transfection Reagents

1. 293T cells (HIV reagent program).
2. Dulbecco's modified Eagle's medium (DMEM) supplemented with 10% fetal bovine serum (FBS) and 1% penicillin/streptomycin (P/S) (DMEM++).
3. Trypsin (0.25%) and ethylenediaminetetraacetic acid (EDTA) (1 mM).
4. Dulbecco's phosphate-buffered saline (DPBS).
5. Opti-MEM™ (Gibco).
6. FuGENE®HD Transfection Reagent (Promega).
7. 6-well tissue culture plates.

2.3 SDS-PAGE and Western Blotting

1. Screw top microcentrifuge tubes.
2. SDS-polyacrylamide gels (Bio-Rad).
3. Transfer materials: Trans-Blot Turbo Transfer System, blotting paper, and PVDF membrane (Bio-Rad).
4. Buffers. 10× SDS-PAGE running buffer (Bio-Rad), 5× transfer buffer (Bio-Rad), 10×-PBS (Bio-Rad), Tween-20, and Li-Cor Blocking Buffer (Li-Cor).
5. Ethanol.
6. 4× Laemmli sample buffer.
7. Chameleon® Duo Pre-Stained Protein Ladder (Li-Cor).
8. 20% sucrose in DPBS.

9. Cell lytic M (Sigma-Aldrich) + protease inhibitor complex (Roche).
10. Antibodies mouse anti-HIVp24 antibody (AIDS reagent program Cat No. 6458), rabbit anti-GFP antibody (Invitrogen Cat No. A6455), mouse anti-HSP90 antibody (Santacruz Cat No. SC-69703), goat anti-mouse IRDye 800 (Li-Cor Cat No. 925-32210), goat anti-rabbit IRDye 780 (Li-Cor Cat No. 926-68071).
11. Imaging System (Li-Cor Odyssey M).

2.4 Virus Collection and Preparation for Fluorescence Imaging

1. 5-mL syringes.
2. Millex-HV 0.45 uM PVDF-sterilized syringe filters.
3. ibiTreated 8-well μ-slides (ibidi).
4. Polybrene.
5. 1.7-mL microcentrifuge tubes.
6. Centrifuge with swinging plate holders.

2.5 Image Collection

1. Nikon Eclipse TE 2000 microscope.
2. SOLA light engine® light source (Lumencor).
3. 100× 1.49 numerical aperture oil objective (Nikon).
4. NIS-Elements software (Nikon).

2.6 Single-Virion Analysis Via MATLAB and Localize

1. MATLAB(MathWorks).
2. Localize Application [11].
3. Excel (Microsoft) and GraphPad Prism 9.

3 Methods

3.1 Preparation of Virus (Fig. 1)

1. FuGENEHD Transfection. The condition of the 293T cells is critical to the production of quality virus particles and reproducible data. Therefore, 293T cells should be split at a high ratio approximately 1 day prior to plating cells for single-virion analysis experiments. For transfection, 293T cells are plated at 3.3×10^5 cells per well of a 6-well plate in 2 mL of DMEM++.
2. Twenty-four hours later, cells are transfected. Transfection should be carried out according to FuGENEHD manufacturer's protocol at a 1:4 ratio of DNA to FuGENEHD reagent diluted in Opti-MEM™.
3. Cells are then incubated at 37 °C in 5% CO_2 for 24 h. FuGENEHD transfection does not require the replacement of media.
4. One well should contain non-transfected cells to act as a Western blot control and to use for determining a threshold for

particle detection. A typical transfection for single-virion analysis uses a combined total of 1.8 μg of DNA including Gag-BSL (550 ng), Gag-CeFP-BSL (500 ng), and Bgl-YFP (750 ng), or MS2-YFP (750 ng) (*see* **Notes 1–3**). Bgl-YFP and MS2-YFP contain nuclear localization signals at their C-terminus (*see* **Note 4**). Untagged ("dark") Gag must account for at least 50% of the total transfected Gag to maintain normal virus particle assembly and morphology (*see* **Note 5**). As shown in Fig. 1a, BSL is located downstream from the *gag* gene, in the *pol* gene, to ensure that only unspliced HIV-1 RNA is labeled. The coat protein of RNA bacteriophage MS2 fused to YFP is used as a negative control in this experiment. MS2 coat protein recognizes MS2 stem loops (MSL), which are not present in the RNA expressed by Gag-BSL and Gag--CeFP-BSL. Therefore, co-transfection of MS2-YFP with Gag-BSL and Gag-CeFP-BSL will produce virus particles labeled with only CeFP from Gag incorporation and the HIV-1 RNA will not be labeled. In addition to serving as a negative control, virus particles generated from MS2-YFP cotransfection can also be used to set a threshold for image analysis of the YFP channel as described below.

5. Harvesting Virus from Supernatant. Twenty-four hours after transfection, cells should produce sufficient levels of virus particles within the supernatant. Longer incubation times are not recommended as they tend to increase the presence of cell debris and background signal. Supernatant from transfected and non-transfected samples is harvested using a 5-mL syringe and passed through a 0.45-μm syringe filter to remove cellular debris (Fig. 1b).
6. Viral supernatants can be divided into microcentrifuge tubes for storage at −80 °C or directly used for imaging and Western blot analysis. Once frozen, each sample should be thawed only once to maintain the integrity of the virus particles.

3.2 Western Blotting of Cell and Virus Lysates

It is advisable to confirm the expression of each Gag construct in cells and whether the protein is assembled into virions when setting up or modifying the system.

3.2.1 Preparation of Cell Samples

1. Following the removal of supernatant/virus particles, cells are washed with 500 μL DPBS followed by direct lysis in wells with 250 μL of Cell Lytic M + PIC.
2. Lysates collected in Eppendorf tubes are clarified by centrifugation at maximum speed for 5 min and stored at −80 °C until ready to use for Western blot analysis.

3.2.2 Preparation of Virus Samples

1 mL of filtered supernatant is loaded carefully over a 100 μL 20% sucrose/DPBS cushion in screw-top microcentrifuge tubes. Virus particles are concentrated by centrifugation at 15,000 rpm for 2 h at 4 °C. The supernatant is removed, and virus particles are lysed in ~80 μL of Cell Lytic M. Samples can be stored at −80 °C until use.

3.2.3 SDS-PAGE and Electrotransfer

10 μg of cell or virus lysate is heated at 99 °C for 5 min with the appropriate volume of 1× protein loading dye and loaded onto an SDS-PAGE gel. This is run in 1× SDS-PAGE running buffer at 200 V for 50 min at room temperature, then transferred with 1× transfer buffer to ethanol-charged PVDF membrane for 10 min at 25 V using the Bio-Rad Trans-Blot Turbo Transfer System, and subsequent membrane is incubated with Li-Cor blocking buffer at RT for 1 h. All 1× buffers are made according to the manufacturer's instructions.

3.2.4 Immunoblotting

The membrane is incubated sequentially with mouse anti-p24, rabbit anti-GFP, and rabbit anti-HSP90 antibodies at 1:5000, 1: 5000, and 1:10,000, respectively, in Li-Cor blocking buffer + 2% Tween for 2 h at room temperature followed by three 5-min washes with PBS+ 1% Tween. Secondary anti-rabbit and anti-mouse fluorescent antibodies are incubated at 1:10,000 for an additional hour, washed as mentioned above, and imaged on a Li-Cor Odyssey M. Analysis of cell and virus lysate from co-transfections with Gag-BSL, Gag-CeFP-BSL, and Bgl-YFP or MS2-YFP indicate equal expression of Gag and RNA-labeling constructs in both samples (Fig. 2a). No detectable Gag is present in non-transfected cell lysate or supernatant. As expected, only Bgl-YFP is present in virus particles as the fusion protein is bound to BSL-containing HIV-1 RNA (Fig. 2a).

3.3 Preparation of Chamber Slides for Microscopy

The ibidi 8-chamber slides have a maximum volume of 300 μL. Therefore, 5–10 μL of filtered virus particles are diluted in 189–194 μL of DPBS and 1 μL of polybrene (final concentration 50 μg/mL) (*see* **Note 6**). Slides are secured to swinging plate-size inserts and centrifuged at 1200 × *g* for 1 h at 18 °C. Alternatively, slides can be placed at 4 °C overnight to allow virus particles to settle if centrifugation of the slides is not possible.

3.4 Fluorescence Microscopy of Single Virions

Epifluorescence microscopy images are captured using an inverted Nikon Eclipse TE 2000 microscope and a 100× 1.49 numerical aperture oil objective, with a SOLA light engine® light source for illumination. Perfect focus is used to minimize changes in the focal plane. Digital images are acquired using a Hamamatsu ORCA-ERA camera and NIS-Elements software with the excitation and emission filter sets 427/10 nm and 472/30 nm for CeFP and 504/12 nm and 542/27 nm for YFP. Comparable imaging systems can be used. However, appropriate controls must be used to

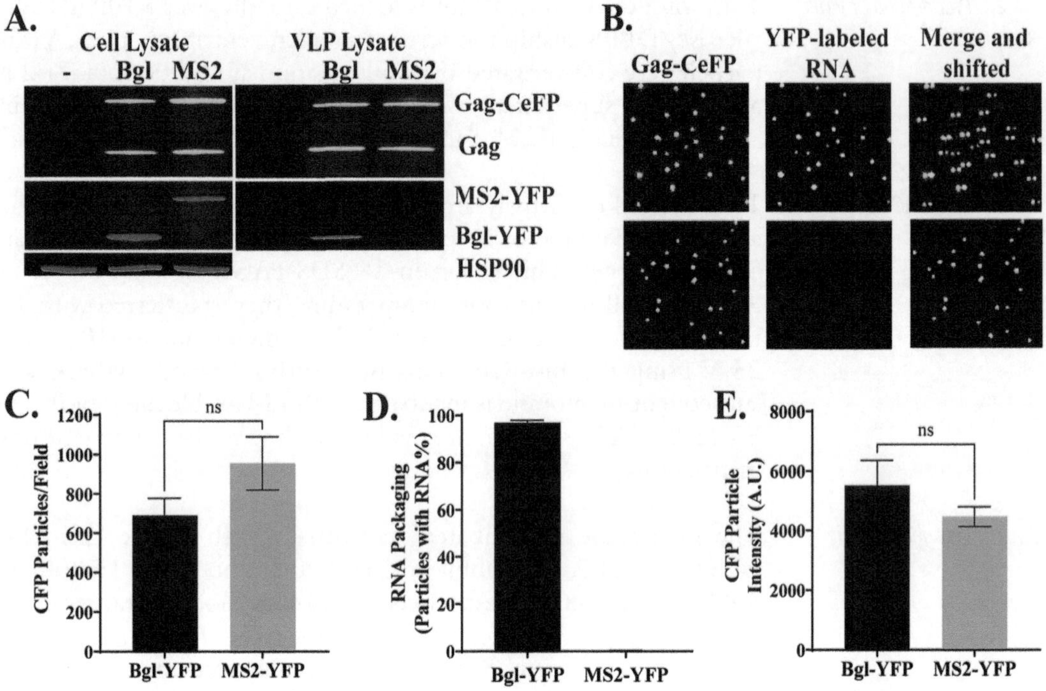

Fig. 2 Representative data from a single-virion analysis experiment. (**a**) Western blot analysis of cell (left panels) and virus lysate (right panels) from untransfected cells (−) or cells co-transfected with Gag-expressing constructs and Bgl-YFP or MS2-YFP. Gag and Gag-CeFP were probed with anti-p24 antibodies and are indicated as bands at ~55 kDa and 83 kDa, respectively. Bgl-YFP and MS2-YFP fusion proteins were detected with anti-GFP antibodies and correspond to bands at ~34 kDa and 43 kDa, respectively. HSP90 was used as a loading control for the cell lysate. (**b**) Representative fluorescence microscopy images of virus particles produced from cells co-transfected with Gag-BSL, Gag-CeFP-BSL, and Bgl-YFP (Top) or MS2-YFP (Bottom). Individual CeFP (Column 1) and YFP (Column 2) channels are shown; to illustrate co-localization, an overlayed image that has been shifted by ten pixels is shown. (**c**) Quantification of the number of CeFP particles per field, (**d**) Proportion of CeFP particles containing HIV-1 RNA with YFP signal, and (**e**) CeFP particle intensity were determined by MATLAB and Localize. An unpaired *t*-test with Welch's correction was used to determine whether samples exhibit significant differences; ns, no significant difference

determine whether a system is compatible with single-virion analysis. For example, the use of positive control virus particles compared to buffer-only samples to verify the sensitivity of the microscope. To ensure that an accurate representation of the virus particle population is imaged, a minimum of 1000 virus particles from several locations within each well are analyzed per experimental condition. It may be necessary to use various concentrations of virus particles to ensure the particles in a field of view are evenly distributed and not too close together. The optimum number of particles per field will depend on the microscope setup and camera and should be determined empirically. In the example shown (Fig. 2b), virus particles produced from cells co-transfected with Gag-BSL, Gag-CeFP-BSL, and Bgl-YFP produce particles labeled with both CeFP and YFP, whereas virus particles from MS2-YFP co-transfection are labeled with only CeFP.

3.5 MATLAB Analysis of Images

The custom MATLAB program used to analyze single-virion images is designed for Nikon image files (.nd2). The MATLAB program will need to be modified if utilizing alternative image file formats. The previously described Localize application [11] is used for the identification of individual fluorescent signals, whereas the MATLAB program streamlines the image analysis for multiple images containing up to three channels. Both programs are available upon request.

3.5.1 Determining RNA and Protein Packaging

1. A custom MATLAB program, Particle_colocalization, was created to determine the co-localization of diffraction-limited fluorescent particles (e.g., virus particles). The program requires a single input, termed "pixelDistance," which represents the maximum pixel distance between two signals to be considered co-localized. We typically use a maximum distance of three pixels, which corresponds to ~0.39 μm based on our microscope setup. The program will prompt the user to select a folder containing the .nd2 files. Following selection, the program will automatically determine the number of channels and convert the .nd2 files to Tagged Image File (TIF) stacks for each channel. For example, two-color images will result in the creation of two TIF files corresponding to the two channels. The program will prompt the user to analyze the images using Localize.
2. To identify the diffraction-limited fluorescent particles using Localize, each TIF stack must be converted to an IPL file. The IPL file for each channel can then be opened and an appropriate threshold for particle identification can be determined. This threshold should be set based on a sample from non-transfected cells where no fluorescent particles are expected. Alternatively, virus particles containing only one fluorescent label, such as the MS2-YFP sample with virus particles containing only CeFP label, can be used to determine an appropriate threshold for the YFP channel. Once the appropriate threshold for each channel has been determined, this threshold should be used for all replicates. This is followed by the localization of particles resulting in the creation of a .loc file containing the coordinates for each identified particle for each image in the image stack.
3. After all IPL files have been analyzed, the subsequent .loc files can be imported individually into MATLAB following the prompts from the Colocalize program. Two outputs, termed "cellsDataSummary" and "cellsData," are generated upon completion of the MATLAB program.

4. The summary sheet contains the co-localization data between the two channels for each image. Columns 1 and 2 contain the number of particles that were only detected in channel 1 or 2, respectively, whereas column 3 contains the number of particles that contained both signals (channels 1 and 2). Rows represent individual images. Raw data including the spatial information regarding individual virus particles are contained within "cellsData."
5. The co-localization summary can be exported to Excel for further analysis. In the example shown (Fig. 2c), total CeFP particles per field were similar between cells co-transfected with Gag constructs and Bgl-YFP or MS2-YFP. Here, the RNA packaging efficiency was calculated in Excel and plotted in GraphPad by determining the percentage of CeFP virus particles that contained YFP-labeled RNA divided by the total number of CeFP particles (Fig. 2d) for each set of images in a replicate.

3.5.2 Determining Fluorescence Intensity of Particle Components

1. Following the determination of particle co-localization another MATLAB program can be implemented to combine the particle fluorescence intensity data, which were also analyzed using Localize and output to "cellsData," from multiple images for the same sample. This is accomplished using a second custom MATLAB program: Particle_intensities. This program requires three input variables.
2. First, the user must create a new variable called "sampleRows" and indicate the range of images taken for a single sample in a two-column format where the first column is the starting image for a sample, and column 2 is the final. The number of rows should match the sample number. For example, an experiment with three samples and ten images for each sample would be listed as 2–11, 12–21, and 22–31 in columns 1 and 2 of rows 1–3, respectively.
3. The second input is the "cellsData" output from the Particle_colocalization analysis.
4. The third input is the number of channels that were analyzed (1, 2, or 3 for single, dual-, and three-color images, respectively).

3.5.3 An Output Datasheet Will Be Created

1. The sample group will be listed in the first column and the intensity data listed in the second. For each sample, all of the particle intensities of each particle are contained within the cells in the second column. The number of columns in these cells will be determined by the number of channels analyzed, and the particle intensities will be separated into columns corresponding to their co-localization. For a two-color image

with CeFP (channel 1) and YFP (channel 2): Columns 1 and 2 contain the individual intensity of particles that are not co-localized with each other, i.e., CeFP only or YFP only, respectively. Column 3 contains the CeFP intensity of particles that are co-localized, i.e., CeFP and YFP positive, whereas the fourth column is the YFP intensity corresponding to those co-localized particles. Several parameters can be examined from this data. For example, the median CeFP intensity of all particles (Fig. 2e) or just those containing RNA.

2. We have previously determined by Western blot and ELISA that the median fluorescence intensity corresponding to tagged Gag correlates directly with the amount of protein packaged. Therefore, this is a useful tool to examine relative protein incorporation.

4 Notes

One great advantage of the single-virion analysis is its flexibility. This assay can easily be adapted to address a multitude of experimental questions by exchanging or combining different fluorescently tagged proteins, either viral or host. Additional unique stem loops can be added into the DNA constructs to tag different RNAs and examine the packaging of different RNAs simultaneously or alter HIV-1 RNA packaging through noncanonical RNA:protein interactions. However, modification of the system does require optimization as described below.

1. RNA Packaging Mutants. The ability of mutant RNA to be packaged into particles can be easily carried out by altering the Gag- and Gag-CeFP-encoding constructs. These studies can also be extended to other viruses such as HIV-2 wherein the role of individual guanosine residues in the 5′UTR of HIV-2 RNA has been examined [5]. When performing mutational analysis of HIV-1 RNA, it is necessary to make matching mutations within all HIV-1 RNA-expressing constructs. It is interesting to note that overexpression of Gag can potentially mask RNA packaging defects. Therefore, Gag expression must be considered to ensure that the experimental system is not saturated. This could be mitigated by limiting the total Gag-expressing DNA transfected into cells.
2. HIV-1 Viral Protein Mutants. Although Gag is the most abundantly packaged viral protein, alternative viral protein targets amenable to fluorescent labeling include an accessory protein, Vpr, and Vpr fusions with integrase, as well as specific domains of Gag such as MA and CA [12–14]. To examine the specific domains of Gag essential for efficient HIV genome packaging,

we adapted the single-virion analysis system and separated the particle-assembly and RNA-packaging properties of Gag. We replaced NC with a leucine zipper motif (Fig. 3b); the resulting mutant Gag polyprotein can assemble into particles that do not contain viral RNAs. We then introduced a Gag-BglG fusion protein that packaged RNA through BglG:BSL interactions. Using this noncanonical packaging system, we were able to assess the RNA packaging ability of Gag mutants that typically could not assemble into particles and removed the complication of the mutation disrupting the interactions between the NC domain and packaging signal [15].

3. Nonviral Protein Packaging. It is important to consider the number of targeted proteins that are packaged into a virus particle and the limit of fluorescence detection of the microscope when selecting a protein to tag fluorescently. HIV-1 Gag is a particularly attractive target as each viral particle contains thousands of Gag polyproteins [16]. However, several nonviral proteins are known to be incorporated into HIV-1 virions including host restriction factors A3G and A3F. It is possible to fluorescently tag APOBEC3 proteins in a similar manner to Gag and study the packaging of various APOBEC3 proteins into virus particles (Fig. 3c) [12, 17]. Previous studies have utilized a small peptide, S15-mCherry, to label the surface of virus particles [18]. This peptide could be implemented to label virus particles as well.

4. Customization of Fluorophore and Stem–Loop Combinations. Typically, our approach utilizes Gag that is C-terminally tagged with CeFP. Alternative placements of fluorophores, such as between MA and CA, and substitution of CeFP with other fluorescent proteins have been done successfully [12, 15]. We illustrate the single-virion analysis system here using the BglG protein recognition of BSL. RNA labeling systems based on bacteriophage MS2 and PP7 coat proteins and their corresponding RNA stem loops have been well-established [19, 20], and we have successfully used these systems as well [2, 15]. The existing multiple pairs of RNA-binding protein and stem–loop sequences provide flexibility in the system, and it is possible to tag a single RNA with two labels, or two distinct RNAs with different fluorophores (Fig. 3a) [2, 21]. It is important to note that introducing many sets of stem loops may impact gene expression and virus particle production.

5. Optimization of Plasmid Ratios. Minor adjustments in ratios of plasmids may be necessary for individual experimental conditions, but it is recommended that the amount of unlabeled Gag has to be at least 50% of the total Gag (labeled and unlabeled). This allows the viral assembly to occur normally and facilitates the production of particles with morphology similar to that of

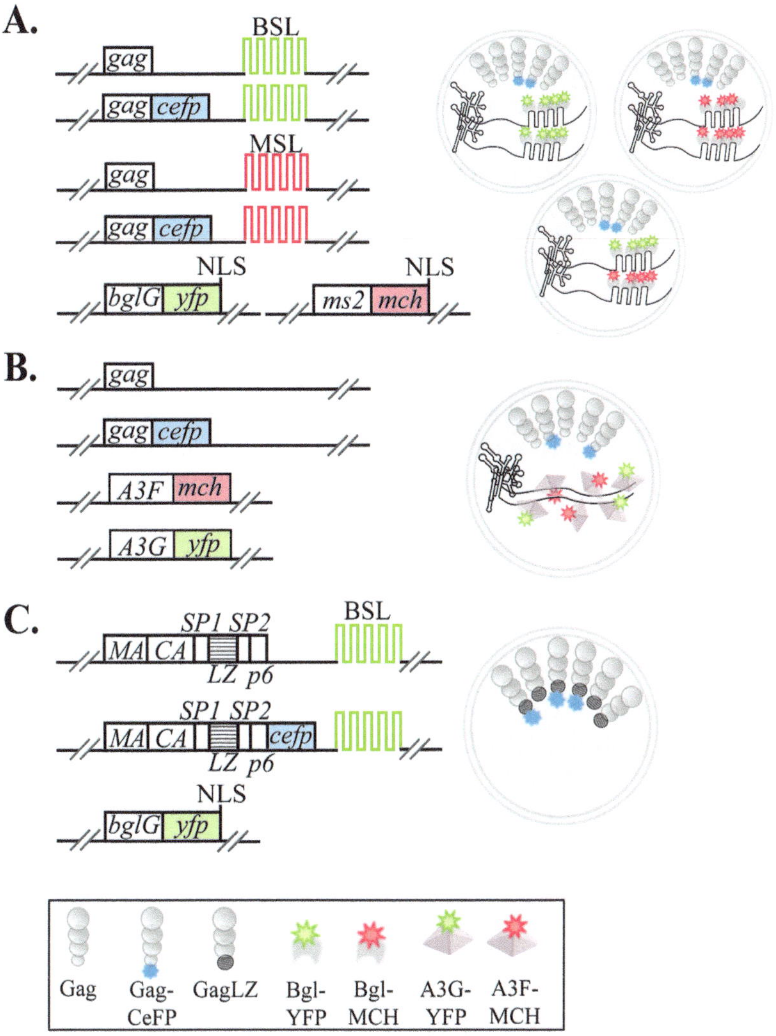

Fig. 3 Extensions of the single-virion analysis system. (**a**) Utilization of two different stem loops with distinct fluorescent fusion proteins allows for the detection of different RNA species within packaged particles. (**b**) Detection of host A3G and A3F proteins in HIV-1 particles. (**c**) Replacement of the NC domain of Gag with a LZ motif results in particles lacking labeled RNA. CeFP, YFP, and mCherry (MCH) fusion proteins are shown as blue, green, and pink, respectively. Stem loops corresponding to Bgl and MS2 coat proteins are indicated by green and pink structures in the constructs shown on the left. Expected virus particles are depicted on the right, and the legend at the bottom indicates viral components

wild-type immature viruses [3, 4]. Additionally, the fluorescent protein-tagged RNA-binding protein must be expressed at a high level to efficiently label RNA. It is critical to ensure that positive control samples do not have significant proportions of particles lacking HIV-1 RNA signal (CeFP only particles) or HIV-1 Gag signal (YFP only particles). For example, if expressing Gag and HIV-1 RNA from independent promoters with vastly different expression levels, it is possible to saturate the system with Gag, which will assemble into virus particles even without viral RNA present, leading to the appearance of HIV-1 RNA lacking particles. Alternatively, if a fluorescently tagged Gag is not efficiently expressed and incorporated into particles, this can result in particles with only HIV-1 RNA signals when the fluorescently tagged Gag is below the limit of detection. Therefore, it is critical to carefully select and optimize any system with a wild-type control prior to experimentation with mutants.

6. Imaging of Samples with Low Virus Production. The volume of virus particles centrifuged onto ibidi slides can be increased to allow for the examination of samples with low numbers of particles. Alternatively, virus particles can be concentrated by ultracentrifugation prior to slide preparation. Alternative slides can be used to accommodate larger sample volumes, but the amount of virus used for visualization will need to be optimized.

Acknowledgements

This work is supported by the Intramural Research Program, National Cancer Institute, National Institutes of Health (NIH); Innovation funds from the Office of AIDS Research, NIH to WSH and to VKP; and the Supplemental funds from the Center for Cancer Research to WSH and to VKP.

References

1. Freed EO (2015) HIV-1 assembly, release and maturation. Nat Rev Microbiol 13(8): 484–496. https://doi.org/10.1038/nrmicro3490
2. Chen J, Nikolaitchik O, Singh J, Wright A, Bencsics CE, Coffin JM, Ni N, Lockett S, Pathak VK, Hu WS (2009) High efficiency of HIV-1 genomic RNA packaging and heterozygote formation revealed by single virion analysis. Proc Natl Acad Sci USA 106(32): 13535–13540. https://doi.org/10.1073/pnas.0906822106
3. Larson DR, Johnson MC, Webb WW, Vogt VM (2005) Visualization of retrovirus budding with correlated light and electron microscopy. Proc Natl Acad Sci USA 102(43): 15453–15458. https://doi.org/10.1073/pnas.0504812102
4. Jouvenet N, Bieniasz PD, Simon SM (2008) Imaging the biogenesis of individual HIV-1 virions in live cells. Nature 454(7201): 236–240. https://doi.org/10.1038/nature06998
5. Umunnakwe CN, Duchon A, Nikolaitchik OA, Rahman SA, Liu Y, Chen J, Tai S, Pathak VK, Hu WS (2021) Specific guanosines in the HIV-2 leader RNA are essential for efficient viral genome packaging. J Mol Biol 433(2): 166718. https://doi.org/10.1016/j.jmb.2020.11.017
6. Liu Y, Nikolaitchik OA, Rahman SA, Chen J, Pathak VK, Hu WS (2017) HIV-1 sequence necessary and sufficient to package non-viral RNAs into HIV-1 particles. J Mol Biol 429(16):2542–2555. https://doi.org/10.1016/j.jmb.2017.06.018
7. Moore MD, Nikolaitchik OA, Chen J, Hammarskjold ML, Rekosh D, Hu WS (2009) Probing the HIV-1 genomic RNA trafficking pathway and dimerization by genetic recombination and single virion analyses. PLoS Pathog 5(10):e1000627. https://doi.org/10.1371/journal.ppat.1000627

8. Dilley KA, Nikolaitchik OA, Galli A, Burdick RC, Levine L, Li K, Rein A, Pathak VK, Hu WS (2017) Interactions between HIV-1 Gag and viral RNA genome enhance virion assembly. J Virol 91(16):e02319-16. https://doi.org/10.1128/JVI.02319-16
9. Desimmie BA, Burdick RC, Izumi T, Doi H, Shao W, Alvord WG, Sato K, Koyanagi Y, Jones S, Wilson E, Hill S, Maldarelli F, Hu WS, Pathak VK (2016) APOBEC3 proteins can copackage and comutate HIV-1 genomes. Nucleic Acids Res 44(16):7848–7865. https://doi.org/10.1093/nar/gkw653
10. Chen J, Grunwald D, Sardo L, Galli A, Plisov S, Nikolaitchik OA, Chen D, Lockett S, Larson DR, Pathak VK, Hu WS (2014) Cytoplasmic HIV-1 RNA is mainly transported by diffusion in the presence or absence of Gag protein. Proc Natl Acad Sci USA 111(48):E5205–E5213. https://doi.org/10.1073/pnas.1413169111
11. Zenklusen D, Larson DR, Singer RH (2008) Single-RNA counting reveals alternative modes of gene expression in yeast. Nat Struct Mol Biol 15(12):1263–1271. https://doi.org/10.1038/nsmb.1514
12. Li C, Burdick RC, Nagashima K, Hu WS, Pathak VK (2021) HIV-1 cores retain their integrity until minutes before uncoating in the nucleus. Proc Natl Acad Sci USA 118(10):e2019467118. https://doi.org/10.1073/pnas.2019467118
13. McDonald D, Vodicka MA, Lucero G, Svitkina TM, Borisy GG, Emerman M, Hope TJ (2002) Visualization of the intracellular behavior of HIV in living cells. J Cell Biol 159(3):441–452. https://doi.org/10.1083/jcb.200203150
14. Burdick RC, Delviks-Frankenberry KA, Chen J, Janaka SK, Sastri J, Hu WS, Pathak VK (2017) Dynamics and regulation of nuclear import and nuclear movements of HIV-1 complexes. PLoS Pathog 13(8):e1006570. https://doi.org/10.1371/journal.ppat.1006570
15. Duchon A, Santos S, Chen J, Brown M, Nikolaitchik OA, Tai S, Chao JA, Freed EO, Pathak VK, Hu WS (2021) Plasma membrane anchoring and Gag:Gag multimerization on viral RNA are critical properties of HIV-1 gag required to mediate efficient genome packaging. mBio 12(6):e0325421. https://doi.org/10.1128/mbio.03254-21
16. Carlson LA, Briggs JA, Glass B, Riches JD, Simon MN, Johnson MC, Muller B, Grunewald K, Krausslich HG (2008) Three-dimensional analysis of budding sites and released virus suggests a revised model for HIV-1 morphogenesis. Cell Host Microbe 4(6):592–599. https://doi.org/10.1016/j.chom.2008.10.013
17. Burdick RC, Hu WS, Pathak VK (2013) Nuclear import of APOBEC3F-labeled HIV-1 preintegration complexes. Proc Natl Acad Sci USA 110(49):E4780–E4789. https://doi.org/10.1073/pnas.1315996110
18. Campbell EM, Perez O, Melar M, Hope TJ (2007) Labeling HIV-1 virions with two fluorescent proteins allows identification of virions that have productively entered the target cell. Virology 360(2):286–293. https://doi.org/10.1016/j.virol.2006.10.025
19. Bertrand E, Chartrand P, Schaefer M, Shenoy SM, Singer RH, Long RM (1998) Localization of ASH1 mRNA particles in living yeast. Mol Cell 2(4):437–445. https://doi.org/10.1016/s1097-2765(00)80143-4
20. Chao JA, Patskovsky Y, Almo SC, Singer RH (2008) Structural basis for the coevolution of a viral RNA-protein complex. Nat Struct Mol Biol 15(1):103–105. https://doi.org/10.1038/nsmb1327
21. Dilley KA, Ni N, Nikolaitchik OA, Chen J, Galli A, Hu WS (2011) Determining the frequency and mechanisms of HIV-1 and HIV-2 RNA copackaging by single-virion analysis. J Virol 85(20):10499–10508. https://doi.org/10.1128/JVI.05147-11

Chapter 7

Correlative Imaging to Detect Rare HIV Reservoirs and Associated Damage in Tissues

Silvana Valdebenito, David Ajasin, Brendan Prideaux, and Eliseo A. Eugenin

Abstract

Correlative light-electron microscopy (CLEM) has evolved in the last decades, especially after significant developments in sample preparation, imaging acquisition, software, spatial resolution, and equipment, including confocal, live-cell, super-resolution, and electron microscopy (scanning, transmission, focused ion beam, and cryo-electron microscopy). However, the recent evolution of different laser-related techniques, such as mass spectrometry imaging (MSI) and laser capture microdissection, could further expand spatial imaging capabilities into high-resolution OMIC approaches such as proteomic, lipidomics, small molecule, and drug discovery. Here, we will describe a protocol to integrate the detection of rare viral reservoirs with imaging mass spectrometry.

Key words Confocal, Electron microscopy, Mass spectrometry imaging, Colocalization, AIDS, HIV-1, Cure, Eradication

1 Introduction

In recent years, correlative light-electron microscopy (CLEM) and other imaging technologies have evolved into multiple combinations among light microscopy, confocal, super-resolution, and electron microscopy [1–6]. These developments have resulted in significant advances in spatial biology and marked the beginning of spatial OMIC studies.

Currently, in the HIV field, identification of viral reservoirs is limited to culture-based assays and reactivation reagents, including quantitative viral outgrowth assays (QVOA), MOLT4/CCR5 viral outgrowth assay (VQA), and quantitative PCR such as droplet digital PCR and additional technologies such as TILDA, iCARED, and Simoa [7–16]. Also, several companies and groups have developed experimental and commercial products based on microscopy techniques to detect DNA and RNA probes with outstanding

Vinayaka R. Prasad and Ganjam V. Kalpana (eds.), *HIV Protocols*, Methods in Molecular Biology, vol. 2807, https://doi.org/10.1007/978-1-0716-3862-0_7,

resolution and precision. However, tissue or cell-related multiplex technologies are extremely limited due to sensitivity, specificity, and resolution (e.g., Akoya Biosciences or Nanostring). Only recently, there have been several reports describing the evolution of microscopy techniques to detect and quantify rare viral reservoirs, residual viral replication, and associated tissue damage in people living with HIV (PLWH), even under the current ART era [17–24]. These approaches have been applied to detect mucosal HIV reservoirs and subsequently use the microscopy information or coordinates to perform laser capture microdissection further to characterize their proteomic/small molecule/lipid profile as well as the presence of pathogens in the same tissues at the same coordinates [25–28].

The future of correlative microscopy points toward a world of automation and expansion into emerging technologies such as MSI and targeted isolation of labeled cells or structures. The advances in these technologies will result in microscopy and multi-OMICs moving together. The correlative platform is challenging but achievable; however, the future of correlative microscopy and OMICs will result in large amounts of spatial data that require specific algorithms for acquisition and analysis, even for rare cell types such as viral reservoirs.

The protocol described below includes sample preparation for confocal to visualized FISH and immunofluorescence (HIV DNA FISH, HIV mRNA FISH, and immunostaining for viral proteins and cellular markers in this order) in correlation with mass spectrometry imaging (MSI) for further OMIC approaches. Here, we need to emphasize the importance of choosing the proper protocol for sample preparation that ensures either the best ultrastructural maintenance or the preservation of samples' antigenicity or probe binding, as well as appropriate standard correlative techniques.

2 Materials

The tissues used in these experiments must be unfixed, fresh, or frozen for lipid detection. There are two potential protocols to address the experiments described. First, serial sections to perform fluorescence or non-fluorescence-based correlative microscopy and MSI, and second, MSI first and then microscopy analysis using the same tissue slice (Fig. 1a, b), respectively (*see* **Note 2**). We will focus on the former of these here as this approach allows more data to be obtained than the latter.

2.1 Buffers and Solutions

1. PNA, BNA, Stellaris, or RNAscope probes.
2. PNA ISH detection kit (Dako, K5201).
3. 0.5 M EDTA, pH 8.0.
4. Fish gelatin 45% (Sigma, G7765).

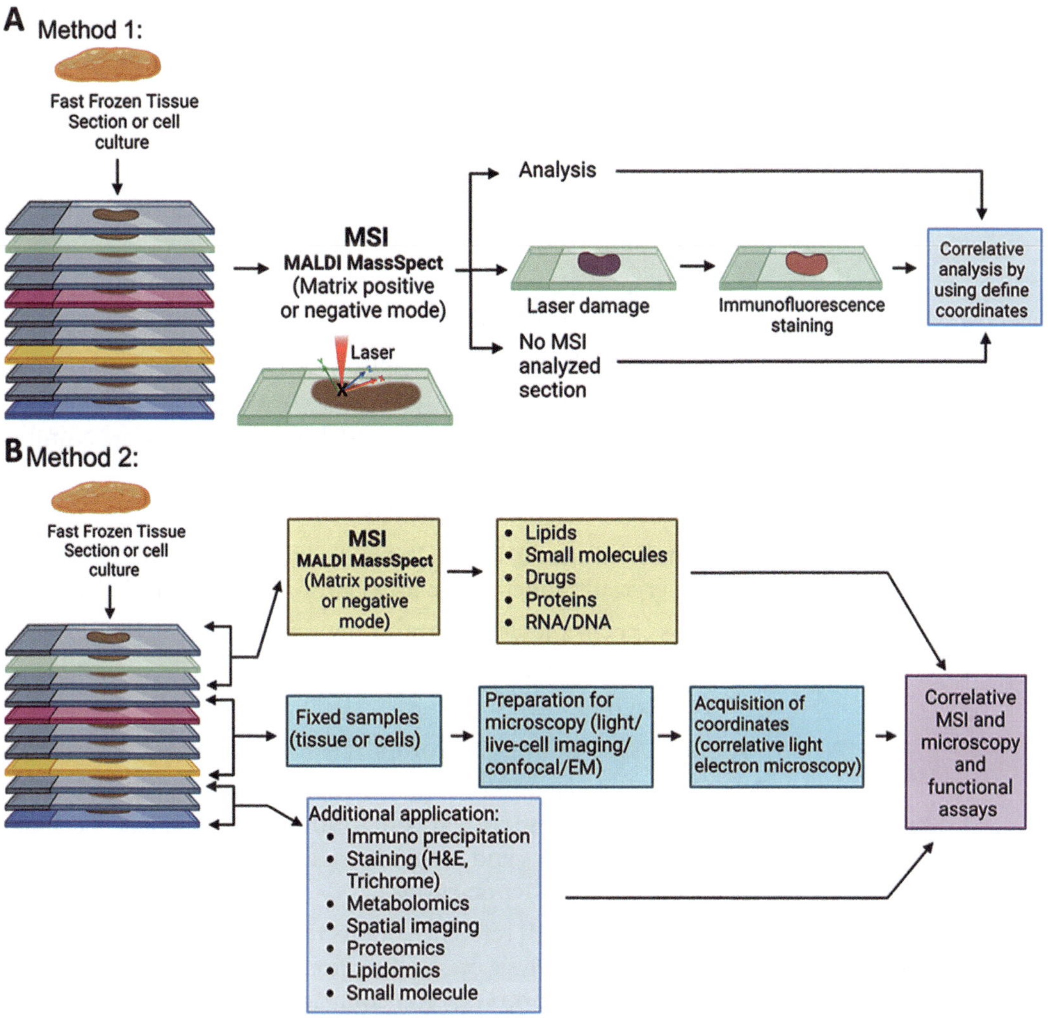

Fig. 1 Novel methodologies to establish correlative experiments. Correlative microscopy and mass spectrometry imaging (MSI) can be used for subsequent targeted OMICs approaches. (**a**) Method 1 involves the direct immunostaining of the same sample subjected to MSO. (**b**) Method 2 corresponds to a serial section method to perform at least three different analyses in the same sample, including staining, MSI, and additional applications such as laser capture and OMICs approaches

5. Albumin from bovine serum fraction V.
6. Horse serum (Sigma, H1138).
7. HIV-p24 antibody (HIV reagent program).
8. Rabbit anti-human CD3.
9. Rat anti-human CD4.
10. Rabbit anti-human Iba-1.
11. Hydrophobic barrier pen (ACD Bio, catalog number 310018).
12. Positive control probe (ACD Bio, catalog number 313901).

13. Negative control probe (ACD Bio, catalog number 310043).
14. For the mass spectrometry reagents (*see* Subheading 3.11).
15. Proteinase K (20 U/mL).
16. Dako Pen (Dako Catalog number S2002).

2.2 Equipment

1. Confocal or epifluorescence microscope (Nikon A1, Japan, or Leica SP8, Germany).
2. Imaging workstations (NIS elements, Nikon, Tokyo, Japan, or Zeiss connection, Germany).
3. HybEZ hybridization system (HybEZ, 110 V AC, Cat No. 310010).
4. HybEZ humidity control tray with lid (HybEZ, Cat. Number 310012).
5. Incubators (Lab Companion SI 600).
6. Trim20 diamond knife (Fisher Scientific, Cat. 50-190-0152) and cryo ultramicrotome for cryoEM (Leica EM UC7).
7. Mass spectrometer (Thermo Scientific Q-Extractive HF).
8. Sprayer (TM-Sprayer, HTX Imaging, HTX Technologies LLC).
9. A cryostat and ultramicrotome (Leica CM 1860 UV).
10. MSI and tissue/cell slides like (Alkaline earth boro-aluminosilicate glass, Part # CB-90IN-SN111, Delta Technologies Limited, and regular tissue/cell slide (Catalog # 12-550-15, Fisher Scientific).
11. Tissue slices with reference points to determine coordinates for subsequent analysis or serial sections for tomography or other 3D reconstructions (several companies supply them).
12. A live-cell or fluorescence microscope including light, confocal, super-resolution, TIRF, or laser capture (Nikon Eclipse T1 or $T1_2$) for acquisition and analysis.
13. Transmission and/or scanning electron microscope (Zeiss Sigma EV500 or FIB).

2.3 Probes and Matrix

1. FISH probes: To dissolve DNA and mRNA probes (testing, negative and positive control probes), resuspend them in formamide at a concentration of 100 μM. A detailed protocol has been recently published extensively for staining and troubleshooting. The current chapter will provide only minor details, but we invite the readers to the staining protocol [20]. All probes are stored at −20 °C.

   The matrix must be prepared fresh: For positive mode, prepare 20 mg/mL of 2′,4′,6′-trihydroxyacetophenone monohydrate (Sigma catalog number T64602) dissolved in 40%

methanol (Fisher Scientific A456-4), 40% acetonitrile (Fisher A955-4), 20% ultrapure water (Fisher W6-4), and 0.1% trifluoroacetic acid (Alfa Aesar catalog number LO6374), for negative mode, prepare 5 mg/mL 1,5-diaminonaphthalene (Sigma catalog number D21200) dissolved in 70% acetone (Fisher A949-4) and 30% ultrapure water (Fisher W6-4).

3 Methods

Carry out all procedures at the stipulated temperature.

3.1 Sample Fixation

For microscopy-based imaging, there is a need for fixation (tissues/cells; *see* Fig. 1). Depending on the protocol selected in Fig. 1, tissues (from humanized mice, monkeys, or humans) or cells can be fixed in multiple fixatives, including formalin, aldehydes, glutaraldehyde, ethanol, or acetone. Some require further permeabilization with Triton X-100 or IGEPAL CA-630 (detergents to enable probes, antibodies, or dyes to cross the membranes). However, for MSI, the sample must be fresh, unfixed, or with minimal fixation to maintain lipids in their natural state (cryostat or tissue slices or cells).

3.2 Sample Preparation

As indicated above, two potential approaches (Fig. 1) can be used. Approach 1: Unfixed tissue(s) is cut and mounted at 10–20 μm thickness using Leica CM 1860 cryostat. Here, the MSI experiment and analysis are first performed. Subsequently, the slide could be used for other fluorescent imaging studies after the MSI matrix has been washed off. This method allows for identifying and examining the various components of interest in the same tissue slice. However, this method has some disadvantages related to laser-induced tissue damage. The laser-induced tissue damage can be prevented by reducing the laser repetition rate. Approach 2: It focuses on the use of serial sections of the unfixed tissues cut at 10–20 μm for MSI and subsequent sections for fixation and immunostaining. The main point here is to keep the same (or close) coordinates to enable the correlative section between the different techniques. We will focus on approach 2.

For MSI sections, after sectioning, tissues are placed in a vacuum unit for 15 min and then transferred to a −80 °C freezer in well-labeled small Ziplock bags for storage. Prior to MALDI-MSI analysis, the mounted tissue section(s) are removed from the −80 °C freezer and allowed to equilibrate to room temperature for 15 min in the vacuum unit. For positive mode ionization, 20 mg/mL of 2′,4′,6′-trihydroxyacetophenone monohydrate is dissolved in 40% methanol, 40% acetonitrile, 20% ultrapure water, and 0.1% trifluoroacetic acid. For the negative mode of ionization, prepare 5 mg/mL 1,5-diaminonaphthalene dissolved in 70%

acetone and 30% Ultrapure water. The main difference between the positive ion mode ionization is the protonation, and the negative mode ionization is based on deprotonation.

3.3 Matrix Application

For the ionization, the right matrix is applied to the tissue by automated air spray deposition using the TM-Sprayer (HTX Technologies LLC, Chapel Hill, NC). The nozzle temperature is set to 60 °C, the flow rate is 50 μL/min, and 20 passes over the tissue are performed. MALDI-MSI analysis is performed using Q Exactive HF Hybrid Quadrupole–Orbitrap Mass Spectrometer (Thermo Fisher Scientific) equipped with an elevated pressure matrix-assisted laser/desorption ionization (EP MALDI) source integrating an Nd: YAG laser (Spectroglyph LLC, Kennewick, WA). Other equipment and companies also can be used to perform these analyses, such as Bruker.

3.4 Data Collection for MSI

Data are acquired at 40 μm^2 lateral resolution in negative ion mode at a laser power of 7 μJ. External mass calibration can be performed using calibration solutions (A39239, Thermo Fisher Scientific). In the case of lipid analysis, the accurate mass-measured lipid peaks are identified by matching reference lipids in the LIPID MAPS and Human Metabolome Database within a ±0.002 Da mass tolerance window. Thermo RAW format data files are converted to imzML using Image Insight software (Spectroglyph LLC, Kennewick, WA). Data visualization is performed as indicated below. Further analysis can be achieved with other MSI software like SCiLS.

3.5 Serial Sections and Detection of Viral Reservoirs in Tissues and Rare Cells

Depending on the cell/tissue type and size, identifier (DNA, mRNA, miRNA, protein, or drug), and the abundance of the cell type analyzed, probably multiple serial sections could be required. Regarding tissue-associated HIV reservoirs [19, 20, 22, 24], we suggest 10–15 slides with tissue sizes 2.5 × 3 cm. Viral reservoirs can be distributed between a few to hundreds of micrometers apart and, in blood, have been calculated to be one positive cell in 10^6–10^7 uninfected cells. However, in tissues, the numbers and rates are even lower. Also, consider that a particular cell can be present (depending on the size and sectioning) in different Z planes. Thus, the coordinates by microscopy could provide spatial resolution and deconvolution to reconstruct and identify rare viral reservoirs and subcellular structures in parallel with OMIC analysis, such as MSI (see example in Fig. 2 and [19]).

For viral reservoirs in large tissues such as the brain, tissue pieces are serially sectioned (cortex and hippocampal section sizes were 6.7 ± 3.03 cm^2, 11–12 serial sections of 10 μm thickness each). Large tissue pieces are crucial in some tissues with a low abundance of viral reservoirs. To identify and establish the location of viral reservoirs, cells with integrated proviral DNA and positive for viral mRNA and HIV-p24 proteins versus the total number of cells (DAPI positive) are the targets to identify [19, 20, 24]. Using

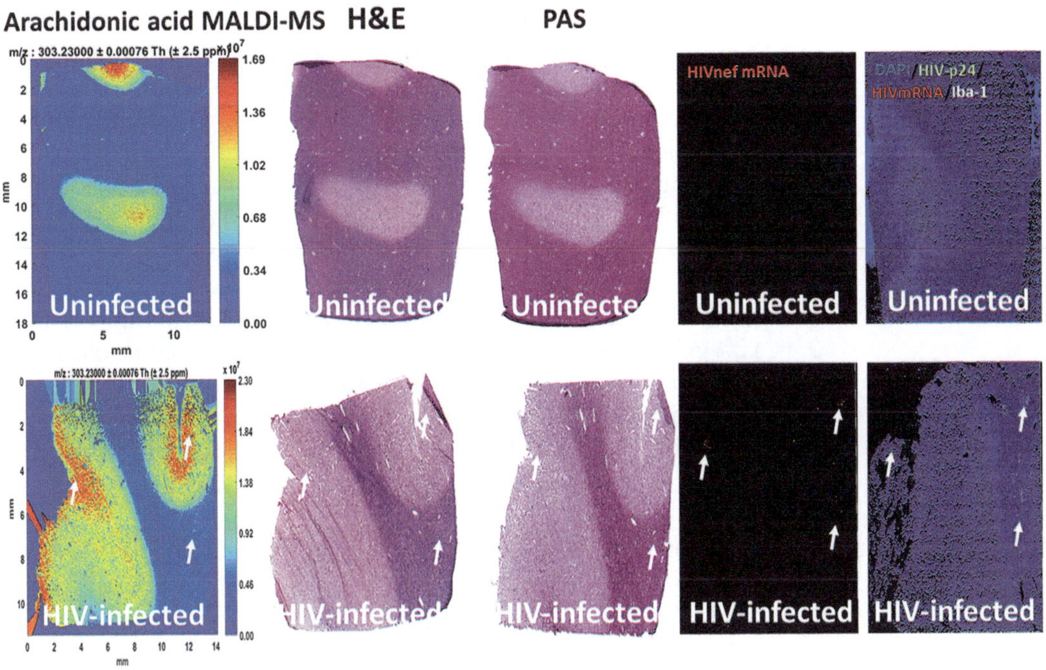

Fig. 2 (Using the method in Fig. 1b), MSI, H&E, PAS, and confocal microscopy for rare viral reservoirs within an individual's brain under ART without systemic viral replication detected. The top row represents an example of arachidonic acid distribution in an uninfected human brain, and the lower row represents the same lipid distribution in HIV-infected brain tissue obtained from an individual without systemic replication. Serial sections were stained for H&E and PAS to observe structures, and subsequent sections were stained for HIVnef RNA (red staining) to detect the presence of HIV mRNA, HIV-p24 (green staining) to identify HIV protein, and Iba-1 (white staining, microglia/macrophage marker) to represent cellular markers and DAPI (to identify nuclei). No staining for HIV RNA or protein was found in uninfected brains, as expected. in an HIV-infected brain, arachidonic acid levels are significantly higher. Arrows represent myeloid cells expressing HIVnef RNA and HIV-p24. An application of this method allows us to correlate the presence of viral components in cells with H&E and PAS alterations and changes in the relative levels of arachidonic acid in the white and gray matter

these imaging technologies, the location of viral reservoirs can be determined, and spatial colocalization (or proximity) of these viral reservoirs to lipid signature changes or EM profiles can be identified depending on the application selected (see examples of the workflow in Fig. 1a or b). For instance, examining spatial colocalization of viral reservoirs and inflammatory lipid signatures could reveal novel findings [29, 30]. One such inflammatory lipid is arachidonic acid, which is mostly found to be linked with phospholipids. Arachidonic acid is a precursor to the prostaglandin and leukotriene pathways, and it has been shown to be associated with HIV infection [31, 32]. An example of the application of MSI and viral reservoir localization is shown in Fig. 2; the arrows denote areas with viral reservoirs as assessed by confocal microscopy, HIV–DNA, and the MSI provides the dysregulation of arachidonic acid, and an inflammatory lipid is further catabolized into various pro-inflammatory metabolites such as prostaglandin E2 and

Leukotrienes, around the viral reservoirs in large pieces of human brains (Fig. 2). Thus, the integration of microscopy and untargeted analysis provides a new dimension in spatial resolution.

This approach is not just limited to tissues. Both adherent and suspension cells can be examined using the same approaches. For adherent cells, coat the MSI slide with poly-Lysine, add cells with O rings to hold cells in a designated part of the slide, and incubate at 37 °C. Subject cells to planned experiments, wash, and coat with matrix. For suspension cells, like peripherical blood monocyte cells (PBMCs), embed cells in gelatin to facilitate sectioning. Briefly, a 12% gelatin solution is prepared; add cells to the gelatin solution in a cryostat tray; and put it at −80 °C to form the block with embedded cells. Move the frozen cryostat tray to the cold cryostat for sectioning. Mount slices of 10–20 μm thick sections to the MSI slides and coat them with designated matrix [33].

3.6 Translation into EM

After the identification and recording of the coordinates of the rare viral reservoirs by microscopy, as shown in Fig. 2, a serial section can be fixed for TEM [34] or SEM imaging [35, 36] or to trim/flatten their surfaces before the analysis with integrated systems such as FIB-SEM [35] and SBF-SEM [37–39]. Identifying the coordinates of rare cells is essential to identify the area of interest and target the EM approach into single cells (see correlative methods below). *See* notes for caution with EM reagent.

3.7 EM Analysis

Upon identification of the areas of interest, a targeted tissue/cell preparation or sectioning is performed using a glass or diamond knife mounted on an ultramicrotome, which allows for obtaining thin sections down to 20 nm. Identifying the coordinates by confocal or live-cell imaging is essential to focus the search on a specific or associated structure to later over-impose light and EM imaging. Cutting the entire tissue is unnecessary but is preferred (depending on the grid size); the localization can be easily identified using staining. For TEM analysis, section thickness can span between 40 nm and 200–500 nm: Thinner sections provide a lower contrast; conversely, the electron beam is not efficiently transmitted through ticker sections [40]. For SEM imaging, instead, sections can be approximately 150 nm thin, flattened on silicon wafers that are placed on SEM supports [35, 36]. Similarly, frozen samples can be cryosectioned and imaged with cryo-electron microscopes, ensuring all these processes are carried out below the devitrification temperature [41]. This last method is preferred to perform correlative microscopy.

3.8 Labeling for CLEM

Labeling can be carried out before the sample embedding [42] or after, directly over sections collected on TEM grids [43] or SEM wafers [44, 45] (*see* **Note 4**). Labeling can occur before the embedding procedure at two different steps: before or after the fixation (*see* **Note 3**). When the labeling is performed on living cells, in vivo imaging is possible [42, 46]. An exciting alternative to the in vivo detection of markers is to perform the labeling on samples that are mildly fixed right after the FM/LM imaging [46, 47]. This protocol has the remarkable advantage of preserving either the ultrastructure or the antigenicity, but also the potential applications for MSI and laser capture. As controls, uninfected tissues or cryoEM can be used.

3.9 Post-Embedding Labeling

Cryosectioned and resin-embedded sections, collected over TEM grids [43] or SEM wafers [44], can be immunolabeled with antibodies conjugated to electron-dense nanoparticles, visible with EMs (e.g., gold nanoparticles–nanogold) or probes/antibodies in serial sections.

3.10 Immunofluorescence for Both Approaches (Serial Sections or After MSI)

Cells are grown, or tissues are mounted on coverslips or MSI slices. The protocols and flow have been recently described [48]. As described above, keeping the spatial coordinates is essential to identify cell types or organelles in specific cell types for subsequent MSI or EM analysis. *See* notes.

Follow these steps:

1. Fix sections with fresh 4% paraformaldehyde (PFA) for 20 min.
2. Wash with 1× PBS for 5 min.
3. Incubate in 100% absolute ethanol (Fisher catalog number A409-4).
4. Air-dry slides for 5 min.
5. Incubate slides in 0.1% Triton X-100 for 1 min.
6. Wash slides in 1× PBS three times (5 min/wash).
7. Incubate slides in 0.3% Sudan Black in 70% ethanol in a glass beaker with continuous stirring with a stirring bar for 2 h. *Note: Ensure this is done in the dark.*
8. Wash slides in 70% ethanol twice (5 min/wash).
9. Incubate slides in fresh sodium borohydride solution (1 mg/mL in 1× PBS, pH 7.5) for 15 min. *Note: Prepare solution and incubate samples on ice.*
10. Wash slides with 1× PBS two times (5 min/wash).
11. Using a Dako pen, draw to encircle the tissue or cells on the slides and wait till it is dry. *(Make sure drawn lines are visible.)*
12. Wash slides with deionized water for 5 min.
13. Prepare two humidity chambers (one for room temperature and the other for incubator).
14. Place the slides in the room-temperature humidity chamber.
15. Prepare proteinase K solution in 1× TBS (20 μg/mL).
16. Add proteinase L solution onto the tissue sections until they are covered.

17. Incubate at room temperature for 10 min.
18. Wash slides in deionized water for 3 min.
19. Immerse slides in 95% ethanol for 20 s.
20. Air-dry slides for 5 min.
21. Follow the kit manufacturer's instructions for DNA and/or RNA FISH probe staining.
22. For immunostains, block tissues/cells with blocking buffer overnight at 4 °C in a humidity chamber.
23. Add primary antibodies (at validated dilutions/concentrations) and incubate at 4 °C overnight.
24. Wash three times in 1× TBS (5 min/wash).
25. Add secondary antibodies (at validated dilutions/concentrations) and incubate in a humidity chamber at room temperature for 4 min.
26. Wash five times in 1× TBS (5 min/wash).
27. Add DAPI containing mounting media and place cover glass on each slide.

3.11 Mass Spectrometry Imaging as a Spatial/Untargeted Method to Identify New Disease Biomarkers in Viral Reservoirs and Surrounding Uninfected Cells

MSI is a powerful technique for visualizing the spatial distribution of biomolecules in situ, and recent advances in microscopy laser capture microdissection and OMIC analysis further strengthen the techniques. Usually, a typical MSI of tissues from HIV-infected individuals or uninfected controls is performed by rastering an ionization beam across a tissue surface at defined x and y coordinates, thereby desorbing/ionizing analytes that are subsequently extracted into the mass spectrometer for analysis. Software is used to reconstruct an image from the dataset in which each pixel consists of a mass spectrum. Images can be plotted for any selected peak in the acquired mass spectrum, with the relative abundances displayed as a false-color image across the analyzed region of interest (ROI). Now, we can select specific ROI that targets particular cell populations and areas of interest. Because the software generates an image, we can identify the cells or area of the tissue analyzed by microscopy, as indicated in Fig. 2, where the HIV viral reservoir is identified, and the lipid signatures close to it are identified. Overall, the combination of microscopy with MSI or laser capture and subsequent OMIC approaches (proteomics, lipidomics, metabolomics, and small molecule quantification) could provide new insight into the spatial distribution.

Multiple paths for mass spectrometry imaging (MSI) are possible. As shown in Fig. 1, fresh or frozen tissues are to be used in these experiments, and the experiment can be done either alone (Fig. 1a) or in conjunction with other applications (Fig. 1b). Note: MSI requires specific slides (*see* Subheading 2). Examine the tissue size to determine if there is a need to cut the tissue further within acceptable size. The upper limit of tissue size that can fit on the MSI slide is (2 cm × 2 cm), and the mounted tissues must be positioned in the middle of the MSI slide.

For protocols described in Fig. 1,

- Mount tissues on the mobile stage knob with O.C.T compound (Tissue-Tek, Sakura Finetek, USA).
- Using refrigerated cryostat mounted with appropriate microtome blade (e.g., Diamond DT315D50), cut tissues (within 7–20 um thickness) and mount on the mass spectrometry slide (alkaline earth boro-aluminosilicate glass, Part # CB-90IN-SN111, Delta Technologies Limited) and regular microscope slide (Catalog # 12-550-15, Fisher Scientific) with little to no ruffle.
- Determine the matrix to coat the tissue slice. This is dependent on whether the MSI experiment is targeted or bulk MSI. This will be imperative to identify the specific matrix to be used. For positive mode, prepare 20 mg/mL of 2′,4′,6′--trihydroxyacetophenone monohydrate (Sigma catalog number T64602) dissolved in 40% methanol (Fisher Scientific A456-4), 40% acetonitrile (Fisher A955-4), 20% ultrapure water (Fisher W6-4), and 0.1% trifluoroacetic acid (Alfa Aesar catalog number LO6374), and for negative mode, prepare 5 mg/mL 1,5-diaminonaphthalene (Sigma catalog number D21200) dissolved in 70% acetone (Fisher A949-4) and 30% ultrapure water (Fisher W6-4).
- Vortex the matrix and put it in an ultrasonic bath for further dissolution.
- Set up the sprayer (TM-Sprayer, HTX Imaging, HTX Technologies LLC) with 50% methanol (continuous flow at 0.05 mL/min flow rate), ultrapure compressed nitrogen gas flow through, and inject prepared matrix.
- Spray the tissues on the mass spectrometry slides using a preset spraying protocol with the designated matrix (mentioned earlier).
- Collect sprayed tissues. Note: Sprayed tissues can be examined with a slide scanner to determine the matrix coverage of the tissues.
- Store the sprayed slides at −80 °C overnight.
- Run tissues in the mass spectrometry device with the stated m/z range (e.g., Thermo Scientific Q Exactive HF) (*see* **Note 1**).
- Collected spectral and position files are exported and analyzed with various analysis software (e.g., SCiLS). An example of such analysis is shown in Fig. 2, comparing the levels of arachidonic acid in both uninfected and HIV-infected human brain tissues.

3.12 Spatial Resolution Between Microscopy and MSI

The combination of microscopy imaging capabilities and MSI could "break" the limited resolution of MSI. MSI is limited to the size of the recorded pixel. Currently, achieving an excellent spatial resolution by MSI has been a challenge due to the diameter of the ionization beam that determines the spatial resolution. Moreover, by adjusting the optics of the ion source, a resolution of 1.4 μm to 600 nm has been described [49]. Thus, subcellular distribution is achievable and could provide an outstanding resolution with the combination of microscopy techniques [50–55]. Currently, there are several commercial software to perform the experiments proposed, including Zen (Zeiss), NIS (Nikon), Image J (https://imagej.net/software/fiji/), ICY-CLEM (https://icy.bioimageanalysis.org/plugin/ec-clem/), IMOD (https://bio3d.colorado.edu/imod/), and ilastik (www.ilastik.org) (*see* **Note 8**). However, many others can be customized according to the user's needs. Additional resources can be found at http://microscopyu.com/; http://zeiss-campus.magnet.fsu.edu/; http://www.leica-microsystems.com/science-lab/; http://olympusmicro.com/; http://www.microscope-microscope.org/; or http://www.fei.com/Education-Resources/. This software's main goal is to acquire pictures and record information about specific cellular locations and depths in correlation with the tissue organization and structure (*see* **Note 5**).

3.13 Correlation Strategies

As mentioned above, one of the main challenges in correlative microscopy is visualizing the structures and making them recognizable with all the imaging techniques used in the selected correlative approach. Below, we will overview the main strategies commonly used for correlative microscopy. These strategies include morphological, tissue slice coordinates or segmentation, superimposed into the EM [56, 57], and labeling (nanoparticles of gold). However, always consider that your sample can shrink, challenging your ability to determine the exact location [58–62]. The software that enables the user to achieve these correlations is described below (*see* **Note 6**).

3.14 Tools for Correlation

Among the strategies for correlating images acquired with different microscopes, using tools such as patterned TEM grids or gridded coverslips is particularly useful, making the correlation easier to track, including identifying the tissue and cell(s) that need to be analyzed. Still, also, the use of different kinds of grids can expedite the correlation. Finder grids are conventional grids for TEM with a peculiar identifier that can be used for the orientation within a sample at low and higher magnifications. DNA probes, antibodies, or MSI in serial sections can obtain some of these identifiers. Further, tissues and cells can be observed in the grid to confirm location. Image registration can be manually performed or using specific software for the coordinates conversion between LM and EM modes. A great resource is using gridded coverslips, either plastic or glass. Cells/tissues can be grown, immunolabeled (pre- or post-embedding), and processed directly over those substrates for resin embedding. After its polymerization, the embedded sample resin block will be removed from the coverslip, and the gridded pattern will remain printed over the block, making any trimming, sectioning, and imaging directed to the region of interest marked with its relative coordinates based on MSI or confocal microscopy to rare cells of interest. Further, staining with gold markers to identify cells or structures helps drive the correlation of data acquired with different instruments, especially when finer-scale registration is needed.

An excellent method for correlation is using gold particles that align the images (gold particles, 10 nm). The presence of these molecules enables good tomographic and correlative analyses. Nevertheless, the need for normalization and proper controls (negative and positive) is essential (*see* **Note 7**).

3.15 Software-Assisted Strategies

During the last decade, many research groups and companies have been developing configuration management workflows, software, and scripts for the automated registration of images acquired with different instruments [63, 64]. Unsupervised and supervised algorithms are widely employed for data clustering and classification [65–69]. Unsupervised algorithms do not require previous knowledge; thus, untargeted approaches are useful for exploratory data analysis, such as in biomarker discovery. Supervised algorithms are highly effective in predicting patterns and features of an imaging dataset. Contrary to unsupervised methods, supervised study requires a specified set of spectra to be known in advance and thus relies on targeted classification. Machine learning algorithms such as support vector machines (SVM) and random decision forests (RDF) use the predefined set of spectra to determine correlating structures in the data. In our MSI analysis, we used SCiLS processing software for MSI datasets and support data from all mass spectrometer manufacturers, including Thermo (*see* **Note 6**).

Additionally, the software enables the reconstruction, visualization, and analysis of 3D MALDI imaging data sets. Numerous databases may be used to assist with lipid identification from both complete and MS/MS scan mode data. Examples include LIPID MAPS Structure Database (The LIPID MAPS Lipidomics Gateway, http://www.lipidmaps.org/) and Metabolomics Workbench Metabolite Database (http://www.metabolomicsworkbench.org/), both University of California (San Diego, USA), as well as the Human Metabolome database and METLIN (Scripps Research Institute, LA Jolla, USA). All data were generally expressed as mean ± standard deviation (S.D.). Differences among groups were analyzed using a t-test with Welch's correction or the Holm–Sidak method (alpha = 0.05). The level of significance was accepted at $p \leq 0.05$. GraphPad Prism 8 software was used for the statistical analyses performed.

4 Notes

1. **The mass spectrometry machine's parameters can vary from experiment to experiment.**
2. Before starting an experiment, coordinating the software and the sections must be perfectly aligned to prevent miscalculations.
3. For fixations with glutaraldehyde, store the solution at −20 °C. Maintaining this stock for a long time (more than a month) will increase autofluorescence.
4. The use of glass and grids with high performance is essential.
5. Data handling and storage are another key concern in micro-copy and high-throughput MSI, mainly when dealing with high spatial and mass resolution data that can reach several hundred gigabytes per tissue section. Physical transfer of large imaging datasets between multiple study sites can prove highly cumbersome and time-consuming. To address this, cloud-based storage can be developed to rapidly share datasets between study sites, allowing remote collaborations between the different equipment used.
6. Analysis: With advances in instrumentation, the complexity of datasets generated in MSI has increased tremendously, demanding advanced statistical methods for reliable interpretation and prospective biomarker discovery. Unsupervised and supervised algorithms are widely employed for data clustering and classification. Unsupervised algorithms do not require previous knowledge; thus, untargeted approaches are useful for exploratory data analysis, such as in biomarker discovery. In MSI, principal component analysis (PCA) can be applied to reduce the dimensionality of the data set to the most significant *m/z* distributions. Hierarchical cluster analysis (HCA), another classical unsupervised tool, is used to bin together spatially resolved spectra based on their spectral similarity, allowing for the identification of colocalized signals in imaging datasets. Supervised algorithms are highly effective in predicting patterns and features of an imaging dataset. Contrary to unsupervised methods, supervised analysis requires a specified set of spectra to be known in advance and thus relies on targeted classification. Machine learning algorithms such as support vector machines (SVM) and random decision forests (RDF) use the predefined set of spectra to determine correlating structures in the data.
7. Normalization and proper negative and positive controls are critically important for the interpretation of correlative microscopy, including tissue and subcellular localization such as

nuclei, cytoplasm, or cell-type staining. In the case of MSI, some artifacts may originate from matrix application, matrix crystal heterogeneity, or ion source contamination. Among the few different normalization strategies for MSI datasets, total ion current (TIC) normalization is the most commonly implemented algorithm. In this approach, all mass spectra are divided by their TIC so that all ranges in a dataset have the same integrated area under the spectrum, based on the assumption that there are a comparable number of signals in each spectrum. These limits signal comparison to similar types of tissue.

8. Depending on the approach (*see* Fig. 1 and example in Fig. 2), using custom-made or commercial instruments speeds up the correlative analysis and reduces mistakes due to problematic manual alignments. For example, Zeiss made available the Shuttle & Find module that can run with ZEN Connect and AxioVision (old version of Zen) software for automated relocation via overview images or fiducial calibrations and alignment of ROIs between light and electron microscopes [70]. When these workflows are not available, several software packages can be used for manual or automatic correlation, such as Amira (Thermo Fisher), Imaris (Oxford Instruments), and open-source plugins for ImageJ/Fiji (NIH). Thus, the possibilities are growing and depend on the level of expertise of each group or the company that provides the interface.

References

1. Ando T, Bhamidimarri SP, Brending N, Colin-York H, De Collinson L, de Jonge N, Pablo PJ, Debroye E, Eggeling C, Franck C, Fritzsche M, Gerritsen H, Giepmans BNG, Grunewald K, Hofkens J, Hoogenboom JP, Janssen KPF, Kaufman R, Klumpermann J, Kurniawan N, Kusch J, Liv N, Parekh V, Peckys DB, Rehfeldt F, Reutens DC, Roeffaers MBJ, Salditt T, Schaap IAT, Schwarz US, Verkade P, Vogel MW, Wagner R, Winterhalter M, Yuan H, Zifarelli G (2018) The 2018 correlative microscopy techniques roadmap. J Phys D Appl Phys 51(44):443001
2. Umakoshi T, Fukuda S, Iino R, Uchihashi T, Ando T (2020) High-speed near-field fluorescence microscopy combined with high-speed atomic force microscopy for biological studies. Biochim Biophys Acta Gen Subj 1864(2):129325
3. Guizetti J, Mäntler J, Müller-Reichert T, Gerlich DW (2010) Correlative time-lapse imaging and electron microscopy to study abscission in HeLa cells. Methods Cell Biol 96:591–601
4. Muller-Reichert T, Gross H (1996) Microscopic analysis of DNA and DNA-protein assembly by transmission electron microscopy, scanning tunneling microscopy and scanning force microscopy. Scanning Microsc Suppl 10:111–20; discussion 120-1
5. Müller-Reichert T, Srayko M, Hyman A, O'Toole ET, McDonald K (2007) Correlative light and electron microscopy of early Caenorhabditis elegans embryos in mitosis. Methods Cell Biol 79:101–119
6. Muller-Reichert T, Verkade P, Preface. (2014) Correlative light and electron microscopy II. Methods Cell Biol 124:xvii–xviii
7. Baxter AE, Niessl J, Fromentin R, Richard J, Porichis F, Charlebois R, Massanella M, Brassard N, Alsahafi N, Delgado GG, Routy JP, Walker BD, Finzi A, Chomont N, Kaufmann DE (2016) Single-cell characterization of viral translation-competent reservoirs in HIV-infected individuals. Cell Host Microbe 20(3):368–380
8. Baxter AE, Niessl J, Fromentin R, Richard J, Porichis F, Massanella M, Brassard N, Alsahafi N, Routy JP, Finzi A, Chomont N, Kaufmann DE (2017) Multiparametric

characterization of rare HIV-infected cells using an RNA-flow FISH technique. Nat Protoc 12(10):2029–2049

9. Fujinaga K, Cary DC (2020) Experimental systems for measuring HIV latency and reactivation. Viruses 12(11):1279
10. Gil C, Garcia MT, Garcia F, Miró JM, Agüero F, Alós L, Zamora L, Capón A, Costa J, Pumarola T, Gatell JM (2011) Evaluation of the Roche COBAS(R) TaqMan (R) HIV-1 test for quantifying HIV-1 RNA in infected cells and lymphoid tissue. J Virol Methods 174(1–2):69–76
11. Henrich TJ, Gallien S, Li JZ, Pereyra F, Kuritzkes DR (2012) Low-level detection and quantitation of cellular HIV-1 DNA and 2-LTR circles using droplet digital PCR. J Virol Methods 186(1–2):68–72
12. Mehta K, Gohil Y, Mishra S, D'Silva A, Amanullah A, Selvam D, Pargain N, Nala N, Sanjeeva GN, Ranga U (2021) An improved Tat/Rev induced limiting dilution assay with enhanced sensitivity and breadth of detection. Front Immunol 12:715644
13. Palmer S (2013) Advances in detection and monitoring of plasma viremia in HIV-infected individuals receiving antiretroviral therapy. Curr Opin HIV AIDS 8(2):87–92
14. Powell L, Dhummakupt A, Siems L, Le Singh D, Duff Y, Uprety P, Jennings C, Szewczyk J, Chen Y, Nastouli E, Persaud D (2021) Clinical validation of a quantitative HIV-1 DNA droplet digital PCR assay: applications for detecting occult HIV-1 infection and monitoring cell-associated HIV-1 dynamics across different subtypes in HIV-1 prevention and cure trials. J Clin Virol 139:104822
15. Zhang Y, Planas DR, Marchand L, Massanella M, Chen H, Wacleche VS, Gosselin A, Goulet JP, Filion M, Routy JP, Chomont N, Ancuta P (2020) Improving HIV outgrowth by optimizing cell-culture conditions and supplementing with all-trans retinoic acid. Front Microbiol 11:902
16. Eriksson S, Graf EH, Dahl V, Strain MC, Yukl SA, Lysenko ES, Bosch RJ, Lai J, Chioma S, Emad F, Abdel-Mohsen M, Hoh R, Hecht F, Hunt P, Somsouk M, Wong J, Johnston R, Siliciano RF, Richman DD, O'Doherty U, Palmer S, Deeks SG, Siliciano JD (2013) Comparative analysis of measures of viral reservoirs in HIV-1 eradication studies. PLoS Pathog 9(2):e1003174
17. Castellano P, Prevedel L, Eugenin EA (2017) HIV-infected macrophages and microglia that survive acute infection become viral reservoirs by a mechanism involving Bim. Sci Rep 7(1):12866
18. Castellano P, Prevedel L, Valdebenito S, Eugenin EA (2019) HIV infection and latency induce a unique metabolic signature in human macrophages. Sci Rep 9(1):3941
19. D'Amico D, Di Barone R, Felice V, Ances B, Prideaux B, Eugenin EA (2023) Chronic brain damage in HIV-infected individuals under antiretroviral therapy is associated with viral reservoirs, sulfatide release, and compromised cell-to-cell communication. Cell Mol Life Sci 80(4):116
20. Donoso M, D'Amico D, Valdebenito S, Hernandez CA, Prideaux B, Eugenin EA (2022) Identification, quantification, and characterization of HIV-1 reservoirs in the human brain. Cells 11(15):2379
21. Eugenin EA, Berman JW (2016) Improved methods to detect low levels of HIV using antibody-based technologies. Methods Mol Biol 1354:265–279
22. Ganor Y, Real F, Sennepin A, Dutertre CA, Prevedel L, Xu L, Tudor D, Charmeteau B, Couedel-Courteille A, Marion S, Zenak AR, Jourdain JP, Zhou Z, Schmitt A, Capron C, Eugenin EA, Cheynier R, Revol M, Cristofari S, Hosmalin A, Bomsel M (2019) HIV-1 reservoirs in urethral macrophages of patients under suppressive antiretroviral therapy. Nat Microbiol 4(4):633–644
23. Lutgen V, Narasipura SD, Barbian HJ, Richards M, Wallace J, Razmpour R, Buzhdygan T, Ramirez SH, Prevedel L, Eugenin EA, Al-Harthi L (2020) HIV infects astrocytes in vivo and egresses from the brain to the periphery. PLoS Pathog 16(6):e1008381
24. Valdebenito S, Castellano P, Ajasin D, Eugenin EA (2021) Astrocytes are HIV reservoirs in the brain: a cell type with poor HIV infectivity and replication but efficient cell-to-cell viral transfer. J Neurochem 158(2):429–443
25. Marakalala M, Raju J, Sharma RM, Zhang K, Eugenin YJ, Prideaux EA, Daudelin B, Chen IB, Booty PY, Kim MG, Eum JH, Via SY, Behar LE, Barry CE, Mann M, Dartois V, Rubin EJ (2016) Inflammatory signaling in human tuberculosis granulomas is spatially organized. Nat Med 22(5):531–538
26. Mishra BB, Lovewell RR, Olive AJ, Zhang G, Wang W, Eugenin E, Smith CM, Phuah JY, Long JE, Dubuke ML, Palace SG, Goguen JD, Baker RE, Nambi S, Mishra R, Booty MG, Baer CE, Shaffer SA, Dartois V, McCormick BA, Chen X, Sassetti CM (2017) Nitric oxide prevents a pathogen-permissive granulocytic inflammation during tuberculosis. Nat Microbiol 2:17072
27. Rella CE, Ruel N, Eugenin EA (2014) Development of imaging techniques to study the

pathogenesis of biosafety level 2/3 infectious agents. Pathog Dis 72(3):167–173
28. Subbian S, Eugenin E, Kaplan G (2014) Detection of mycobacterium tuberculosis in latently infected lungs by immunohistochemistry and confocal microscopy. J Med Microbiol 63 (Pt 11):1432–1435
29. Malik S, Theis M, Eugenin EA (2017) Connexin43 containing gap junction channels facilitate HIV bystander toxicity: implications in NeuroHIV. Front Mol Neurosci 10:404
30. Malik S, Valdebenito S, D'Amico D, Prideaux B, Eugenin EA (2021) HIV infection of astrocytes compromises inter-organelle interactions and inositol phosphate metabolism: a potential mechanism of bystander damage and viral reservoir survival. Prog Neurobiol 206:102157
31. Samikkannu T, Rao K, Ding H, Agudelo M, Raymond AD, Yoo C, Nair P (2014) Immunopathogenesis of HIV infection in cocaine users: role of arachidonic acid. PLoS One 9(8): e106348
32. Nokta MA, Hassan MI, Loesch KA, Pollard RB (1995) HIV-induced TNF-alpha regulates arachidonic acid and PGE2 release from HIV-infected mononuclear phagocytes. Virology 208(2):590–600
33. de Boer R, van der Klei IJ (2023) Correlative light- and electron microscopy in peroxisome research. Methods Mol Biol 2643:93–104
34. Parlanti P, Cappello V, Brun F, Tromba G, Rigolio R, Tonazzini I, Cecchini M, Piazza V, Gemmi M (2017) Size and specimen-dependent strategy for x-ray micro-ct and tem correlative analysis of nervous system samples. Sci Rep 7(1):2858
35. Parlanti P, Pal-Ghosh S, Williams A, Tadvalkar G, Popratiloff A, Stepp MA (2020) Axonal debris accumulates in corneal epithelial cells after intraepithelial corneal nerves are damaged: a focused Ion Beam Scanning Electron Microscopy (FIB-SEM) study. Exp Eye Res 194:107998
36. Parlanti P, Cappello V (2022) Microscopes, tools, probes, and protocols: a guide in the route of correlative microscopy for biomedical investigation. Micron 152:103182
37. Pope I, Tanner H, Masia F, Payne L, Arkill KP, Mantell J, Langbein W, Borri P, Verkade P (2023) Correlative light-electron microscopy using small gold nanoparticles as single probes. Light Sci Appl 12(1):80
38. Denk W, Horstmann H (2004) Serial block-face scanning electron microscopy to reconstruct three-dimensional tissue nanostructure. PLoS Biol 2(11):e329
39. Cocks E, Taggart M, Rind FC, White K (2018) A guide to analysis and reconstruction of serial block face scanning electron microscopy data. J Microsc 270(2):217–234
40. Winey M, Meehl JB, O'Toole ET, Giddings TH Jr (2014) Conventional transmission electron microscopy. Mol Biol Cell 25(3):319–323
41. Chlanda P, Sachse M (2014) Cryo-electron microscopy of vitreous sections. Methods Mol Biol 1117:193–214
42. Melo RC, Morgan E, Monahan-Earley R, Dvorak AM, Weller PF (2014) Pre-embedding immunogold labeling to optimize protein localization at subcellular compartments and membrane microdomains of leukocytes. Nat Protoc 9(10):2382–2394
43. Robinson JM, Takizawa T (2009) Correlative fluorescence and electron microscopy in tissues: immunocytochemistry. J Microsc 235(3):259–272
44. Clarkson C, Smeal RM, Hasenoehrl MG, White JA, Rubio ME, Wilcox KS (2020) Ultrastructural and functional changes at the tripartite synapse during epileptogenesis in a model of temporal lobe epilepsy. Exp Neurol 326: 113196
45. Gasaway C, Mastalerz M, Krause F, Clarkson C, Debuhr C (2018) Application of micro-FTIR mapping and SEM to study compositional heterogeneity of siltstones: example from the Late Devonian-Early Mississippian Middle Bakken Member. J Microsc 269(3): 195–211
46. Oliver C, Jamur MC (2010) Fixation and embedding. Methods Mol Biol 588:353–362
47. Van Jiménez N, Donselaar EG, De Winter DA, Vocking K, Verkleij AJ, Post J (2010) A. *Gridded Aclar: preparation methods and use for correlative light and electron microscopy of cell monolayers, by TEM and FIB-SEM.* J Microsc 237(2):208–220
48. Simrén J, Elmgren A, Blennow K, Zetterberg H (2023) Fluid biomarkers in Alzheimer's disease. Adv Clin Chem 112:249–281
49. Niehaus M, Soltwisch J, Belov ME, Dreisewerd K (2019) Transmission-mode MALDI-2 mass spectrometry imaging of cells and tissues at subcellular resolution. Nat Methods 16(9): 925–931
50. Bien T, Bessler S, Dreisewerd K, Soltwisch J (2021) Transmission-mode MALDI mass spectrometry imaging of single cells: optimizing sample preparation protocols. Anal Chem 93(10):4513–4520
51. Chan Y, Pathmasiri H, Pierre-Jacques KC, Cologna D, Gao SMR (2023) Gel-assisted mass spectrometry imaging. bioRxiv

52. Elhanani O, Keren L, Angelo M (2022) High-dimensional tissue profiling by multiplexed ion beam imaging. Methods Mol Biol 2386:147–156
53. Maimó-Barceló A, Garate J, Bestard-Escalas J, Fernández R, Berthold L, Lopez DH, Fernández JA, Barceló-Coblijn G (2019) Confirmation of sub-cellular resolution using oversampling imaging mass spectrometry. Anal Bioanal Chem 411(30):7935–7941
54. Meibom A, Plane F, Cheng T, Grandjean G, Haldimann O, Escrig S, Jensen L, Daraspe J, De Mucciolo A, Bellis D, Rädecker N, Martin-Olmos C, Genoud C, Comment A (2023) Correlated cryo-SEM and CryoNanoSIMS imaging of biological tissue. BMC Biol 21(1):126
55. O'Neill KC, Dueñas ME, Larson E, Forsman TT, Lee YJ (2022) Enhancing metabolite coverage for matrix-assisted laser desorption/ionization mass spectrometry imaging through multiple on-tissue chemical derivatizations. Methods Mol Biol 2437:197–213
56. Murphy GE, Narayan K, Lowekamp BC, Hartnell LM, Heymann JAF, Subramaniam J (2011) Correlative 3D imaging of whole mammalian cells with light and electron microscopy. J Struct Biol 176(3):268–278
57. Cao T, Zach C, Modla S, Powell D, Czymmek K, Niethammer M (2014) Multimodal registration for correlative microscopy using image analogies. Med Image Anal 18(6):914–926
58. Mohammadian S, Agronskaia AV, Blab GA, van Donselaar EG, de Heus C, Liv N, Klumperman J, Gerritsen HC (2020) Integrated super resolution fluorescence microscopy and transmission electron microscopy. Ultramicroscopy 215:113007
59. Mohammadian S, Fokkema J, Agronskaia AV, Liv N, de Heus, Van C, Donselaar E, Blab GA, Klumperman J, Gerritsen HC (2019) High accuracy, fiducial marker-based image registration of correlative microscopy images. Sci Rep 9(1):3211
60. Bykov YS, Cohen N, Gabrielli N, Manenschijn H, Welsch S, Chlanda P, Kukulski W, Patil KR, Schuldiner M, Briggs JAG (2019) High-throughput ultrastructure screening using electron microscopy and fluorescent barcoding. J Cell Biol 218(8):2797–2811
61. Bykov YS, Cortese M, Briggs JA, Bartenschlager R (2016) Correlative light and electron microscopy methods for the study of virus-cell interactions. FEBS Lett 590(13):1877–1895
62. Schorb M, Gaechter L, Avinoam O, Sieckmann F, Clarke M, Bebeacua C, Bykov YS, Sonnen AF, Lihl R, Briggs JAG (2017) New hardware and workflows for semi-automated correlative cryo-fluorescence and cryo-electron microscopy/tomography. J Struct Biol 197(2):83–93
63. Baatsen P, Gabarre S, Vints K, Wouters R, Vandael D, Goodchild R, Munck S, Gounko NV (2021) Preservation of fluorescence signal and imaging optimization for integrated light and electron microscopy. Front Cell Dev Biol 9:737621
64. Munck S, Swoger J, Coll-Lladó M, Gritti NV, Velde G (2021) Maximizing content across scales: moving multimodal microscopy and mesoscopy toward molecular imaging. Curr Opin Chem Biol 63:188–199
65. Last MGF, Voortman LM, Sharp TH (2023) Measuring cryo-TEM sample thickness using reflected light microscopy and machine learning. J Struct Biol 215(2):107965
66. Opstad IS, Godtliebsen G, Ströhl F, Myrmel T, Ahluwalia BS, Agarwal K, Birgisdottir ÅB (2022) Three-dimensional structured illumination microscopy data of mitochondria and lysosomes in cardiomyoblasts under normal and galactose-adapted conditions. Sci Data 9(1):98
67. Rohde F, Braumann UD, Schmidt M (2020) *Correlia:* An ImageJ plugin to co-register and visualise multimodal correlative micrographs. J Microsc
68. Schmidt M, Rohde F, Braumann UD (2021) Visualization and co-registration of correlative microscopy data with the ImageJ plugin Correlia. Methods Cell Biol 162:353–388
69. Weiner E, Pinskey JM, Nicastro D, Otegui MS (2022) Electron microscopy for imaging organelles in plants and algae. Plant Physiol 188(2):713–725
70. Blazquez-Llorca L, Hummel E, Zimmerman H, Zou C, Burgold S, Rietdorf J, Herms J (2015) Correlation of two-photon in vivo imaging and FIB/SEM microscopy. J Microsc 259(2):129–136

Part II

Biomolecular Condensates

Chapter 8

Characterization of Nuclear HIV-Induced Membraneless Organelles Through Fluorescence Microscopy

Viviana Scoca and Francesca Di Nunzio

Abstract

The postnuclear entry steps of HIV-1 involve reverse transcription, uncoating, and integration into the host genome. The differential regulation of these steps has a significant impact on HIV overall replication, including integration site selection and viral gene expression. Recently, another important phenomenon has been uncovered as part of HIV interplay with the nuclear environment, specifically involving the cleavage and polyadenylation specific factor 6 (CPSF6) protein. This phenomenon is the formation of nuclear HIV-induced membraneless organelles (HIV-1 MLOs). In this article, we will describe the methods used to assess the composition and liquid–liquid phase separation (LLPS) properties of these organelles using fluorescence microscopy. The study of HIV-1 MLOs represents a new frontier that may reveal previously unknown key players in the fate of HIV-infected cells.

Key words HIV-1, CPSF6, HIV-1 MLOs, LLPS, Fluorescence microscopy, Time-lapse microscopy, FRAP

1 Introduction

The investigation of early replication steps of HIV-1 is crucial to gain a better understanding of viral mechanisms, including reverse transcription, uncoating, nuclear import, and integration. Recent advancements in microscopy technologies for live and fixed cells have revolutionized the established knowledge of HIV molecular biology, particularly with regard to the discovery of reverse transcription occurring in the nuclear compartment [1–5]. This alternative location of reverse transcription, previously thought to only take place in the cytoplasm, has been supported by evidence from different laboratories [2–5]. Furthermore, a comprehensive investigation of HIV infection in non-mitotic and death-resistant macrophage cells [6] has revealed differences in viral replication kinetics [4, 7] and viral behavior within this particular cell type.

Vinayaka R. Prasad and Ganjam V. Kalpana (eds.), *HIV Protocols*, Methods in Molecular Biology, vol. 2807,
https://doi.org/10.1007/978-1-0716-3862-0_8,

Notably, in macrophage cells, it has been observed that viral complexes accumulate in specific regions of the nucleus [4, 8]. These compartments also harbor nuclear factors such as CPSF6, which interacts with HIV capsid [9–11], and nuclear speckle factors such as SC35 and SON [3, 4, 8]. Through the characterization of these viral/host clusters, their biomolecular condensate nature has been uncovered, leading to their designation as HIV-induced membraneless organelles (HIV-1 MLOs) [5, 12].

Membraneless organelles are present in cells under both homeostatic and stress conditions, and their assembly often involves a process called liquid–liquid phase separation (LLPS) [13–16]. LLPS occurs through the interaction of proteins that contain intrinsically disordered regions (IDRs) [17], such as CPSF6 [18, 20]. Here, we describe the materials and methods to characterize HIV-1 MLOs and explore their LLPS behaviors during viral postnuclear entry steps in macrophages [5, 19, 20]. The experimental techniques involve immunostaining of viral and host components, coupled with RNA FISH to visualize the HIV genome. Studies conducted on fixed cells allow for the examination of the components and morphology of HIV-1 MLOs (Fig. 1). Additionally, the generation of cells expressing fluorescent CPSF6 enables the monitoring of CPSF6 protein dynamics and diffusion within the MLOs that contain viral components, such as the integrase (IN), using time-lapse microscopy (Fig. 1).

2 Materials

2.1 Cell Culture and Differentiation (See Note 1)

1. HEK293T human embryonic kidney cells (ATCC) to produce lentiviral vectors (LVs) and HIV.
2. THP-1 monocytic cell line (ATCC).
3. Phorbol 12-myristate 13-acetate (PMA) to differentiate THP-1 (Sigma).
4. Dulbecco's modified Eagle's medium (DMEM), for HEK293T, supplemented with 10% FBS and 1% penicillin–streptomycin (100 U/mL).
5. Roswell Park Memorial Institute (RPMI) 1640 Medium, for THP-1, supplemented with 10% fetal bovine serum (FBS) and 1% penicillin–streptomycin solution (100 U/mL).

2.2 Plasmids and Vector Production

1. HIV-1ΔEnvIN-HA plasmid encodes for the genome of the single-round HIV-1 LAI (BRU). IN protein is fused to the HA tag [21, 22] for IN detection via immunostaining.

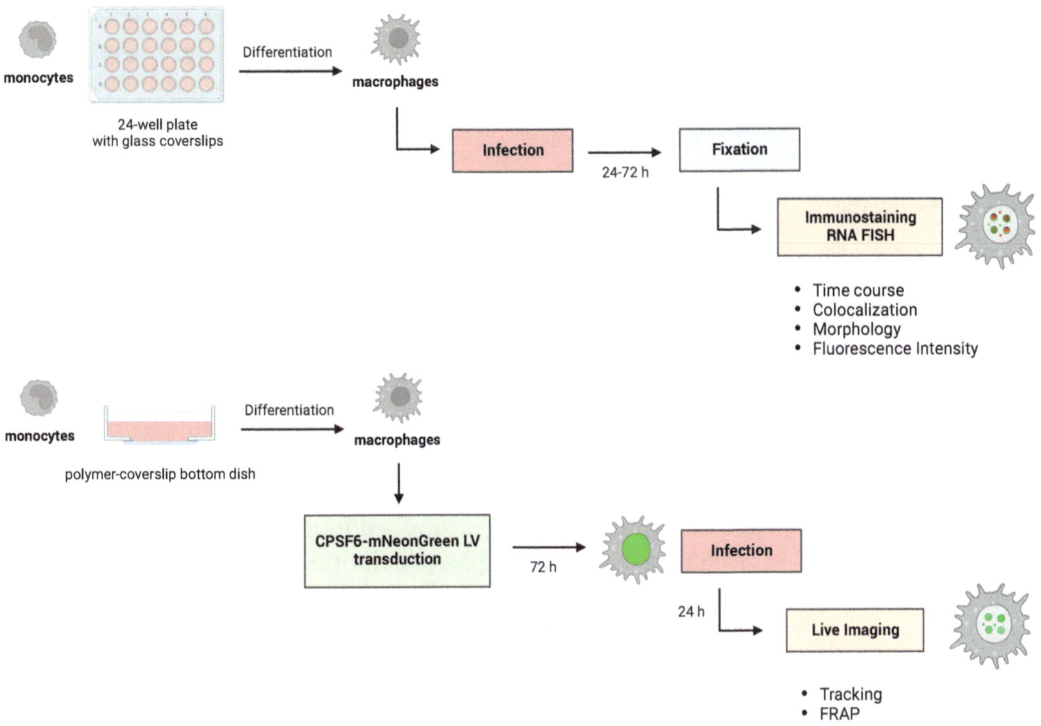

Fig. 1 Overview of HIV-1 MLO characterization via fluorescence imaging. For fluorescence microscopy, the cells are plated on glass coverslips. The protein staining and the probe hybridization to HIV sequences allow viral and host component detection, as well as the study of HIV-1 MLO morphology. For live imaging, the cells are plated on a polymer coverslip bottom dish. The stable expression of fluorescent CPSF6 permits the monitoring of CPSF6 proteins and HIV-1 MLOs in live

2. $SIV_{MAC-251}$ Vpx plasmid [23].
3. Gag-IN-Ruby (GIR) plasmid [24, 25].
4. pSICO-CPSF6-mNeonGreen encoding fluorescent CPSF6 (Addgene plasmid #16758) [26].
5. Packaging plasmid (gag-pol-tat-rev) (TheraVectys, Institut Pasteur).
6. pHCMV-VSV-G envelope plasmid.
7. 1 M calcium chloride solution.
8. 2× HEPES-buffered saline, PH 7.0.
9. Nuclease-free water.
10. SFCA membrane filters, 0.45 μm, for syringes.

2.3 Fixed-Cell Imaging

1. Coverslips, 12 mm, #1.5H.
2. 4% paraformaldehyde in PBS.
3. Dark humid chamber.

4. Glass slides.
5. ProLong Diamond Antifade Mountant.
6. Nuclease-free water.
7. Hoechst 33342.

2.3.1 Protein Staining

1. 0.15% glycine in PBS.
2. 0.5% Triton X-100 in PBS.
3. 1% BSA in PBS.
4. Anti-CPSF6 antibody (Novus Biologicals).
5. Anti-HA antibody (Roche).
6. Anti-rabbit Alexa 488 antibody.
7. Anti-rat Alexa 647 antibody.

2.3.2 Immuno-RNA FISH

1. Blocking/permeabilization solution (1% BSA, 2 mM vanadyl ribonucleoside complexes solution (Sigma), and 0.3% Triton X-).
2. Wash Buffer A: 20% Stellaris Wash Buffer A (LGC Biosearch Technologies), 10% deionized formamide, and 70% nuclease-free water.
3. Hybridization Buffer: 10% deionized formamide and 90% Stellaris Hybridization Buffer (LGC Biosearch Technologies).
4. Stellaris Wash Buffer B (LGC Biosearch Technologies).
5. HIV-1 *pol* primary probes (Table 1).
6. Fluorescent secondary probe (Cy5/AA TGC ATG TCG ACG AGG TCC GAG TGT AA/Cy5Sp/).
7. NEB 3.0 buffer.

2.3.3 Live Microscopy

1. Polymer-coverslip bottom μ-dish 35 mm (ibidi).
2. Hoechst 33342.

3 Methods

3.1 Viral and Vector Production

Lentiviral vectors (LVs) and HIV-1 viruses are produced via plasmid transfection in HEK293T cells through calcium chloride coprecipitation.

1. Plate six to seven million cells in 10-cm cell culture dishes.
2. The day after, prepare plasmid mix as follows (volumes per dish):
   - 125 μl of 1 M calcium chloride solution
   - For VSV-G HIV-1ΔEnvIN-HA + Vpx viruses: 10 μg HIV-1ΔEnvIN-HA plasmid, 2.5 μg of pHCMV-VSV-G

Table 1
Probes used for RNA FISH against HIV-1 pol

Probe name	Sequence
HIV1-01	GGG GAT TGT AGG GAA TTC CAA ATT CCT GCT TTT ACA CTC GGA CCT CGT CGA CAT GCA TT
HIV1-02	CTT TTA GCT GAC ATT TAT CAC AGC TGG CTA TTA CAC TCG GAC CTC GTC GAC ATG CAT T
HIV1-03	GTG TGC TGG TAC CCA TGC CAG ATA GAC TTA CAC TCG GAC CTC GTC GAC ATG CAT T
HIV1-04	AAT ACT GGA GTA TTG TAT GGA TTT TCA GGC CCT TAC ACT CGG ACC TCG TCG ACA TGC ATT
HIV1-05	TTT TAC TGG TAC AGT CTC AAT AGG GCT AAT GGT TAC ACT CGG ACC TCG TCG ACA TGC ATT
HIV1-06	TAT GTT GAC AGG TGT AGG TCC TAC TAA TAC TGT TAC ACT CGG ACC TCG TCG ACA TGC ATT
HIV1-07	CTA ATC CTC ATC CTG TCT ACT TGC CAT TAC ACT CGG ACC TCG TCG ACA TGC ATT
HIV1-08	CAA TCA TCA CCT GCC ATC TGT TTT CCA TTT ACA CTC GGA CCT CGT CGA CAT GCA TT
HIV1-09	TTT CCA AAG TGG ATT TCT GCT GTC CCT GTA TTA CAC TCG GAC CTC GTC GAC ATG CAT T
HIV1-10	TTG TGG ATG AAT ACT GCC ATT TGT ACT GCT GTT ACA CTC GGA CCT CGT CGA CAT GCA TT
HIV1-11	TTA AGA TGT TCA GCC TGA TCT CTT ACC TGT TTA CAC TCG GAC CTC GTC GAC ATG CAT T
HIV1-12	TAC AGT CTA CTT GTC CAT GCA TGG CTT CTT ACA CTC GGA CCT CGT CGA CAT GCA TT
HIV1-13	TCA TGT TCA TCT TGG GCC TTA TCT ATT CCT TAC ACT CGG ACC TCG TCG ACA TGC ATT
HIV1-14	TGT CAG TTA GGG TGA CAA CTT TTT GTC TTC CTT TAC ACT CGG ACC TCG TCG ACA TGC ATT
HIV1-15	TGC TCC TAC TAT GGG TTC TTT CTC TAA CTT TAC ACT CGG ACC TCG TCG ACA TGC ATT
HIV1-16	TCT GTT AGT GCT TTG GTT CCT CTA AGG AGT TTT TAC ACT CGG ACC TCG TCG ACA TGC ATT
HIV1-17	CTG TAT GTC ATT GAC AGT CCA GCT GTC TTT TTT ACA CTC GGA CCT CGT CGA CAT GCA TT
HIV1-18	TGG CAG CAC TAT AGG CTG TAC TGT CCT TAC ACT CGG ACC TCG TCG ACA TGC ATT

(continued)

Table 1
(continued)

Probe name	Sequence
HIV1-19	TCT GAT GTT TTT TGT CTG GTG TGG TAA GTC CCT TAC ACT CGG ACC TCG TCG ACA TGC ATT
HIV1-20	CCT CAA CAG ATG TTG TCT CAG CTC CTC TTA CAC TCG GAC CTC GTC GAC ATG CAT T
HIV1-21	ATT GCT GGT GAT CCT TTC CAT CCC TGT TAC ACT CGG ACC TCG TCG ACA TGC ATT
HIV1-22	TTT CTT TTT TAA CCC TGC GGG ATG TGG TAT TCT TAC ACT CGG ACC TCG TCG ACA TGC ATT
HIV1-23	TTT AAC TTT TGG GCC ATC CAT TCC TGG CTT ACA CTC GGA CCT CGT CGA CAT GCA TT
HIV1-24	CCC TAT CTT TAT TGT GAC GAG GGG TCG TTG TTA CAC TCG GAC CTC GTC GAC ATG CAT T

plasmid, and 3 μg of SIVMAC-251 Vpx [23]. VSV-G HIV-1ΔEnv + Vpx viruses can also be produced in combination with 7 μg GIR (Gag-IN-Ruby plasmid) [24, 25] fluorescent integrase (IN) detection.

- For CPSF6-mNeonGreen LV, 10 μg of transfer vector (Addgene #167587), 10 μg of packaging plasmid (gag--pol-tat-rev), and 2.5 μg of pHCMV-VSV-G.
- Nuclease-free water to 500 μl of final volume.

3. Add drop-by-drop the plasmid mix to a tube containing HEPES-buffered saline (500 μl per dish), while vortexing the tube.
4. Add drop-by-drop the transfection mix to the cells, making sure to cover the whole area of the dish.
5. Change media 4–5 h post-transfection.
6. Forty-eight hours after transfection, collect the supernatant in 50-mL falcon tubes.
7. Centrifuge at 2500 rpm for 5 min to clarify.
8. Filter the supernatant with SFCA membrane filters, 0.45 μm.
9. Concentrate viral particles with ultracentrifugation at 22,000 rpm, for 1 h at 4 °C.
10. Resuspend viral pellet in desired volume.
11. Store viral aliquots at −80 °C.

Lentiviral vectors and viruses can be titered by qPCR of U5R sequence, 3 days post-transduction of cells of interest.

3.2 Imaging of Endogenous HIV-1 MLOs

To investigate the localization of viral components with respect to host nuclear factors and the formation of HIV-induced membraneless organelles (HIV-1 MLOs), HIV infection can be performed in different target cells, such as HeLa, lymphocytes, and macrophages. HIV-1 MLO frequency of detection and kinetics of formation highly depend on the cell type and multiplicity of infection (MOI).

In macrophage-like cells, HIV-1 MLOs can be easily spotted in the nucleus of infected cells 24 h postinfection (p.i.) using an MOI of 5 and a confocal microscope for their visualization. They appear as quasi-spherical signals in the interchromatin space colocalizing with viral integrase (IN) and incoming viral RNA [5] (Fig. 2).

3.2.1 Viral and Host Factor Detection in HIV-1 MLOs

1. Plate five hundred thousand THP-1 cells in a 24-well plate with 12-mm Ø coverslips on the bottom, making a mix of cells, RPMI, and PMA (160 nM) for differentiation (coverslip should not float and should stay at the bottom of the well).
2. Forty-eight hours post-differentiation, the adherent cells are infected with VSV-G HIV-1ΔEnvIN-HA + Vpx viruses (MOI between 5 and 20) in RPMI (10% FBS, 1% penicillin–streptomycin solution (100 U/mL), and PMA (160 nM)) (*see* **Note 2**). Remove supernatant, wash the well once with RPMI, and aliquot the virus in the well.

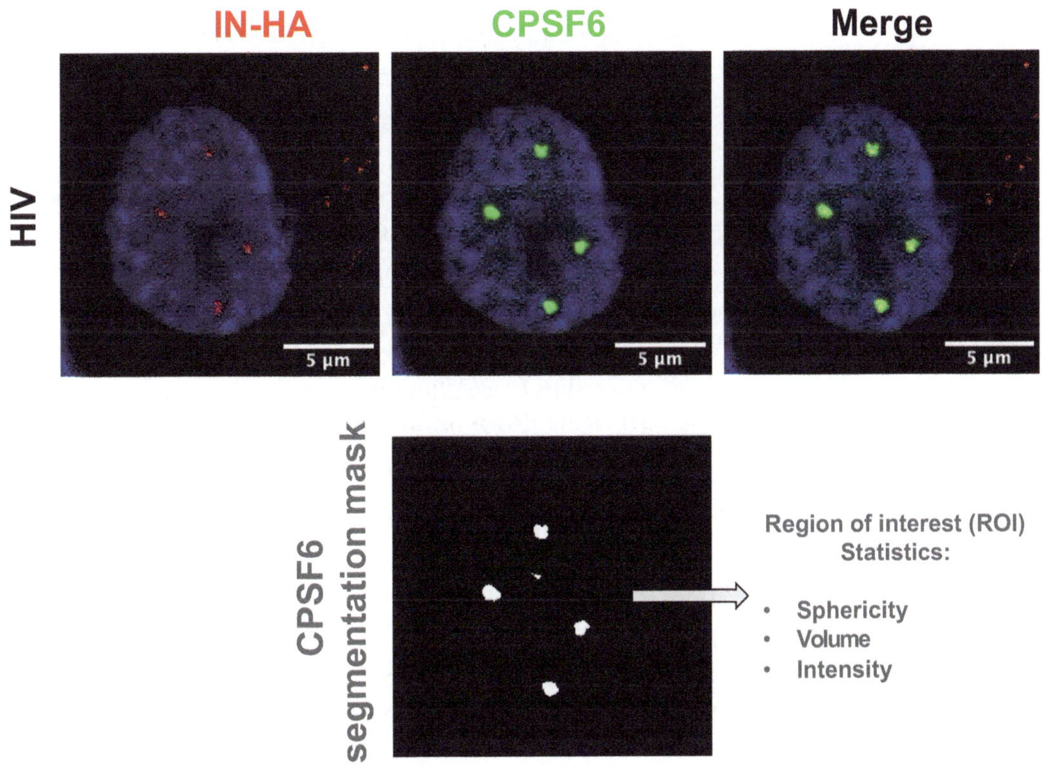

Fig. 2 Analysis of HIV-1 MLOs in fixed cells. Differentiated THP-1 cell, 3 days p.i. with HIV-IN-HA (MOI 5). The panel shows the mask for CPSF6 signal segmentation generated in Icy software via HK-means plugin

3. Infection is stopped by removing supernatant and washing the coverslips twice with PBS. Cells are fixed according to time points of study with 4% paraformaldehyde (PFA) for 15 min.
4. Discard PFA and wash twice with PBS (*see* **Note 3**).

Integrase/CPSF6 Protein Staining

1. Treat the coverslips in the wells with 0.15% glycine for 10 min.
2. Wash once with PBS and permeabilize cells with 0.5% Triton X-100 for 30 min.
3. Wash once with PBS and block with 1% bovine serum albumin (BSA) for 30 min.
4. Make primary antibody mix in 1% BSA at 40 μl per coverslip (anti-HA 1:500 + anti-CPSF6 1:400) (*see* **Note 4**).
5. Place coverslips on parafilm in a dark humid chamber. Pipet a 40 μl drop of antibody mix onto the coverslip.
6. Incubate 1 h at room temperature in the dark.
7. Make secondary antibody mix in 1% BSA at 40 μl per coverslip (for anti-rat for IN-HA 1:100 + anti-rabbit for CPSF6 1:300–1:500).
8. Wash the coverslips 10 times, plunging them in BSA 1% solution.
9. Place coverslips on parafilm in a dark humid chamber. Pipet a 40 μl drop of antibody mix onto the coverslips.
10. Incubate 45 min at room temperature in the dark.
11. Wash the coverslips 10 times, plunging them in 1% BSA solution.
12. Place coverslips on parafilm and pipet a 40 μl drop of Hoechst diluted in water (1:10,000) onto the coverslips.
13. Incubate 5 min at room temperature in the dark.
14. Prepare a drop of ProLong Diamond Antifade Mountant on glass slides.
15. Wash coverslips twice plunging them in water and mount with the cells facing down on the mountant drop. Press gently.
16. Allow mounting media to dry, in the dark, before imaging.

HIV-1 RNA/CPSF6 Immuno-RNA FISH

1. Treat coverslips in the wells with the blocking/permeabilization solution (1% BSA, 2 mM vanadyl ribonucleoside complexes solution (Sigma #94742), and 0.3% Triton X-100).
2. Proceed as in paragraph above for antibody staining (**steps 4–11**) but keep all reagents in RNAse-free condition. Make antibody mix in blocking/permeabilization solution and perform washes in PBS.

3. Fix coverslips with 4% PFA for 10 min (a second time).
4. Wash coverslips twice with PBS and store at 4 °C until RNA FISH reagents are ready.
5. Prepare the fluorophore-conjugated probes for hybridization:
   - Prepare the mix of 24 smiFISH probes (Table 1) at 20 μM.
   - Prepare the alignment mix enough for 1 μl per coverslip.

     Example for 10 μl Mix (About Nine Coverslips)

     - 40 pmol of primary probe (20 μM)—2 μl.
     - 50 pmol of probe—FLAP-Cy5—0.5 μl.
     - 1× NEB 3 buffer (10×)—1 μl.
     - Nuclease-free H2O to 10 μl— 6.5 μl.
   - Mix well and perform alignment cycle in PCR machine (85 ° C 3 min, 65 °C 3 min, 25 °C 5 min, 4 °C).
6. Take the plate with fixed coverslips, remove PBS, and incubate with Wash Buffer A at room temperature up to 5 min (*see* **Note 5**).
7. Add the aligned probe mix to the Hybridization Buffer (1:50) (e.g., 10 μl mix in 500 μl of buffer) and mix well.
8. Place coverslips on parafilm in a dark humid chamber and pipet a 50 μl drop of hybridization mix onto the coverslips.
9. Incubate overnight at 37 °C in the dark.
10. The day after, move coverslips to a clean 24-well plate and add Wash Buffer A.
11. Incubate the plate for 30 min at 37 °C in the dark.
12. Make Hoechst mix in nuclease-free water, 1:5000.
13. Remove Wash Buffer A and add Hoechst to the wells.
14. Incubate 5 min at room temperature in the dark.
15. Remove Hoechst and add Wash Buffer B (prepare according to Stellaris protocol) to the wells.
16. Incubate up to 5 min at room temperature.
17. Prepare a drop of ProLong Diamond Antifade Mountant on glass slides.
18. Wash coverslips twice plunging them in water and mount with the cells facing down on the mountant drop. Press gently.
19. Dry the coverslips in the dark before imaging.

Three-dimensional images can be acquired with a fluorescence microscope or a confocal microscope for a better resolution (e.g., an inverted LSM700 microscope, Zeiss) with a 63× objective (Plan-Apochromat, oil immersion, NA = 1.4).

3.2.2 HIV-1 MLO 3D Analysis

From the acquired images, it is possible to visualize CPSF6 clusters, only in the infected condition, while, in uninfected cells, CPSF6 is randomly distributed inside the nucleus [5]. CPSF6 clusters colocalize with most of the intranuclear HIV integrase protein foci and viral RNA (Fig. 2). The viral RNA in CPSF6 clusters is present under nevirapine treatment, meaning that the incoming particles inside the organelles have not yet started the reverse transcription [5].

To characterize these viral/host clusters, we have analyzed two different time points (24 h and 72 h p.i.). By segmenting CPSF6 signal (Fig. 2), it is possible to retrieve different statistics of the region of interest (ROI) such as the volume, the sphericity, and the signal intensity [5].

With the Icy software [27], CPSF6 cluster segmentation can be performed via HK-Means block [28] (https://icy.bioimageanalysis.org/plugin/hk-means/) followed by the ROI statistics exportation (https://icy.bioimageanalysis.org/plugin/roi-statistics/) (Fig. 2).

3.3 Live Cell Imaging of HIV-1 MLOs

With the aim of studying HIV-1 MLO dynamics, it is necessary to establish a system to track CPSF6 and HIV in living cells. The target cells can be transduced with the fluorescent CPSF6 lentiviral vector and infected with HIV, fluorescent (HIV GIR), or not (*see* Subheading 3.1 for production).

3.3.1 Fluorescent CPSF6 Transduction

1. Plate one to two million THP-1 cells on a polymer-coverslip bottom dish, making a mix of cells in RPMI (10% FBS, 1% penicillin–streptomycin solution (100 U/mL), and PMA (160 nM)).
2. Differentiate THP-1 cells into macrophage-like cells for 48 h.
3. Transduce the adherent cells with CPSF6 lentiviral vector in RPMI (10% FBS, 1% penicillin–streptomycin solution (100 U/mL), and PMA (160 nM)). To avoid excessive protein overexpression and potential formation of aggregates, it is important to use a low MOI, between 0.01 and 0.2.
4. Remove supernatant, wash once with RPMI, and aliquot the viral mix.
5. Let transduction last for 3 days.
6. Remove supernatant and infect cells with HIV-IN-HA or with HIV GIR virus at MOI 10 in RPMI (10% FBS, 1% penicillin–streptomycin solution (100 U/mL), and PMA (160 nM)) (*see* **Note 2**).
7. Proceed to live imaging acquisition at the selected time point p.i. (*see* **Note 6**).

When CPSF6-mNeonGreen THP-1-infected cells are fixed and stained for CPSF6 and for IN-HA, it is possible to observe that almost the totality of HIV-1 MLOs are NeonGreen positive, and the CPSF6-mNeonGreen clusters colocalize with HIV complexes indicated by the IN labeled. This indicates that the exogenous CPSF6 marks the majority of HIV-1 MLOs and that the proteins behave as the endogenous ones [5].

If CPSF6-mNeonGreen THP-1 cells are infected with HIV GIR virus, it is possible to follow, in live, the IN-labeled virus in CPSF6 clusters.

3.3.2 Dynamicity of HIV-1 MLOs

Starting at 24 h postinfection (p.i.), cells that exhibit fluorescent HIV-1 MLOs can be tracked over time. The use of a spinning-disk microscope, such as Nikon Ti2-E microscope, allows for the monitoring of HIV-1 MLO dynamics, revealing their high level of dynamism with fusion and fission events, which are characteristic of biomolecular condensates [5].

3.3.3 Diffusion of CPSF6 Proteins Within HIV-MLOs

Fluorescence recovery after photobleaching (FRAP) can be employed to study the spontaneous assembly of CPSF6 protein and the maintenance of equilibrium within HIV-1 MLO components. This technique allows for the investigation of protein diffusion. Using a spinning-disk microscope with a FRAP module, the stably transduced and infected cells are identified, and the area corresponding to the HIV-1 MLO is irradiated with a 488 laser for a duration of 200 milliseconds (ms) per pixel. This irradiation results in the bleaching of NeonGreen fluorescence within the region (Fig. 3). Subsequently, a time lapse is initiated to monitor the fluorescence restoration of the CPSF6 cluster. This restoration occurs due to the constant exchange of molecules between the interior and exterior of the MLO, becoming detectable as a decrease in fluorescence intensity of the depleted CPSF6 proteins (Fig. 3).

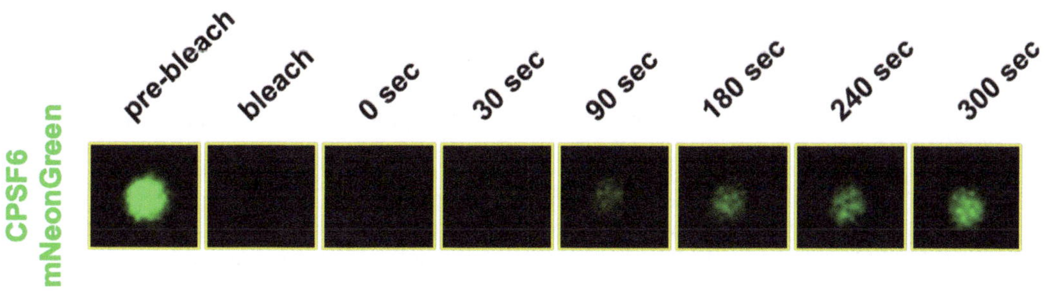

Fig. 3 Fluorescence recovery after photobleaching of CPSF6 in HIV-1 MLO. Frames extracted from fluorescence recovery after photobleaching (FRAP) time lapse in THP-1 cells expressing CPSF6 mNeonGreen, infected with HIV-1 (MOI 10, 3 days p.i)

4 Notes

1. All cells are kept in incubator at 37 °C and 5% CO2.
2. For control conditions, make viral/RPMI/PMA mix adding drugs such as nevirapine (NEV) 10 μM or raltegravir (RAL) 20 μM.
3. The plate with coverslips in PBS can be stored at 4 °C (no longer than 3 days; for the immune-RNA FISH, it is recommended to process the samples the same day).
4. Secondary-only control coverslips are washed and kept in PBS until secondary antibody incubation.
5. Manipulate Wash Buffer A and Hybridization Buffer under chemical hood since they contain formamide.
6. At the time point of acquisition, change medium with fresh RPMI with Hoechst 1:10,000 for nuclei staining.

References

1. Scoca V, Nunzio FD (2021) The HIV-1 capsid: from structural component to key factor for host nuclear invasion. Viruses 13:273
2. Dharan A, Bachmann N, Talley S et al (2020) Nuclear pore blockade reveals that HIV-1 completes reverse transcription and uncoating in the nucleus. Nat Microbiol 5:1088–1095
3. Selyutina A, Persaud M, Lee K et al (2020) Nuclear import of the HIV-1 core precedes reverse transcription and uncoating. Cell Rep 32(13):108201
4. Rensen E, Mueller F, Scoca V et al (2021) Clustering and reverse transcription of HIV-1 genomes in nuclear niches of macrophages. EMBO J 40:e105247
5. Scoca V, Morin R, Collard M et al (2023) HIV-induced membraneless organelles orchestrate post-nuclear entry steps. J Mol Cell Biol 14(11):mjac060. https://doi.org/10.1093/jmcb/mjac060
6. Le Douce V, Herbein G, Rohr O et al (2010) Molecular mechanisms of HIV-1 persistence in the monocyte-macrophage lineage. Retrovirology 7:1–16
7. Arfi V, Rivière L, Jarrosson-Wuillème L et al (2008) Characterization of the early steps of infection of primary blood monocytes by human immunodeficiency virus type 1. J Virol 82:6557–6565
8. Francis AC, Marin M, Singh PK et al (2020) HIV-1 replication complexes accumulate in nuclear speckles and integrate into speckle-associated genomic domains. Nat Commun 11(1):3505
9. Price AJ, Fletcher AJ, Schaller T et al (2012) CPSF6 defines a conserved capsid interface that modulates HIV-1 replication. PLoS Pathog 8:e1002896
10. Bhattacharya A, Alam SL, Fricke T et al (2014) Structural basis of HIV-1 capsid recognition by PF74 and CPSF6. Proc Natl Acad Sci USA 111:18625–18630
11. Buffone C, Martinez-Lopez A, Fricke T et al (2018) Nup153 unlocks the nuclear pore complex for HIV-1 nuclear translocation in nondividing cells. J Virol 92:e00648–e00618
12. Scoca V, Di Nunzio F (2021) Membraneless organelles restructured and built by pandemic viruses: HIV-1 and SARS-CoV-2. J Mol Cell Biol 13:259–268
13. Brangwynne CP, Eckmann CR, Courson DS et al (2009) Germline P granules are liquid droplets that localize by controlled dissolution/condensation. Science 324:1729–1732
14. Banani SF, Rice AM, Peeples WB et al (2016) Compositional control of phase-separated cellular bodies. Cell 166:651–663
15. Alberti S, Gladfelter A, Mittag T (2019) Considerations and challenges in studying liquid-liquid phase separation and biomolecular condensates. Cell 176:419–434
16. Di Nunzio F (2023) Stress-induced condensate switch awakens sleeping viruses. Cell Host Microbe 31:679–680

17. Li P, Banjade S, Cheng HC et al (2012) Phase transitions in the assembly of multivalent signalling proteins. Nature 483(7389):336–340
18. Greig JA, Nguyen TA, Lee M et al (2020) Arginine-enriched mixed-charge domains provide cohesion for nuclear speckle condensation. Mol Cell 77:1237–1250.e1234
19. Ay S, Di Nunzio F (2023) HIV-induced CPSF6 condensates. J Mol Biol 435(16): 168094
20. Di Nunzio F, Uversky VN, Mouland AJ (2023) Biomolecular condensates: insights into early and late steps of the HIV-1 replication cycle. Retrovirology 20(1):1–20
21. Petit C, Schwartz O, Mammano F (1999) Oligomerization within virions and subcellular localization of human immunodeficiency virus type 1 integrase. J Virol 73:5079–5088
22. Petit C, Schwartz O, Mammano F (2000) The karyophilic properties of human immunodeficiency virus type 1 integrase are not required for nuclear import of proviral DNA. J Virol 74: 7119–7126
23. Durand S, Nguyen X-N, Turpin J et al (2013) Tailored HIV-1 vectors for genetic modification of primary human dendritic cells and monocytes. J Virol 87:234–242
24. Hulme AE, Kelley Z, Foley D et al (2015) Complementary assays reveal a low level of CA associated with viral complexes in the nuclei of HIV-1-infected cells. J Virol 89: 5350–5361
25. Dharan A, Talley S, Tripathi A et al (2016) KIF5B and Nup358 cooperatively mediate the nuclear import of HIV-1 during infection. PLoS Pathog 12:e1005700
26. Zhong Z, Ning J, Boggs EA et al (2021) Cytoplasmic CPSF6 regulates HIV-1 capsid trafficking and infection in a cyclophilin A-dependent manner. MBio 12:e03142–e03120
27. De Chaumont F, Dallongeville S, Chenouard N et al (2012) Icy: an open bioimage informatics platform for extended reproducible research. Nat Methods 9(7):690–696
28. Dufour A, Meas-Yedid V, Grassart A et al (2008) Automated quantification of cell endocytosis using active contours and wavelets. In: Proceedings – international conference on pattern recognition. Tampa, FL, USA, pp. 1–4, https://doi.org/10.1109/ICPR.2008.4761748

Chapter 9

Detection of CPSF6 in Biomolecular Condensates as a Reporter of HIV-1 Nuclear Import

Charlotte Luchsinger and Felipe Diaz-Griffero

Abstract

The initial stages of HIV-1 infection involve the transport of the viral core into the nuclear compartment. The presence of the HIV-1 core in the nucleus triggers the translocation of CPSF6/CPSF5 from paraspeckles into nuclear speckles, forming puncta-like structures. While this phenomenon is well-documented, the efficiency of CPSF6 translocation to nuclear speckles upon HIV-1 infection varies depending on the type of cell used. In some human cell lines, only 1–2% of the cells translocate CPSF6 to nuclear speckles when exposed to a 95% infection rate. To address the issue that only 1–2% of cells translocate CPSF6 to nuclear speckles when a 95% infection rate is achieved, we screened several human cell lines and identified a human a cell line in which approximately 85% of the cells translocate CPSF6 to nuclear speckles when 95% infection rate is achieved. This cellular system has enabled the development of a robust fluorescence microscopy method to quantify the translocation of CPSF6 into nuclear speckles following HIV-1 infection. This assay holds the potential to support studies aimed at understanding the role of CPSF6 translocation to nuclear speckles in HIV-1 infection. Additionally, since the translocation of CPSF6 into nuclear speckles depends on the physical presence of the viral core in the nucleus, our method also serves as a reporter of HIV-1 nuclear import.

Key words HIV-1, CPSF6, CPSF5, Nuclear speckles, Nuclear import, Biomolecular condensates, Phase separation, Fluorescence microscopy

1 Introduction

The viral core, consisting of 1500 monomers of the capsid protein (p24), encases the viral RNA genome. This viral core migrates to the nucleus, where it is believed that the RNA undergoes a process of reverse transcription to become viral DNA [1–3]. As reverse transcription takes place, the viral core undergoes uncoating, which involves the dissociation of monomeric capsid proteins from the core. The presence of the viral core in the nucleus triggers the translocation of cleavage and polyadenylation-specific factor 6 (CPSF6) from nuclear paraspeckles to nuclear speckles (NS) [2, 4, 5] (Fig. 1). Nuclear speckles are dynamic,

Vinayaka R. Prasad and Ganjam V. Kalpana (eds.), *HIV Protocols*, Methods in Molecular Biology, vol. 2807,
https://doi.org/10.1007/978-1-0716-3862-0_9,

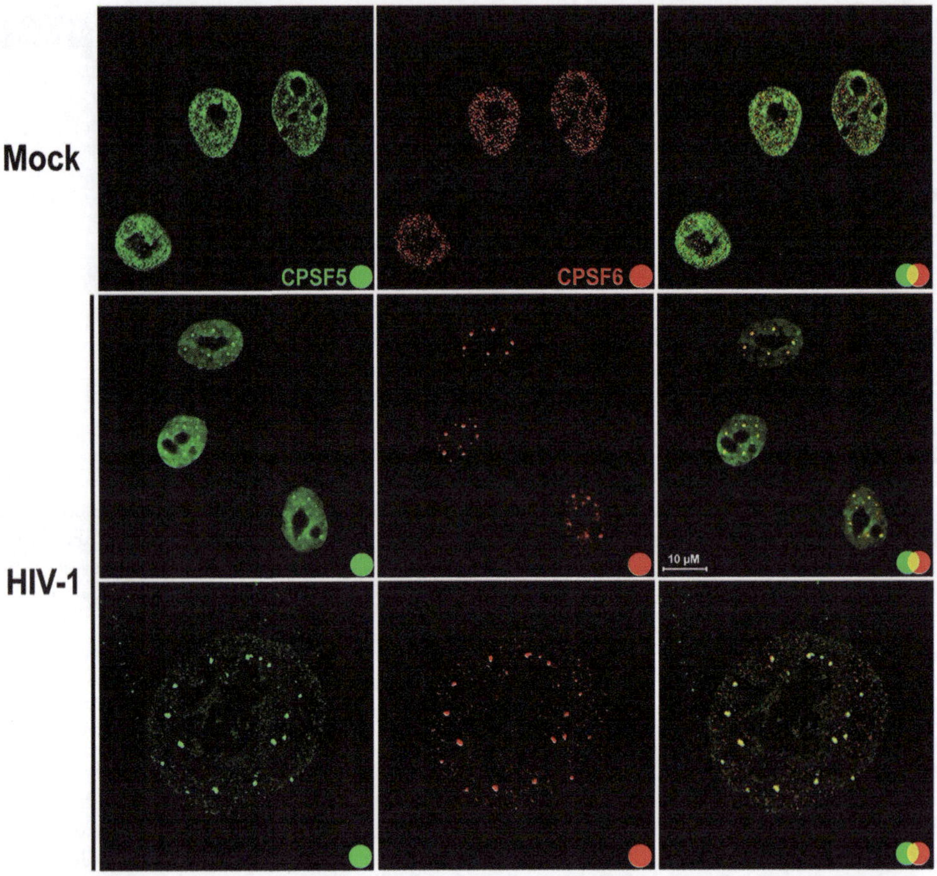

Fig. 1 Translocation of CPSF6 and CPSF5 to biomolecular condensates (nuclear speckles) induced by HIV-1 infection. A549 cells were either infected with HIV-1-Luc at an approximate MOI of 2 or left uninfected (mock) for 24 h. Following this, cells were fixed, permeabilized, and co-immunostained using a rabbit polyclonal antibody targeting CPSF6 alongside a mouse monoclonal antibody against CPSF5. Secondary antibodies used were Alexa-488-conjugated donkey anti-mouse IgG (green channel) and Alexa-594-conjugated donkey anti-rabbit IgG (red channel). Subsequently, stained cells were observed using fluorescence microscopy. The third image in each row was generated by merging the red and green channels, where yellow color indicates the overlapping localization of the red and green channels. These experiments were replicated at least three times, and the provided images represent a typical result. Scale bar = 10 μm. (This figure was taken from our publication [13])

boundary-less structures that are clearly defined entities within the nucleus and behave as biomolecular condensates [6]. NS, which play a crucial role in regulating expression of highly active genes, consist of proteins and RNA that coordinate transcription and splicing processes [7–9]. NS are also known as SC35-rich domains [10] as they are easily visualized by antibodies directed against the spliceosome assembly factor SC35 [11].

The recruitment of CPSF6 and CPSF5 to NS induced by HIV-1 infection requires the presence of an intact core in the nuclear compartment (Fig. 1), as viruses with capsid mutations N74D,

A77V, or N57S are unable to induce the translocation of CPSF6 to NS [2, 4]. Translocation of CPSF6 to NS induced by HIV-1 infection alters the fluorescence microscopy pattern of CPSF6 from diffuse nuclear staining to large CPSF6 condensates or puncta-like structures that are readily recognizable [2, 4] (Fig. 1). The addition of the small pharmacological molecules PF74, GS-CA-1, or lenacapavir during infection inhibits translocation of CPSF6 to NS [2, 4, 12].

Genetic mapping studies to identify viral determinants for the translocation of CPSF6 to NS revealed that viral particles formed by expressing HIV-1 gag-pol pseudotyped with VSV-G are sufficient for inducing CPSF6 translocation to NS [13]. We showed that translocation of CPSF6 to NS does not require reverse transcription, integration, or viral genomic RNA in the particle [13]. In addition, we showed that nuclear speckles containing CPSF6/CPSF5 upon HIV-1 infection behave as biomolecular condensates since the use of osmotic stress and 1,6-hexanediol disrupts these structures [13]. The role of this translocation in HIV-1 replication, if any, is not understood.

Although the translocation of CPSF6 to NS upon HIV-1 infection is widely accepted, its study has been limited by the experimental models available. In some human cell lines, only 1–2% of the infected cells translocate CPSF6 to nuclear speckles upon a total 95% infection rate. This limits the dynamic range of the assay making it difficult to use it as a reliable assay to study translocation of CPSF6 to NS. To address this issue, we performed a screen to identify human cells that will induce translocation of CPSF6 to NS in 80–100% of the infected cells (Fig. 2) [13]. As shown in Fig. 2, we identified the human cell line A549, where 85% of infected cells translocate CPSF6 to NS, as a reliable model for studying CPSF6 translocation to NS upon HIV-1 infection [13]. This method provides a dependable and quantitative assay for measuring the translocation of CPSF6 to NS upon HIV-1 infection. This assay can be used to investigate the significance of CPSF6 translocation to nuclear speckles in the context of HIV-1 infection. This assay can also be used as a reporter for the presence of the HIV-1 core in the nuclear compartment, making it a surrogate assay for HIV-1 nuclear import.

2 Materials

Utilize analytical reagents and ultrapure water (with a resistance of >18 megohms/cm) to prepare all solutions. Store these solutions at room temperature unless specified otherwise. Adhere meticulously to all waste disposal regulations when discarding waste materials, particularly when dealing with HIV-1-infected samples.

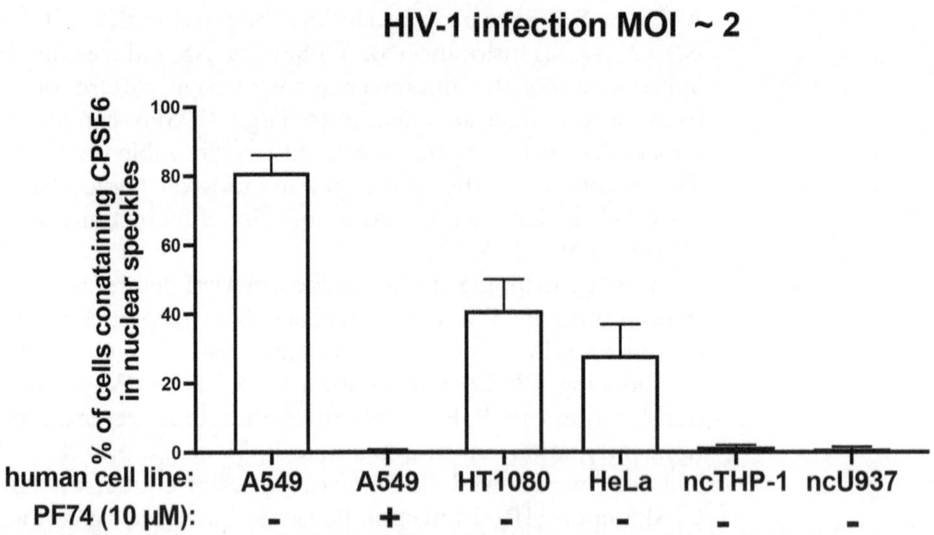

Fig. 2 Percentage of infected cells where translocation of CPSF6 to nuclear speckles upon HIV-1 infection has occurred. A549, HT1080, HeLa, THP-1, and U937 cells were infected with HIV-1-GFP at an approximate MOI of 2 (tittered in human A549 cells), in the presence or absence of 10 μM PF74, for 24 h. For THP-1 and U937 cells, PMA was used to induce differentiation to a non-cycling (nc) state before infection. Following infection, the cells were fixed, permeabilized, and co-immunostained using a rabbit polyclonal antibody targeting CPSF6. The secondary antibody employed was Alexa-594-conjugated donkey anti-rabbit IgG (red channel). Nuclei were stained with DAPI. Stained cells were examined through fluorescence microscopy. The percentage of cells containing CPSF6 in nuclear speckles was determined by visually inspecting 100 cells per sample in three independent experiments (total $n = 240$ cells)

2.1 Cell Culture

1. A549 cells are a line derived from human lung carcinoma (ATCC; Cat# CCL-1885).
2. HEK293/17 cells are derived from human embryo kidney tissue (ATCC; Cat# CRL-11268).
3. Dulbecco's Modified Eagle Medium (DMEM) high glucose supplemented with 10% heat-inactivated FBS, 100 U/mL penicillin, 100 μg/mL streptomycin, 29.2 mg/mL L-glutamine, and 5 μg/mL plasmocin, which is used to prevent mycoplasma contamination (complete DMEM).
4. Dulbecco's phosphate buffered saline (PBS 1×).
5. 0.25% Trypsin, and 2.21 mM EDTA.

2.2 HIV-1 Viruses and Plasmids

1. Pseudotyped HIV-1 viruses with the envelope of VSV-G expressing green fluorescent protein (GFP) or luciferase (Luc) as a reporter were prepared, as described below [14].
2. Polyethylenimine (PEI) 1 μg/mL.
3. Plasmids expressing the following proteins or polyproteins: HIV-1 gag-pol (NL4-3), Tat, Rev, and VSV-G.

4. Plasmids containing a GFP reporter for HIV-1: LTR-GFP-LTR.
5. Methanol-free Paraformaldehyde (PFA).
6. Glycine.

2.3 Immunofluorescence Reagents

1. Washing solution: PBS 1× in ultrapure H_2O.
2. Fixing solution: 4% Paraformaldehyde (PFA) in PBS 1×.
3. Quenching solution: 0.1 M glycine in PBS 1×.
4. Permeabilization solution: 0.5% Triton-X100 in PBS 1×.
5. Blocking solution: 3% bovine serum albumin (BSA) in PBS 1×.
6. Primary antibodies: 1:1000 dilution of rabbit α-CPSF6 (Cat#ab99347, Abcam), mouse α-CPSF5 antibody (Cat#H00011051-M12, Novus Biologicals) and mouse α-SC35 antibody (Cat#ab11826, Abcam) 3% BSA in PBS 1×.
7. Secondary antibodies: 1:1000 dilution of Alexa Fluor 488-conjugated, Alexa Fluor 594-conjugated, or Alexa Fluor 647-conjugated donkey α-rabbit IgG and α-mouse IgG (Cat#A21206; #A21202; #A21207; #A21203, #A31573, Life Technologies), 3% BSA in PBS 1×.
8. Nuclei staining: 1 μg/mL 4′,6-diamidino-2-phenylindole (DAPI) in 3% BSA-PBS 1× (Cat# D1306, Life Technologies).
9. Mounting medium for fluorescence microscopy (Cat#345789, EMD Millipore).
10. Ethanol 70%.

2.4 Acquisition, Processing, and Analysis of Images

1. Fluorescence microscope with filters for 405, 488, and 594 nm wavelengths (Zeiss AxioObserverZ1 Motorized Inverted Fluorescence Microscope). This microscope is equipped with a 63× oil immersion objective (NA 1.4) and a digital camera (AxioCam MRm, Carl Zeiss).
2. A computer with a software with the capacity of performing deconvolution using an acquired point spread function (ZEN 3.3 blue edition software).
3. Adobe Photoshop CS5 software.

3 Methods

This protocol outlines an indirect immunofluorescence method to simultaneously visualize the translocation of CPSF6 and CPSF5 proteins to nuclear speckles during HIV-1 infection (Fig. 1). We will provide a comprehensive description of this procedure using A549 cells, a cellular model that we have previously validated for the quantitative analysis and characterization of CPSF5/CPSF6 condensates [13]. This method offers the advantage of being easily adaptable to various cell types, such as monocytes, T cells, and macrophages.

3.1 Preparation of Cells

1. Clean and sterilize the coverslips by immersing them in 70% ethanol.
2. Eliminate any remaining ethanol from the coverslips using a vacuum and allow them to air dry.
3. Position the sterilized coverslips within a 24-well plate.
4. Detach cells from the stock culture utilizing 0.25% trypsin-2.21 mM EDTA, centrifuge at 2000 rpm for 5 min, resuspend, and manually count the cells using a Neubauer counting chamber.
5. Seed cells into the 24-well plate with the coverslips at a final density of 6×10^5 cells per well in 1 mL of complete DMEM. Gently press the coverslips with a tip to eliminate any bubbles formed between the plate and the coverslip. The density may be adjusted based on the cell line used. The optimal cell culture confluence is approximately 30–40%.
6. Incubate the seeded plates in a 5% CO2 incubator at 37 °C overnight.
7. Proceed with the infection protocol.

3.2 HIV-1 Infection

3.2.1 Production of HIV-1 Stocks

Prepare in advance stock of HIV-1 viruses expressing green fluorescent protein (GFP) or luciferase (Luc) as a reporter, as previously described [14]:

1. Seed 3×10^6 HEK293T/17 cells in a 10 cm dish (approximately 70% confluence). Incubate the seeded plates in a 5% CO2 incubator at 37 °C overnight.
2. Prepare a DNA mixture in 3 mL of serum-free DMEM containing 1 μg of each of the following plasmids: HIV-1 gag-pol, tat, rev, and VSV-G; and 5 μg of plasmids LTR-GFP-LTR or LTR-Luc-LTR [14].
3. Create a PEI mixture in 3 mL of serum-free DMEM containing 45 μg of PEI (DNA:PEI ratio 1:5).
4. Gently combine the PEI mixture with the DNA mixture (6 mL). Mix gently and incubate the transfection mixture for 30 min at room temperature.
5. Carefully add the transfection mixture dropwise to the HEK293T/17 cells. Do not conduct prior washing to prevent cell detachment. Incubate the transfected plates in a 5% CO2 incubator at 37 °C overnight.

6. Replace the media with 6 mL of complete DMEM after removing the previous media. Incubate the cells in a 5% CO_2 incubator at 37 °C for approximately 48 h.
7. Collect the viral supernatant from each plate and consolidate all individual harvests in one main stock. Centrifuge the viral supernatant at 2000 rpm to pellet the 293 cells collected during harvest. Pass the viral supernatant through a 0.2 μm syringe filter.
8. Virus infectivity is tested by challenging A549 cells in a 24-well plate (4×10^4 cells per well). Forty-eight hours post-challenge, infection is determined by measuring the percentage of GFP-positive cells using a flow cytometer.
9. Determine the viral stock titer using the following equation in A549 cells: Infectious units (IU)/mL = (cell number) × (% of GFP-positive cells) × (dilution factor), where the dilution factor = 1000 μL/viral input (μL).
10. Aliquot the viral stock and store it at −80 °C to prevent loss of infectivity.

3.2.2 Infecting the Cells

1. Calculate the volume of HIV-1 stock for an MOI ~2, using the following equation: MOI = [(virus stock IU/mL) × (volume of virus used)]/ (number of cells in infection).
2. Prepare an infection mixture with the virus volume calculated in the previous step in complete DMEM. Preferably, to prepare the infection mixture, use warm medium (37 °C). For a 24-well plate, use 1 mL as final volume per infection.
3. Remove the media from the plates seeded the previous day and wash with PBS 1×. Prevent cells from drying out.
4. Infect cells by adding 1 mL of the warm infection mixture into the cells.
5. Incubate the infected plates in a 5% CO_2 incubator at 37 °C for ~24 h. (Time of infection, *see* **Note 1**).
6. Proceed with staining protocol.

3.3 Indirect Immunofluorescence

3.3.1 Solutions Preparation

Before starting the staining protocol, prepare all essential materials and solutions, which include washing, fixing, quenching, permeabilization, and blocking solutions, along with the primary antibody, secondary antibody, and nucleus staining solution. Prepare enough solutions, accounting for the usage of 1 mL per coverslip, except for the antibodies and nucleus staining solution, which needs smaller volumes.

1. Washing solution: Prepare stock solution of PBS 1× by adding 50 mL 10× PBS to 450 mL ultrapure H_2O in a 500 mL sterile bottle. Mix and adjust pH to 8.0.

2. Fixing solution: Prepare 4% PFA solution by adding 10 mL 16% w/v PFA methanol free to 30 mL PBS 1× in a 50 mL conical tube. Gently rotate the tube for 15 min. Centrifuge at 5000 rpm. Always use fresh fixing solution.
3. Quenching solution: Prepare 0.1 M glycine in PBS 1× solution. To this end, weigh 300 mg of glycine (75.07 g/mol) in a 50 mL conical tube and add 40 mL of PBS 1×. Mix by inversion until completely dissolved.
4. Permeabilization solution: Prepare 0.5% triton X-100 solution by adding 200 μL of triton X-100 to 40 mL PBS 1× in a 50 mL conical tube. Gently rotate the tube for 15 min, avoiding formation of foam.
5. Blocking solution: To prepare 3% BSA in PBS 1× solution, weigh 1.5 g of BSA in a 50 mL conical tube and make up with PBS 1× to 50 mL. Mix by inversion until completely dissolved. Pass solution through 0.2 μm syringe filter using a 20 mL syringe.
6. Primary antibodies: Centrifuge stock antibodies for 5 min. Dilute rabbit α-CPSF6, mouse α-CPSF5, or mouse α-SC-35 to 1:1000 in the above blocking solution to prepare primary antibodies solutions. Mix and centrifuge diluted primary antibodies before use.
7. Secondary antibodies: Centrifuge stock antibodies for 5 min. Dilute donkey α-rabbit and α-mouse conjugated to Alexa 488, 594, or 647 in the above blocking solution to make secondary antibodies solutions. (Choice of the fluorochrome-conjugated to the secondary antibody, *see* **Note 2**). Mix and centrifuge dilution primary antibodies before use.
8. Nucleus staining solution: Centrifuge DAPI 1 mg/mL stock for 5 min. To prepare a 1 μg/mL DAPI solution, add 1 μL of DAPI stock in the above blocking solution. Mix and centrifuge before use. Alternatively, DAPI can be added directly in the dilution of the secondary antibodies.
9. Incubation chamber: A suitable chamber can be created by placing moist paper towel in a plastic box, as it serves as a dark, sealed humidified chamber. Coat one side of a 24 well-plate with a hydrophobic surface (parafilm); slightly stretch it, so it attaches smoothly to the plastic. Place it inside of box.

3.3.2 Immunofluorescence Staining

The entire staining procedure will be conducted in 24-well plates. Use a sterile well within the 24-well plate for all solution incubations. Follow these instructions for handling the coverslips: (1) Lift the coverslip using tweezers; (2) Submerge the coverslips in PBS 1×; (3) Remove excess PBS; (4) Transfer to a 24-well plate containing the respective solution. For antibody incubations, create a droplet with the appropriate dilution on parafilm and position the coverslips with the cells facing downward onto the droplet. Repeat this process throughout the entire protocol. (For guidance on handling coverslips, *see* **Note 3**.)

1. Use tweezers to lift the coverslips out of the 24-well plate (*see* **Note 4**).
2. Thoroughly wash the coverslips twice with PBS 1×. Eliminate excess liquid by touching an edge of the coverslip to absorbent paper. Repeat this process for all subsequent washes in the following steps (*see* **Note 5**).
3. Treat the cells with 1 mL of fixing solution. Incubate for 15 min at room temperature in the dark (cover the plate with aluminum foil). Avoid leaving the cells in the fixing solution for more than 30 min. Wash the coverslip with PBS 1× (this step serves as a stopping point; *see* **Note 6**).
4. Incubate the coverslip with 1 mL of quenching solution for 10 min at room temperature. Wash the coverslip with PBS 1× (optional step; *see* **Note 7**).
5. Permeabilize the cells with 1 mL of permeabilization solution. Incubate for 5 min at room temperature. Wash the coverslip with PBS 1× (*see* **Note 8**).
6. Apply 1 mL of blocking solution to the coverslip. Incubate for 60 min at room temperature. Wash the coverslip with PBS 1×.
7. For single staining, apply 20 μL of rabbit α-CPSF6 and mouse α-CPSF5 at 1:1000 dilutions on a hydrophobic surface of the chamber. For double staining, apply 10 μL of rabbit α-CPSF6 at 1:500 dilution and 10 μL of mouse α-CPSF5 at 1:500 dilution or 10 μL of mouse α-SC-35 at 1:500 dilution. Incubate for 60 min at room temperature. Wash the coverslip with PBS 1× (*see* **Note 9**).
8. Apply 20 μL of Alexa Fluor 594-conjugated donkey α-rabbit IgG or α-mouse IgG at a 1:1000 dilution on a hydrophobic surface of the chamber. For double staining, apply 10 μL of Alexa Fluor 647 (or 488)-conjugated donkey α-rabbit IgG at 1:500 dilution and 10 μL of Alexa Fluor 594-conjugated donkey α-mouse IgG CPSF5 at 1:500. Incubate for 60 min at room temperature. Wash the coverslip with PBS 1× (*see* **Note 10**).
9. Apply 20 μL of the nucleus staining dilution. Incubate for 5 min at room temperature. Wash the coverslip with PBS 1×.
10. Place 10 μL of mounting solution on a microscope slide and position the coverslip with the cells facing the mounting solution. Gently press the coverslip with a tip to eliminate any bubbles. Carefully remove excess mounting media.

11. Store the slides in a dark box at room temperature until the mounting solution has dried. Alternatively, place the slides in an oven at 37 °C for 60 min.
12. Store the slides in a dark box at 4 °C.

3.4 Acquisition, Processing, Quantification, and Analysis of Images

Perform 2-D image data acquisition utilizing a fluorescence microscope (Zeiss AxioObserver.Z1) equipped with filters for wavelengths of 405 nm, 488 nm, and 594 nm, coupled with a digital camera. Capture images in a 512 × 512 resolution, channel by channel, saved in a 16-bit format. Analyze the images using software capable of deconvolution using an acquired point spread function (such as ZEN 3.3 blue edition software). Process image figures using Adobe Photoshop CS5 software (Adobe Systems, Mountain View, CA). Quantify the percentage of infected cells (anticipated around 95%) and the percentage of infected cells containing CPSF6 or CPSF5 in nuclear speckles by visually inspecting a minimum of 100 cells per sample across three independent experiments. Subsequently, calculate the mean and standard deviation values using statistical software (GraphPad Prism 8), followed by conducting a statistical analysis employing an unpaired t-test.

4 Notes

1. Within 24 h, A549 cells exhibit the highest percentage of cells containing CPSF5/CPSF6 condensates. However, with longer infection times (48–96 h), the percentage of CPSF5/CPSF6 condensates per cell decreases, while their size increases [13].
2. The selection of the fluorochrome conjugated to the secondary antibody relies on several factors: (1) the type of HIV-1 infection—cells infected with HIV-1 expressing GFP or Luc as a reporter gene; (2) for double staining—whether it is necessary to simultaneously stain CPSF5 and CPSF6 or CPSF6 and SC35.
3. If a coverslip is accidentally dropped on the bench, follow these steps: (1) Use tweezers to lift the coverslip; (2) Wash the coverslip with PBS 1×; (3) Hold the coverslip against the light and, using a tip, trace a small line on the edge. If the cells are removed where the line was drawn, it indicates the side of the coverslip containing cells.
4. Removing coverslips from the 24-well plate may be challenging. A simple method involves lifting one edge of the coverslip with a tip and then gripping it with tweezers.
5. Proper washing of the coverslip is essential for achieving high-quality staining. To perform an effective wash, use the method

described above and dip the coverslip in PBS 1× for 5 s, repeating at least 10 times.

6. While fixed coverslips can be stored in PBS 1× overnight at 4 °C, it is advisable to complete the entire staining protocol on the same day as the fixation.
7. The use of quenching solution is optional, but it reduces non-specific background by leveraging the ability of glycine to eliminate any free aldehyde groups that may generate autofluorescence.
8. Prolonged exposure to the permeabilization solution should be avoided, as it may lead to cell membrane damage, including nuclear and organellar membranes.
9. Aldehyde fixation can induce autofluorescence, so including a negative control lacking staining is crucial to verify the stain's specificity. Autofluorescence can contribute to background signal and result in false-positive staining.
10. To detect unspecific binding or cross-reaction between the secondary antibodies, incorporate a control that solely employs secondary antibodies. This involves fixing, permeabilizing, blocking, and staining with the secondary antibody but without using the primary antibody.

Acknowledgments

This work was supported by an NIH grant AI087390 to F.D.-G. We are grateful to the NIH AIDS repository for providing plasmids and antibodies.

References

1. Dharan A, Bachmann N, Talley S, Zwikelmaier V, Campbell EM (2020) Nuclear pore blockade reveals that HIV-1 completes reverse transcription and uncoating in the nucleus. Nat Microbiol 5(9):1088–1095. https://doi.org/10.1038/s41564-020-0735-8
2. Selyutina A, Persaud M, Lee K, KewalRamani V, Diaz-Griffero F (2020) Nuclear import of the HIV-1 Core precedes reverse transcription and uncoating. Cell Rep 32(13):108201. https://doi.org/10.1016/j.celrep.2020.108201
3. Burdick RC, Li C, Munshi M, Rawson JMO, Nagashima K, Hu WS, Pathak VK (2020) HIV-1 uncoats in the nucleus near sites of integration. Proc Natl Acad Sci USA 117(10):5486–5493. https://doi.org/10.1073/pnas.1920631117
4. Francis AC, Marin M, Singh PK, Achuthan V, Prellberg MJ, Palermino-Rowland K, Lan S, Tedbury PR, Sarafianos SG, Engelman AN, Melikyan GB (2020) Publisher correction: HIV-1 replication complexes accumulate in nuclear speckles and integrate into speckle-associated genomic domains. Nat Commun 11(1):6165. https://doi.org/10.1038/s41467-020-20152-w
5. Rensen E, Mueller F, Scoca V, Parmar JJ, Souque P, Zimmer C, Di Nunzio F (2021) Clustering and reverse transcription of HIV-1 genomes in nuclear niches of macrophages. EMBO J 40(1):e105247. https://doi.org/10.15252/embj.2020105247

6. Ilik IA, Aktas T (2022) Nuclear speckles: dynamic hubs of gene expression regulation. FEBS J 289(22):7234–7245. https://doi.org/10.1111/febs.16117
7. Ilik IA, Aktas T (2021) Nuclear speckles: dynamic hubs of gene expression regulation. FEBS J. https://doi.org/10.1111/febs.16117
8. Galganski L, Urbanek MO, Krzyzosiak WJ (2017) Nuclear speckles: molecular organization, biological function and role in disease. Nucleic Acids Res 45(18):10350–10368. https://doi.org/10.1093/nar/gkx759
9. Lamond AI, Spector DL (2003) Nuclear speckles: a model for nuclear organelles. Nat Rev Mol Cell Biol 4(8):605–612. https://doi.org/10.1038/nrm1172
10. Spector DL, Landon S, O'Keefe RT (1993) Organization of RNA polymerase II transcription and pre-mRNA splicing within the mammalian cell nucleus. Biochem Soc Trans 21(4): 918–920. https://doi.org/10.1042/bst0210918
11. Fu XD, Maniatis T (1990) Factor required for mammalian spliceosome assembly is localized to discrete regions in the nucleus. Nature 343(6257):437–441. https://doi.org/10.1038/343437a0
12. Selyutina A, Hu P, Miller S, Simons LM, Yu HJ, Hultquist JF, Lee K, KewalRamani VN, Diaz-Griffero F (2022) GS-CA1 and lenacapavir stabilize the HIV-1 core and modulate the core interaction with cellular factors. iScience 25(1):103593. https://doi.org/10.1016/j.isci.2021.103593
13. Luchsinger C, Lee K, Mardones GA, KewalRamani VN, Diaz-Griffero F (2023) Formation of nuclear CPSF6/CPSF5 biomolecular condensates upon HIV-1 entry into the nucleus is important for productive infection. Sci Rep 13(1):10974. https://doi.org/10.1038/s41598-023-37364-x
14. Diaz-Griffero F, Perron M, McGee-Estrada K, Hanna R, Maillard PV, Trono D, Sodroski J (2008) A human TRIM5alpha B30.2/SPRY domain mutant gains the ability to restrict and prematurely uncoat B-tropic murine leukemia virus. Virology 378(2):233–242. https://doi.org/10.1016/j.virol.2008.05.008. S0042-6822(08)00310-3 [pii]

Part III

HIV-1 Nuclear Import and Chromatin Association

Chapter 10

Monitoring HIV-1 Nuclear Import Kinetics Using a Chemically Induced Nuclear Pore Blockade Assay

Adarsh Dharan and Edward M. Campbell

Abstract

To integrate with host chromatin and establish a productive infection, HIV-1 must translocate the viral Ribonucleoprotein (RNP) complex through the nuclear pore complex (NPC). Current assay to measure HIV-1 nuclear import relies on a transient byproduct of HIV-1 integration failure called 2-LTR circles. However, 2-LTR circles require complete or near-complete reverse transcription and association with the non-homologous end joining (NHEJ) machinery in the nucleus, which can complicate interpretation of 2-LTR circle formation as a measure of nuclear import kinetics. Here, we describe an approach to measure nuclear import of infectious HIV-1 particles. This involves chemically induced dimerization of Nup62, a central FG containing nucleoporin. Using this technique, nuclear import of infectious particles can be monitored in both primary and cell culture models. In response to host factor depletion or restriction factors, changes in HIV-1 nuclear import can be effectively measured using the nuclear import kinetics (NIK) assay.

Key words HIV-1, Reverse transcription, Nuclear pore complex, Nuclear import, 2-LTR, Nuclear pore blockade, Lentiviruses

1 Introduction

A hallmark of lentiviruses such as HIV-1 is their ability to translocate through the nuclear pore complex in order to facilitate integration of the viral preintegration complex (PIC) with the host chromatin. It appears that HIV-1 capsid may translocate through the nuclear pore complex (NPC), intact or nearly intact, and involves interaction with several host factors, including nucleoporins (Nups). Multiple models have been proposed to explain HIV-1 nuclear import, but the precise mechanism is still under debate (reviewed in [1, 2]). Nevertheless, HIV-1's interaction with NPC components is critical for this process. The conventional measurement of HIV-1 nuclear import is by measuring a byproduct of viral infection called the 2-LTR circles. These unintegrated viral products are formed by non-homologous end joining (NHEJ) of viral

Vinayaka R. Prasad and Ganjam V. Kalpana (eds.), *HIV Protocols*, Methods in Molecular Biology, vol. 2807, https://doi.org/10.1007/978-1-0716-3862-0_10,

DNA, and requires access to the nuclear environment and completion of viral reverse transcription (RT) [3–5]. By designing primers that specifically detect 2-LTR circle junctions, the amount of these unintegrated viral byproducts can be measured in infected cells by quantitative PCR [6, 7]. Although this technique has been and remains the most common method used to quantify HIV1 nuclear import, there are limitations of this approach. First, it measures a byproduct of infection, which does not integrate or recapitulate actual infection events. Reports indicate that sometimes this technique lacks accuracy in monitoring nuclear import of infectious viral Ribonucleoproteins (RNPs) [3, 8–13]. One study analyzed the discrepancy in 2-LTR circle levels during depletion of a HIV-1 host factor TNPO3 [14–19] and found abundant autointegration products can complicate the standard 2-LTR measurement to simply assess nuclear import [20]. Second, this technique fails to quantify the timing of viral nuclear import or any changes arising from host factor modulation. Monitoring the formation of the 2-LTR products during infection has shown their copy numbers to appear 10–12 h post-infection and peaks at 24 h following infection [7, 21]. This is because the process of 2-LTR formation requires completion of viral RT and data from our laboratory and others have shown that nuclear import of HIV-1 precedes the completion of RT [22, 23]. In relevant target cells of infection, we and others have shown that nuclear import occurs 4–6 h following receptor binding [22, 24]. While 2-LTR circles are generally a reliable technique to quantify viral accumulation in the nucleus, the inability to measure the kinetics of nuclear import and confounding phenotypes induced by changes in nuclear import kinetics following host factor modulation cannot be modeled using this approach.

We developed an inducible NPC blockade system that allows monitoring the nuclear import of infectious HIV-1 particles [22]. This assay stems from a previous study that used the blockade technique to monitor protein transport at the ciliary pore complex [25]. This assay relies on the ability to block active transport through the NPC by inducing the drug-induced dimerization of Nup62, a central pore protein of the NPC. This involves the utilization of a NUP62 construct which is tagged with the modified FKBP domain Fv (DmrB domain) and 2 copies of eGFP (Nup62DG) (Fig. 1a). Dimerization of the FKBP domain is dependent on the presence of a homodimerizing (HD) drug. We generated a lentiviral vector expressing this construct and transduced cells to obtain a cell line stably expressing this construct (Fig. 1b). The blockade potently inhibited HIV-1 infection while HD treatment on cells not expressing the NUP62DG construct showed no difference to control (Fig. 1c). NUP62 blockade did not affect HIV-1 reverse transcription but significantly impaired 2-LTR circle formation (Fig. 1d). By selectively inhibiting the nuclear import of

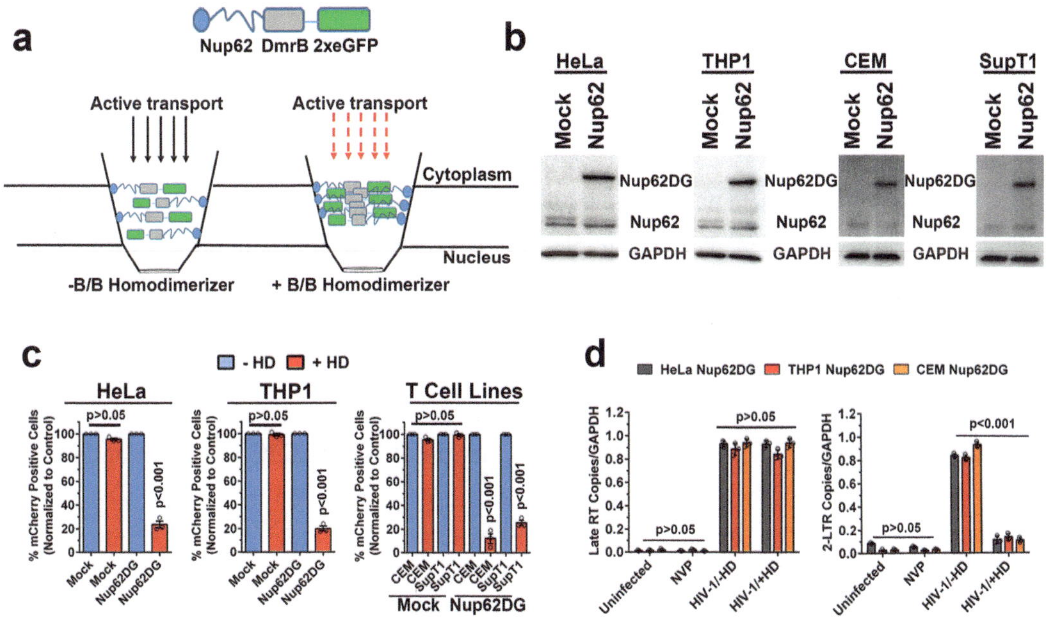

Fig. 1 Inducible nuclear pore blockade to block HIV-1 infection at the nuclear entry stage. (**a**) Schematics of the Nup62 construct fused to DmrB and two copies of eGFP (Nup62DG) used to block active nuclear pore transport after the addition of B/B HD. (**b**) Protein expression in HeLa, THP1, CEM, and SupT1 cells stably expressing the Nup62DG and probed with a Nup62 antibody. (**c**) Mock or Nup62DG-transduced HeLa, THP1-differentiated macrophages, CEM, and SupT1 T cells were synchronously infected with VSV-G pseudotyped R7ΔEnvmCherry in the presence or absence of HD drug for the first 24 h of infection. The HD drug was removed after 24 h, replaced with normal media and infection was assessed 48 h post-synchronized infection by measuring the percentage of mCherry+ cells. The normalized and averaged data (± s.e.m.) from three independent experiments are shown. (**d**) RT–qPCR quantification of reverse transcription and 2-LTR circles in cells expressing Nup62DG, 24 h after HIV-1 infection. Depicted is the mean of biological triplicates (± s.d.). Data shown are representative of three independent experiments. (Reproduced from Ref. [22] based on the permission criteria outlined by Springer Nature Journal)

the HIV-1 genome at various times following a synchronized infection, we used this assay to map the kinetics of viral nuclear import in primary and cell culture models [22]. An advantage of this assay is that the measurement of infectivity following nuclear import necessarily focuses on the infectious fraction of the viral inoculum, whereas cell fractionation and imaging studies generally cannot assert with certainty that the accumulation of viral proteins or genomes in the nucleus represent RNPs that are destined to infect the cell or may be arrested at a post-nuclear import stage of infection. The nature of this assay further allows comparison of nuclear import kinetics between conditions including various HIV-1 capsid mutations or modulation of host factor levels.

2 Materials

2.1 Constructs

1. Lentiviral or retroviral vector expressing the NUP62DG construct. The following protocol describes monitoring HIV-1 nuclear import using the pLVX Nup62DG (Addgene plasmid: 139683) (*see* **Note 1**).
2. Packaging plasmid psPAX2 (Addgene plasmid: 12260) and VSV glycoprotein (Addgene plasmid: 8454) for generation of lentiviral vector in 293Ts.
3. HIV-1 viral plasmids to generate viral particles to perform the NIK assay (*see* **Note 2**).

2.2 Cells and Culture Media

1. 293T cells for production of NUP62DG lentivirus and HIV-1 viral particles.
2. Immortalized cells such as HeLa, THP1 monocytes, Jurkats, and CEM T cells.
3. Primary CD4+ T cells and monocytes isolated from healthy donors using EasySep Human CD4+ T cell enrichment kit or EasySep Human CD14 positive selection kit from STEMCELL Technologies.
4. 293Ts and HeLa cells cultured in High Glucose Dulbecco media containing 10% FBS and antibiotics and monocytes and T cells cultured in RPMI with 10% FBS and antibiotics.

2.3 Consumables and Instruments

1. Tissue culture plates for vector generation (100 mm dishes), transduction (6 well plate), and NIK assay (48 well plate).
2. Temperature-controlled Benchtop centrifuge with plate holders.
3. BD LSRFortessa flow cytometry.

2.4 Chemicals and Reagents

1. B/B Homodimerizer (Takara, 635059) drug (HD) to induce NUP62 dimerization and subsequent nuclear pore blockade.
2. Polyethyleneimine (PEI), MW 25000 transfection reagent (1 mg/mL stock, Polysciences, 23966-1).
3. Human granulocyte macrophage colony stimulating factor and human macrophage colony stimulating factor for monocyte differentiation to macrophages.
4. Human CD3/CD28 T cell activator and Human IL-2 for primary CD4+ T cell activation.

3 Methods

3.1 Lentiviral Vector Generation to Transduce Primary and Cell Culture Model with the hNup62DG Construct

1. The day before lentiviral vector generation, plate 5×10^6 293T cells on a 10 cm dish.
2. On the day of transfection, in a 1.5 mL Eppendorf tube, assemble the transfection mixture in 1 mL of Opti-MEM. Add to the tube 4.5 μg of pLVX NUP62DG, 3.5 μg of psPAX2 and 2 μg of pCMV-VSVg, and 30 μL of PEI transfection reagent.
3. Incubate the transfection mixture for 12 min at room temperature and following the incubation, add the mixture dropwise to the 293T cells.
4. After overnight incubation, replace media with 37 °C warm media.
5. 48 hours following transfection, harvest the 293T supernatant that contain the viral vectors by passing through a 0.45-micron filter.
6. Unconcentrated vector can be used to transduce cell lines such as HeLa, THP1 monocytes, Jurkats, and CEMs (*see* **Note 3**).
7. For transducing primary monocyte derived macrophages (MDMs) and CD4+ T cells, the viral vectors concentrated by an overnight spinoculation at 5000× *g* at 4 °C [26] and resuspended to 1/10th of the initial volume with cell culture media.

3.2 Generation of Cells Expressing the Nup62DG

1. For adherent cell lines such as HeLa, plate 7×10^5 cells on a 6 well plate the day before transducing with the Nup62DG vector. For suspension cells such as T cell lines and primary CD4+ T cells, mix the transducing vector with 2×10^6 cells on the day of transduction on a 6 well plate.
2. Primary CD4+ T cells stimulated for 2 days with 2.5 μg/mL of anti-CD3/CD28 and 10 ng/mL of IL2. Primary monocytes differentiated to macrophages (MDMs) by treating the cell culture media with 50 ng/mL of GM-CSF and 25 ng/mL of MCSF. On day 5 of differentiation, cells are treated overnight with simian immunodeficiency Vpx virus-like particles (*see* **Note 4**).
3. Transduction performed by adding 500–1000 RT units of Nup62DG vector to cells and spinoculating them at 1200× *g* for 2 h at 13 °C using benchtop centrifuges with plate holders [27]. For primary CD4+ T cells, spinoculation performed at 800× *g* for 1 h at 32 °C.
4. Following spinoculation, transfer plates to 37 °C cell culture incubator and incubate overnight with the viral vector (*see* **Note 5**).

5. Next day, replace media with 37 °C warm media and allow transduction to proceed for another 24 h. For MDMs, replace media supplemented with GM-CSF and M-CSF.
6. To generate cell lines stably expressing the NUP62DG, 48–72 h following the addition of vector, replace media with media containing 1 μg/mL of puromycin (*see* **Note 6**).
7. Puromycin selection process not performed on MDMs and CD4+ T cells. Four days following transduction, NIK assay carried out on these transduced cells (*see* **Note 7**).

3.3 HIV-1 Nuclear Import Kinetic (NIK) Assay

1. The day before the NIK assay, plate the NUP62DG stably expressing cells on to multiple wells of a 48 well plate (9×10^4 cells per well). Dividing cells growth arrested by adding 5 μg/mL of Aphidicolin while plating the cells and incubated overnight (16–18 h of incubation).
2. For THP1 monocytes stably expressing the NUP62DG, 1×10^5 cells per well plated on a 48 well plate with 100 ng/mL of PMA to induce monocyte differentiation to macrophages. After 48 h of PMA treatment, cells allowed to rest in media without PMA for another 48 h before HIV-1 infection.
3. On the day of the assay, cells were infected with NL4.3 or R7-based HIV-1 virus harboring a mCherry reporter and infection synchronized by spinoculation at 1200× *g* for 2 h at 13 °C.
4. Following spinoculation, replace media with fresh media. During this media change, replace media in duplicate infected wells with media containing 1.5 μM of the B/B homodimerizing (HD) drug. This time point (0 h) will be the longest duration of nuclear pore blockade (Fig. 2a).
5. In order to induce nuclear pore blockade at different times post-infection, repeat media changes with the homodimerizing drug (*see* **Note 8**) on different HIV-1-infected wells (Fig. 2a).
6. To monitor the changes in NIK, we added the homodimerizing drug every hour during the first 8 h of infection and following this, NPC blockade was done at 10, 12, and 14 h following synchronized infection (*see* **Note 9**).
7. All HD containing media are replaced 24 h following synchronized infection with normal media, regardless of the time point at which HD was added (*see* **Note 10**).
8. Infection allowed to continue for another 48 h. In the case of MDM, infection proceeded for 72 h following media change.
9. Following this incubation period, cells harvested and fixed in 1% paraformaldehyde and reporter gene expression (mCherry in the NL4.3 or R7 HIV-1 vector) as a measure of viral infection analyzed by flow cytometry (*see* **Note 11**). Analysis

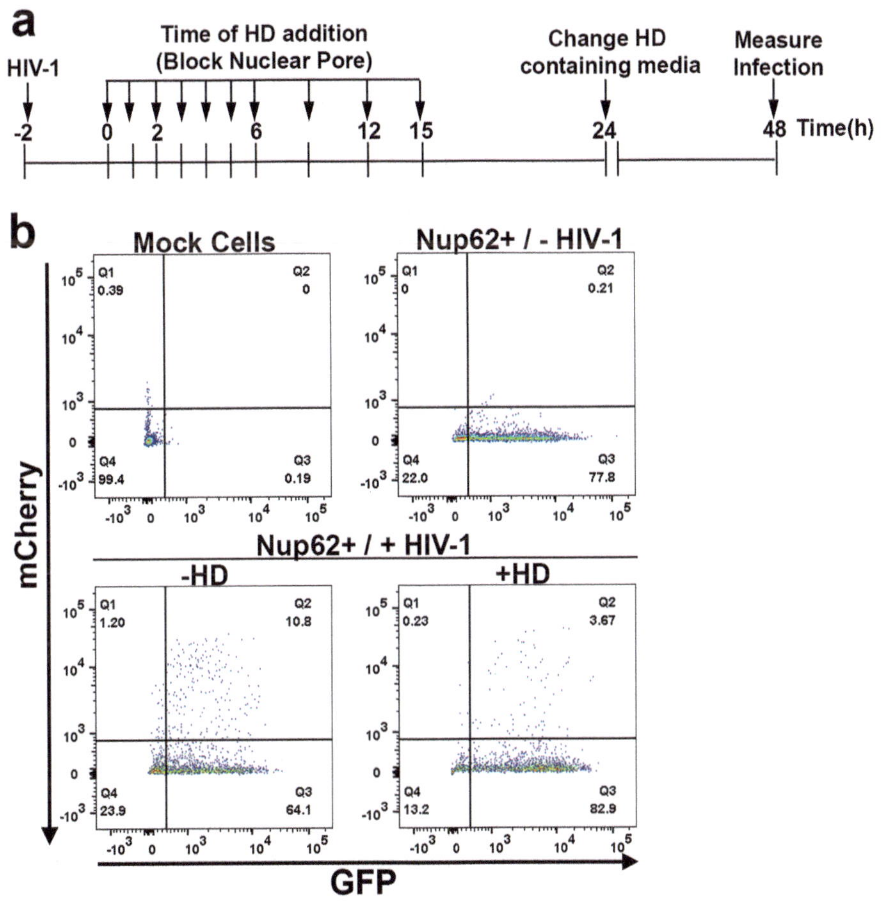

Fig. 2 Schematics of HIV-1 nuclear import kinetic (NIK) assay and gating strategy used for analysis. (**a**) Schematic of how NIK was monitored in different cells expressing the NUP62DG construct. (**b**) Shown is an example of THP1 differentiated macrophages expressing the Nup62DG infected with VSVg-R7ΔEnvmCherry. Data representative of three independent experiments. Analysis included gating for live and single cells. Percent mCherry positive cells calculated as the ratio of double positive population (Q2) to the total amount of Nup62GFP+ population (Q2 + Q3). (Reproduced from Ref. [22] based on the permission criteria outlined by Springer Nature Journal)

performed only on the GFP expressing cells and infection (mCherry) monitored in this GFP+ population (Fig. 2b, Q2).

3.4 Monitoring HIV-1 Nuclear Import Kinetics

1. Percent-infected cells calculated as the ratio of double positive cells (Fig. 2b, Q2) to the total amount of NUP62 GFP+ population (Fig. 2b, Q2 + Q3).
2. To obtain the nuclear import kinetics, raw infectivity values (Fig. 3, upper panel) normalized by setting 0% as the value obtained with the 0-h time point of HD addition (24 h of HD treatment) and 100% to the value obtained from the vehicle control (-HD) (*see* **Note 12**).

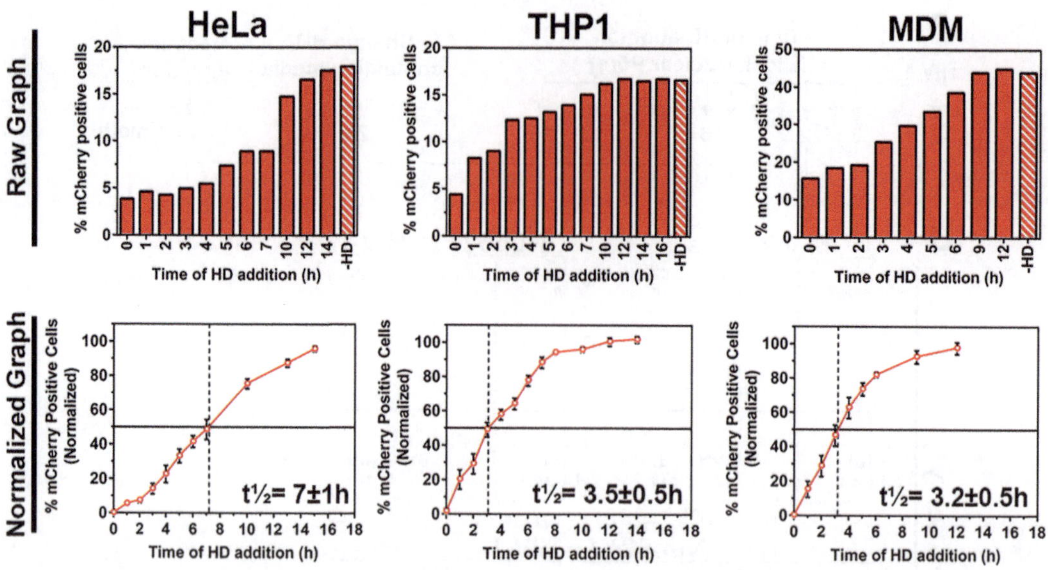

Fig. 3 Monitoring HIV-1 NIK in primary and cell culture models. Monitoring NIK assay in HeLa, THP1 differentiated macrophages, and monocyte derived macrophage cells. Cells were synchronously infected with VSV-G-R7ΔEnvmCherry for 2 h at 13 °C before adding the HD drug to block nuclear pore import. The HD drug was added at different time points post-infection. Cells were incubated with HD until 24 h post-infection and then washed and replaced with normal media. Cells were collected and infection was monitored 48–72 h later by measuring the percentage of mCherry+ cells by flow cytometry. Bar graph depicts raw values and normalized graph (line graphs) attained by setting the baseline (0%) as the maximum inhibition attained by blocking the nuclear pore for 24 h (0 h time point) and maximum infection (100%) as the level of infection attained with no HD treatment (dashed bars, bar graph). Data shown are representative of three independent experiments (± s.e.m.). (Reproduced from Ref. [22] based on the permission criteria outlined by Springer Nature Journal)

3. Following normalization, half time of HIV-1 nuclear import (t½) extrapolated from the X-axis of the normalized graph as the time point when half of the viral inoculum became insensitive to the nuclear pore blockade (Fig. 3, lower panel).
4. Under conditions of host factor depletion or different HIV-1 capsid mutants, NIK monitored by infecting NUP62DG cells with HIV-1 viruses generated with different CA mutants or by depleting the specific host factor by CRISPR technology or siRNA.

4 Notes

1. For this study, the NUP62DG construct cloned into the pLVX lentiviral backbone from Clontech. Other lentiviral and retroviral vectors can be used to express this transgene. Make sure the transgene expresses two copies of GFP as our data shows one copy of GFP is not sufficient to induce NUP62

dimerization and more than two copies results in overcrowding at the NPC (data not shown).

2. HIV-1 vectors containing WT or capsid mutants generated based on individual lab protocols.
3. A real-time PCR-based assay that detects the RT activity of the harvested viral vectors used to titrate the virus [28]. Vectors can be frozen at −80 °C. Avoid multiple freeze thaws. For high transduction efficiency, it is recommended to make a fresh vector before use.
4. Simian Immunodeficiency virus (SIV) containing the accessory protein Vpx used to degrade SAMHD1 levels in cells that improves lentiviral transduction efficiency.
5. We observe a loss in cell viability with longer incubation.
6. With a high transduction efficiency, puromycin selection can help generate cell lines within a weak to perform the NIK assay. We do not advise to freeze the cells following selection as the NUP62DG expression in stably selected cells is downregulated following freeze thaw.
7. We observe a significant loss in cell viability with puromycin selection on these cells. During the NIK assay, we monitor HIV-1 infection in the subset of cells that harbor GFP from the NUP62DG vector transduction.
8. All time points are done in duplicate.
9. Our study has shown nuclear import of WT HIV-1 virus occurs within the first 8 h in HeLa cells and more rapidly (4–6 h) in other relevant target cells. Depending on experimental conditions such as different capsid mutants or changes following host factor depletion, the time points of homodimerizing drug addition should be optimized.
10. To ensure the blockade efficiency of the NUP62DG, collect cells after 24 h of HD treatment (0-h time point) and harvest DNA to quantify 2-LTR circles. A successful blockade would result in a marked decrease in 2-LTR circles compared to vehicle control (-HD).
11. We have also used HIV-1 particles harboring Firefly luciferase reporter gene to monitor infection.
12. A drop in blockade efficiency is observed depending on the duration and passage number the cells are kept in culture. When inhibition to HIV-1 infection following a 24-h NUP62 blockade falls below 75%, the cells are discarded and we prepare fresh transduced cell population. We do not keep the NUP62DG cells for more than 6 weeks in culture.
13. Spinoculation at 13 °C is required during infection as it allows for a synchronized infection that is necessary to interpret NIK

assays. It may be dispensable for generating cells expressing the NUP62-DG construct and assays not involving an assessment of nuclear import kinetics, such as measuring CA mutant sensitivity.

Acknowledgments

This work was supported by NIH grants AI139009 and AI162694 to E.M.C.

References

1. Dharan A, Campbell EM (2022) Teaching old dogmas new tricks: recent insights into the nuclear import of HIV-1. Curr Opin Virol 53:101203. https://doi.org/10.1016/j.coviro.2022.101203
2. Shen Q, Wu C, Freniere C et al (2021) Nuclear import of HIV-1. Viruses 13(11). https://doi.org/10.3390/v13112242
3. Kilzer JM, Stracker T, Beitzel B et al (2003) Roles of host cell factors in circularization of retroviral dna. Virology 314(1):460–467
4. Li L, Olvera JM, Yoder KE et al (2001) Role of the non-homologous DNA end joining pathway in the early steps of retroviral infection. EMBO J 20(12):3272–3281
5. Miller MD, Farnet CM, Bushman FD (1997) Human immunodeficiency virus type 1 preintegration complexes: studies of organization and composition. J Virol 71(7):5382–5390
6. Bukrinsky M, Sharova N, Stevenson M (1993) Human immunodeficiency virus type 1 2-LTR circles reside in a nucleoprotein complex which is different from the preintegration complex. J Virol 67(11):6863–6865
7. Butler SL, Hansen MS, Bushman FD (2001) A quantitative assay for HIV DNA integration in vivo. Nat Med 7(5):631–634
8. Bona R, Baroncelli S, D'Ettorre G et al (2013) Effects of raltegravir on 2-long terminal repeat circle junctions in HIV type 1 viremic and aviremic patients. AIDS Res Hum Retrovir 29(10):1365–1369. https://doi.org/10.1089/AID.2013.0047
9. Leavitt AD, Robles G, Alesandro N et al (1996) Human immunodeficiency virus type 1 integrase mutants retain in vitro integrase activity yet fail to integrate viral DNA efficiently during infection. J Virol 70(2):721–728
10. Mannioui A, Nelson E, Schiffer C et al (2005) Human immunodeficiency virus type 1 KK26-27 matrix mutants display impaired infectivity, circularization and integration but not nuclear import. Virology 339(1):21–30. https://doi.org/10.1016/j.virol.2005.05.023
11. Shun MC, Raghavendra NK, Vandegraaff N et al (2007) LEDGF/p75 functions downstream from preintegration complex formation to effect gene-specific HIV-1 integration. Genes Dev 21(14):1767–1778. https://doi.org/10.1101/gad.1565107
12. Sowd GA, Serrao E, Wang H et al (2016) A critical role for alternative polyadenylation factor CPSF6 in targeting HIV-1 integration to transcriptionally active chromatin. Proc Natl Acad Sci USA 113(8):E1054–E1063. https://doi.org/10.1073/pnas.1524213113
13. Wang H, Jurado KA, Wu X et al (2012) HRP2 determines the efficiency and specificity of HIV-1 integration in LEDGF/p75 knockout cells but does not contribute to the antiviral activity of a potent LEDGF/p75-binding site integrase inhibitor. Nucleic Acids Res 40(22):11518–11530. https://doi.org/10.1093/nar/gks913
14. Cribier A, Segeral E, Delelis O et al (2011) Mutations affecting interaction of integrase with TNPO3 do not prevent HIV-1 cDNA nuclear import. Retrovirology 8:104. https://doi.org/10.1186/1742-4690-8-104
15. De Iaco A, Luban J (2011) Inhibition of HIV-1 infection by TNPO3 depletion is determined by capsid and detectable after viral cDNA enters the nucleus. Retrovirology 8:98. https://doi.org/10.1186/1742-4690-8-98
16. Konig R, Zhou Y, Elleder D et al (2008) Global analysis of host-pathogen interactions that regulate early-stage HIV-1 replication. Cell 135(1):49–60. https://doi.org/10.1016/j.cell.2008.07.032
17. Logue EC, Taylor KT, Goff PH et al (2011) The cargo-binding domain of transportin 3 is required for lentivirus nuclear import. J Virol

85(24):12950–12961. https://doi.org/10.1128/JVI.05384-11

18. Valle-Casuso JC, Di Nunzio F, Yang Y et al (2012) TNPO3 is required for HIV-1 replication after nuclear import but prior to integration and binds the HIV-1 core. J Virol 86(10): 5931–5936. https://doi.org/10.1128/JVI.00451-12
19. Zhou L, Sokolskaja E, Jolly C et al (2011) Transportin 3 promotes a nuclear maturation step required for efficient HIV-1 integration. PLoS Pathog 7(8):e1002194. https://doi.org/10.1371/journal.ppat.1002194
20. De Iaco A, Santoni F, Vannier A et al (2013) TNPO3 protects HIV-1 replication from CPSF6-mediated capsid stabilization in the host cell cytoplasm. Retrovirology 10:20. https://doi.org/10.1186/1742-4690-10-20
21. Bejarano DA, Puertas MC, Borner K et al (2018) Detailed characterization of early HIV-1 replication dynamics in primary human macrophages. Viruses 10(11):620. https://doi.org/10.3390/v10110620
22. Dharan A, Bachmann N, Talley S et al (2020) Nuclear pore blockade reveals that HIV-1 completes reverse transcription and uncoating in the nucleus. Nat Microbiol 5(9): 1088–1095. https://doi.org/10.1038/s41564-020-0735-8
23. Selyutina A, Persaud M, Lee K et al (2020) Nuclear import of the HIV-1 core precedes reverse transcription and uncoating. Cell Rep 32(13):108201. https://doi.org/10.1016/j.celrep.2020.108201
24. Burdick RC, Li C, Munshi M et al (2020) HIV-1 uncoats in the nucleus near sites of integration. Proc Natl Acad Sci USA 117(10): 5486–5493. https://doi.org/10.1073/pnas.1920631117
25. Takao D, Dishinger JF, Kee HL et al (2014) An assay for clogging the ciliary pore complex distinguishes mechanisms of cytosolic and membrane protein entry. Curr Biol 24(19): 2288–2294. https://doi.org/10.1016/j.cub.2014.08.012
26. Qing E, Hantak MP, Galpalli GG et al (2020) Evaluating MERS-CoV entry pathways. Methods Mol Biol 2099:9–20. https://doi.org/10.1007/978-1-0716-0211-9_2
27. O'Doherty U, Swiggard WJ, Malim MH (2000) Human immunodeficiency virus type 1 spinoculation enhances infection through virus binding. J Virol 74(21):10074–10080
28. Vermeire J, Naessens E, Vanderstraeten H et al (2012) Quantification of reverse transcriptase activity by real-time PCR as a fast and accurate method for titration of HIV, lenti- and retroviral vectors. PLoS One 7(12):e50859. https://doi.org/10.1371/journal.pone.0050859

Chapter 11

Biochemical Detection of Capsid in the Nucleus During HIV-1 Infection

Felipe Diaz-Griffero

Abstract

Recent evidence has shown that uncoating and reverse transcription precede nuclear import. These recent breakthroughs have been made possible through the development of innovative biochemical and imaging techniques. This method outlines the biochemical assay used for detecting the presence of the HIV-1 core in the nuclear compartment. In this procedure, human cells are infected with HIV-$1_{NL4\text{-}3}$, with or without the inclusion of PF74, a small molecule that inhibits core entry into the nuclear compartment. Subsequently, cells are separated into cytosolic and nuclear fractions. To assess whether the capsid protein has reached the nuclear compartment, cytosolic and nuclear fractions are subjected to Western blot analysis, utilizing antibodies specific to the HIV-1 capsid protein p24. To validate the true origin of these fractions, Western blot analysis employing antibodies against cytosolic and nuclear markers are also performed. In summary, this assay provides a reliable and efficient means to detect the presence of the HIV-1 capsid protein in the nucleus during infection under various conditions.

Key words HIV-1, Core, Capsid protein, p24, Nuclear import, Cellular fractionation

1 Introduction

Recent findings have suggested that HIV-1 replication involves the entry of the HIV-1 core into the nuclear compartment [1–3]. Contrary to previous beliefs that the HIV-1 core in the cytoplasm undergoes uncoating, which involves the dissociation of monomeric capsid proteins from the core, recent developments have revealed that HIV-1 uncoating and reverse transcription precede the nuclear import of the HIV-1 core [1–3]. This breakthrough has become possible due to the development of novel biochemical and imaging techniques. This protocol will elaborate on the biochemical method employed for detecting the capsid protein in the nuclear compartment during HIV-1 infection.

The biochemical method described here has yielded insights into uncoating, reverse transcription, and nuclear import, revealing several key findings [3]: (1) Disrupting the stability of the HIV-1

Vinayaka R. Prasad and Ganjam V. Kalpana (eds.), *HIV Protocols*, Methods in Molecular Biology, vol. 2807,
https://doi.org/10.1007/978-1-0716-3862-0_11,

core, using rhesus TRIM5α or TRIMCyp [4, 5], hinders the entry of the HIV-1 core into the nuclear compartment. (2) Conversely, reinforcing the stability of the HIV-1 core, through the expression of cytoplasmic CPSF6 or the restriction factor MxB [6–8], facilitates the core's entry into the nuclear compartment. (3) Genetic or pharmacological inhibition of HIV-1 reverse transcription does not prevent the entry of the HIV-1 core into the nuclear compartment. (4) The majority of early and late reverse transcription products are localized in the nuclear fraction during HIV-1 infection. (5) Most of the capsid protein reaching the nuclear compartment during HIV-1 infection results from the import of capsid in higher-order complexes.

Our method involves the separation of HIV-1-infected cells into CYTOSOLIC and NUCLEAR fractions (Fig. 1a, b). Subsequently, these fractions will be examined for the presence of the capsid protein (Fig. 1c). To verify the precise origin of each fraction, we will conduct Western blotting using anti-Nopp40 antibodies as a nuclear marker and anti-GAPDH antibodies as a cytosolic marker (Fig. 1c). This analysis will determine the localization of the capsid protein p24 within the nuclear compartment. As a control, this assay incorporates the use of the small molecule inhibitor PF74 [9], which prevents the entry of the HIV-1 core into the nuclear compartment [3].

2 Materials

2.1 Buffers

Prepare all buffers using Millipore water and analytical grade reagents.

1. Dulbecco's phosphate-buffered saline (PBS) 1X: To prepare 1000 mL, dissolve 8 g of NaCl, 0.2 g of KCl, 1.44 g of Na_2HPO_4, and 0.24 g of KH_2PO_4 in 800 mL of Millipore water. Adjust pH to 7.4 using a pH meter, and the total volume to 1000 mL using Millipore water.
2. Whole cell extract (WCE): 50 mM Tris pH 8.0, 280 mM NaCl, 10% glycerol, 0.5% NP-40, 5 mM $MgCl_2$, 250 units/mL Benzonase, 10 μg/mL ethidium bromide, 1X protease inhibitor.
3. Lysis buffer: 10 mM Tris pH 6.8, 1 mM DTT, 1 mM $MgCl_2$, 10% sucrose, 100 mM NaCl, 0.5% NP-40, 1X protease inhibitor.
4. Nuclear washing buffer: 10 mM Tris, pH, 6.8, 1 mM DTT, 1 mM $MgCl_2$, 10% sucrose, 100 mM NaCl, 1X protease inhibitor.
5. 5X Laemmli buffer: 10% SDS, 50% glycerol, 25% 2-mercaptoethanol, 0.02% bromophenol blue, and 0.3125 M Tris HCl, pH 6.8.

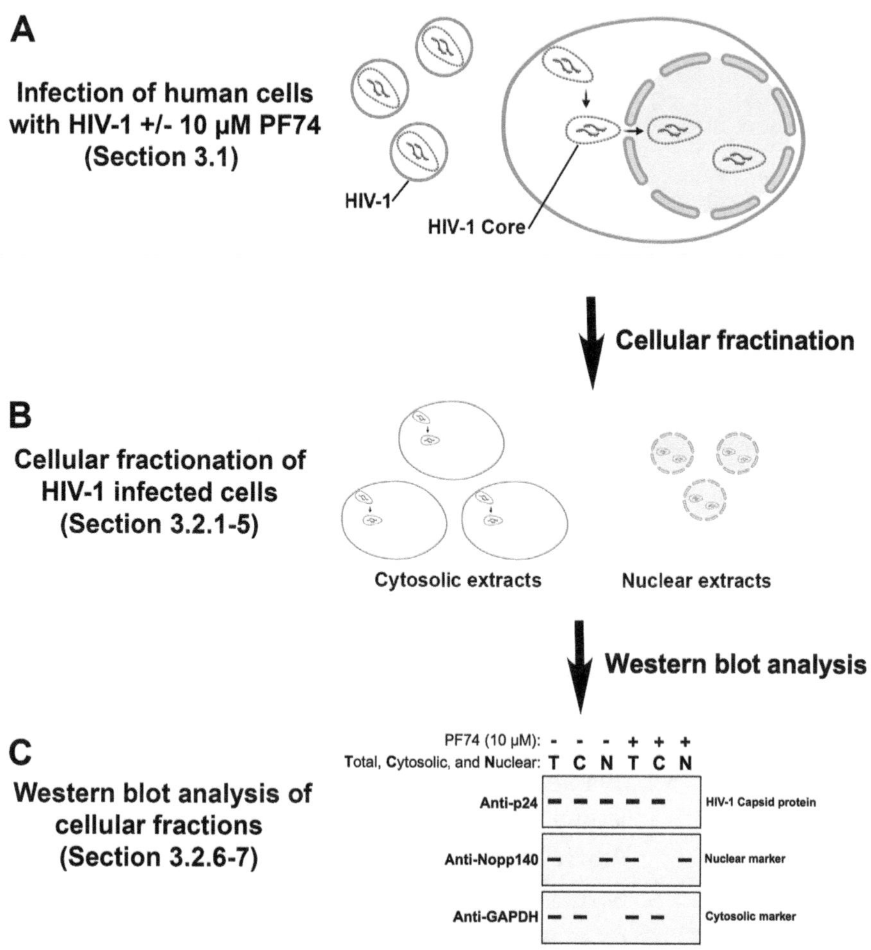

Fig. 1 Biochemical detection of capsid in the nucleus during HIV-1 infection. Human cells are infected with HIV-1 viruses using an MOI = 2 for different times ranging from 1 to 48 h in the presence or absence of 10 μM PF74 (**a**). PF74 blocks the entry of the HIV-1 core into the nucleus. Subsequently cells are separated into cytosolic and nuclear fractions (**b**). **T**otal, **C**ytosolic, and **N**uclear extracts are analyzed by Western blotting using anti-HIV-1 capsid antibodies (**c**). To identify the bona fide origin of cellular extracts, cytosolic and nuclear extracts are analyzed by Western blotting using antibodies against cytosolic(anti-GAPDH) and nuclear proteins (anti-Nopp140)

2.2 Cell Culture and Cell Lines

1. DMEM high glucose. Corning Cat# MT10-017-CV.
2. Trypsin 0.25%. Corning Cat#25-053-CI.
3. Human cell line A549 (ATCC CCL-185).
4. Fetal bovine Serum (FBS) heat inactivated. Gibco Cat# 16140-071.
5. DMEM complete media: DMEM containing 10% FBS, 1% pen-strep, and 2 mM L-glutamine.

2.3 Small Molecule Inhibitors

1. Protease inhibitors: Complete EDTA-free protease inhibitor cocktail. Millipore Sigma Cat# 11873580001.
2. PF74. Sigma-Aldrich Cat# SML0835: prepare as 10 mM stock solution in DMSO.

2.4 Viruses

1. Preparation of HIV-1_{NL4-3} viruses pseudotyped with the envelope of VSV-G: human 293 T cells are transfected with 1 μg each of expression constructs encoding Tat, Rev., and HIV-1 Gag-Pol and 7.5 ug of LTR-GFP-LTR. 24 hours post-transfection, the transfection media is replaced with fresh DMEM complete media, as previously described [3]. 48 hours post-transfection, the media containing the virus is collected and aliquoted. Viruses were titered by measuring the percentage of GFP-positive cells (infected cells) of serially diluted viruses on human A549 cells, as previously described [3].

2.5 Antibodies

1. Mouse monoclonal anti-HIV-1 p24 (183-H12-5C). NIH AIDS repository Cat#3537.
2. Mouse monoclonal anti-glyceraldehyde-3-phosphate dehydrogenase antibody (GAPDH). Invitrogen Cat# AM4300.
3. Rabbit polyclonal anti-alpha tubulin. Invitrogen Cat#PA5-294444.
4. Rabbit polyclonal anti-Nopp140. Gift of U. Thomas Meier, Albert Einstein College of Medicine.
5. Goat anti-mouse IRDye 680LT. LI-COR Cat# 925-68020.
6. Goat anti-rabbit IRDye 680LT. LI-COR Cat# 926-68021.
7. Goat anti-mouse IRDye 800LT. LI-COR Cat# 926-32210.
8. Goat anti-rabbit IRDye 800LT. LI-COR Cat# 925-32211.

2.6 Western Blot Buffers

1. Western blot transfer buffer 1X: 25 mM Tris and 192 mM glycine and 20% methanol.
2. Western blot blocking buffer: 1X PBS, 0.1% Tween 20, and 5% non-fat dry milk.
3. Western blot washing buffer: 1X PBS and 0.1% Tween 20.

3 Methods

3.1 Infection

1. Seeding of cells for infection. Seed two 10 cm plates per every point you have in your experiment (2.5×10^6 A549 cells per plate). For this, detach the cells using 0.25% trypsin treatment, determine the number of cells, and resuspend the cells to a concentration of 2.5×10^6 cells in 10 mL of complete growth

media and add 10 mL per each 10 cm plate. Incubate cells at 37 °C with 5% CO_2 for at least 6 h to allow cells to attach.

2. Infection of cells by HIV-1 (Fig. 1a). Human A549 cells are challenged using HIV-1 at a MOI of 2 for different lengths of time. Although this protocol is described here using human A549 cells, it has also been used with Jurkat, MOLT-3, HeLa, HT1080, and dog Cf2Th cells [3]. To prepare the viral mixture, take 15 mL tube per every point. Put 1 mL of complete growth media into the tube. Add 10 μL of DMSO or 10 μL of 10 mM PF74 (to achieve the final concentration of 10 μM PF74). Gently vortex the tube. Add complete growth media and the virus (final volume should be 10 mL). Replace the old media from the plates using 5 mL of your viral mixture per plate. Incubate at 37 °C with 5% CO_2 for desired time. The assay works using replication competent HIV-1_{NL4-3}, or HIV-1_{NL4-3} Δenv pseudotyped with the envelope of vesicular stomatitis virus (VSV-G). Virus challenges have been performed from 1 to 48 h. Capsid protein is detected in the nuclear fraction 1-h post-infection [3].

3.2 Cellular Fractionation

1. Harvest cells for fractionation. Aspirate medium from both plates. Add 1 mL of 0.25% Trypsin per plate and incubate 2–3 min at 25 °C until cells are completely detached from the plate (check for cell detachment using a low magnification light microscope). You can shake the plates to ensure the complete detachment of the cells (*see* **Note 1**). Transfer all the cells to a 50 mL Falcon tube. To achieve this, add to the first plate 7 mL of DMEM medium without FBS, collect the cells, transfer the whole mixture into the second plate, collect the cells, and transfer the whole mixture into the 50 mL Falcon tube on ice. To ensure the complete transfer of cells repeat these 2 more times, so the final volume in the tube would be 23 mL (*see* **Note 2**).

2. Wash cells. Centrifuge cells at 200× *g* for 7 min at 4 °C. Discard the supernatant and keep the cell pellet. Whenever the cells are not being centrifuged you should keep them on ice. Harvested cells are washed twice using ice-cold 1X PBS. For each wash, resuspend cell pellet in 10 mL of ice-cold 1X PBS and add 40 mL of ice-cold PBS. Centrifuge at 200× *g* for 7 min at 4 °C. Supernatant is discarded and cell pellet is resuspended in 1 mL of ice-cold 1X PBS (*see* **Note 3**).

3. Preparation of the **TOTAL** fraction (Fig. 1b). 1/10 aliquot of the cell suspension (100 μL) is transferred to 1.5 mL tube and centrifuged (200× *g* for 7 min at 4 °C), the supernatant is discarded, and the cell pellet can be processed immediately or can be frozen and kept at −20 °C for 24 h or −80 °C for a longer time. To prepare the total fraction, resuspend the cell

pellet in 38 μL of WCE, which contains 10 μg/mL of ethidium bromide (*see* **Note 4**), 1X protease inhibitor), incubate for 1 h on ice, centrifuge at 20,000× *g* for 1 h at 4 °C, then 34.2 μL of the resulting supernatant mix with 10.8 μL of 5X Laemmli buffer, and use 40 μL of this mixture to measure the **TOTAL** amount of capsid by Western blotting (*see* **Step 6**).

4. Preparation of the **CYTOSOLIC** fraction (Fig. 1b). The remaining cell suspension 9/10 (900 μL) from **step 3** is centrifuged (800× *g* for 7 min at 4 °C), supernatant is discarded, and the cell pellet is resuspended in 315 μL of lysis buffer and incubated for 5 min on ice. Subsequently, the sample is centrifuged at 664× *g* for 2 min at 4 °C. The resulting supernatant and pellet correspond to **CYTOSOLIC** and **NUCLEAR** fractions, respectively (*see* **Note 5**). 1/9 aliquot of the **CYTOSOLIC** fraction (35 μL) is mixed with 15 μL of 5X Laemmli buffer and 40 μL of this mixture is used as Cytosolic fraction for detection by Western blotting (*see* **Step 6**).
5. Preparation of **NUCLEAR** fraction (Fig. 1b). The resultant pellet from **step 4** corresponds to the cellular nuclei. The cellular nuclei are washed twice using 1 mL of nuclear washing buffer by gently inverting the tube 2–3 times. The sample is then centrifuged at 660× *g* for 2 min at 4 °C, and the supernatant is discarded. The nuclear pellet is resuspended in 315 μL of extraction buffer (10 mM Tris, pH = 6.8, 1 mM DTT, 1 mM $MgCl_2$, 10% sucrose, 400 mM NaCl, 1X protease inhibitor), and incubated on ice for 10 min. Subsequently, the sample is centrifuged at 5200× *g* for 2 min at 4 °C. 1/9 aliquot of the supernatant (35 μL) is mixed with 15 μL of 5X Laemmli buffer and 40 μL of this mixture are used as **NUCLEAR** fraction (*see* **Step 6**).
6. Western blot analysis of **TOTAL**, **CYTOSOLIC,** and **NUCLEAR** fractions (Fig. 1c). Loading 40 μL in the Western blot as specified above corresponds to 8% of the total, cytosolic, and nuclear fractions. Western blot analysis will be performed using anti-p24, anti-Nopp140, anti-α-tubulin, or anti-GAPDH antibodies. Detection of Nopp140 is used as a nuclear marker since Nopp140 is a nucleolar phosphoprotein, but other nuclear markers can be used (including but not limited to Lamin A/C, Lamin B1, Histone H1, Histone H3). α-tubulin and GAPDH are used as a cytosolic marker (*see* **Note 6**). After TOTAL, CYTOSOLIC, and NUCLEAR fractions are prepared as described (*see* **Steps 3**, **4**, and **5** of this protocol), they are stable at −20 °C for at least a month. The day you perform the Western Blot, completely thaw the samples and boil them at 100 °C for 3 min. For electrophoresis we use NuPAGE 10% Bis-Tris gels. Load 40 μL of each fraction per well and 2 μL of Protein Standard on a separate well of the

same gel. Run the gel in 1X PAGE running buffer at 25 mA per gel for 2–3 h at 25 °C or until the Bromophenol Blue dye riches the end of the gel. Transfer the proteins to a Nitrocellulose membrane in 1X Western blot transfer buffer at 100 Volts for 1 h at 4 °C or on ice. After the transfer, block the membrane in Western blot blocking buffer for 0.5 h at 25 °C or at 4 °C for 16 h. Incubate with antibody against p24 (1:1000 dilution in Western blot blocking buffer) for 1.5 h at 25 °C or at 4 °C overnight. Wash the membrane in Western blot washing buffer 3 times for 10 min. Incubate with secondary antibody (Goat anti-Mouse IRDye 680LT, 1:10,000 dilution in Western blot blocking buffer) for 45 min at 25 °C. Wash the membrane in Western blot washing buffer 3 times for 10 min. Dry the membrane with a paper towel and scan using the LI-COR Odyssey Infrared Imager or any other fluorescence scan. After that re-hydrate the membrane in Western blot washing buffer for at least 30 min at 25 °C and incubate with the nuclear marker antibody (anti-Nopp140, 1:5000 dilution in Western blot blocking buffer). As a secondary antibody, we use Goat anti-Rabbit IRDye 680LT, 1:10,000 dilution in Western blot blocking buffer. After scanning the membrane, re-hydrate the membrane again and incubate with antibody against α-tubulin (1:8000 dilution in Western blot blocking buffer). As a secondary antibody, we use Goat anti-Rabbit IRDye 800CW (1: 10,000 dilution in Western blot blocking buffer). As alternative anti-GAPDH antibodies can be used (1:5000 dilution in Western blot blocking buffer and Goat anti-Mouse IRDye 800CW 1:10,000 dilution in Western blot blocking buffer) (*see* **Notes 7** and **8**).

7. Quantification of capsid protein in the **NUCLEAR** fraction. To quantify the amount of capsid in the NUCLEAR fraction, we calculate the ratio of capsid protein in the NUCLEAR fraction to the CYTOSOLIC fraction for three independent experiments and perform statistical analysis by applying the unpaired t test.

8. Expected Outcomes of control experiments. It is expected that the small molecule PF74 prevents the entry of the HIV-1 core, which is made of capsid protein, into the nuclear compartment [3]. Although here we describe PF74, the small molecule BI-2 that works in a similar way when compared to PF74 can also be used [3, 10, 11]. It is also expected that there is no cross-contamination between NUCLEAR and CYTOSOLIC fractions meaning that we will not observe a positive signal for GAPDH or α-tubulin in the nuclear fraction when performing Western blot analysis and vice versa. Limitations of the assay are described in **Note 9**.

4 Notes

1. To prevent cell damage during trypsin treatment, it is essential to incubate cells with trypsin for the minimum duration necessary for them to detach from the plate. Extended incubations with trypsin can lead to cell damage and the potential cross-contamination of cytosolic and nuclear fractions.
2. Ensure the collection of all cells from the plate, as this significantly influences the amount of capsid protein detected in the cytosolic fractions.
3. While performing washes, it is crucial to thoroughly eliminate any remnants of virus and trypsin from your cell mixture, as failing to do so will impact the outcome of your experiment.
4. Ethidium bromide is used to unwind genomic DNA an improve lysis.
5. Make sure that the amount of lysis buffer you use is enough to lyse your cell pellet. Incomplete lysis will result in contamination of your nuclear fraction with non-lysed cells. We recommend using 5× cell pellet volumes of the lysis buffer. To determine the correct volume of lysis buffer, you can make a pilot experiment testing several volumes of lysis buffer and checking by Western blot your fractions for nuclear and cytoplasmic markers.
6. It's essential to ensure that the quantity of lysis buffer you employ is sufficient to lyse your cell pellet completely. Insufficient lysis may lead to the contamination of your nuclear fraction with non-lysed cells. We recommend using a volume of lysis buffer equal to 5 times the cell pellet's volume. To determine the correct volume of lysis buffer, you can conduct a pilot experiment, testing various volumes of lysis buffer and constant number of cells, and subsequently assessing your fractions for nuclear and cytoplasmic markers using Western blot analysis.
7. After you detected the signal from your Western blot, you can keep the membrane dry at 25 °C for no longer than 2 weeks before proceeding with the next antibody.
8. It is important that the cytosolic and nuclear markers are similar in amount when comparing the effect of different treatments to the amount of capsid protein in the nucleus.
9. A limitation of this methodology is the fact that our analysis does not provide information regarding the oligomerization state of capsid in the nucleus. After HIV-1-infected cells are separated into cytosolic and nuclear fractions, proteins are extracted with a lysate buffer containing detergent, which disrupts the oligomerization state of the HIV-1 capsid rendering the assay unable to provide information regarding the oligomeric state of the capsid protein during infection.

Acknowledgments

This work was supported by an NIH grant AI087390 to F.D.-G. We are grateful to the NIH AIDS repository for providing plasmids and antibodies.

References

1. Burdick RC, Li C, Munshi M, Rawson JMO, Nagashima K, Hu WS, Pathak VK (2020) HIV-1 uncoats in the nucleus near sites of integration. Proc Natl Acad Sci USA 117(10): 5486–5493. https://doi.org/10.1073/pnas.1920631117
2. Dharan A, Bachmann N, Talley S, Zwikelmaier V, Campbell EM (2020) Nuclear pore blockade reveals that HIV-1 completes reverse transcription and uncoating in the nucleus. Nat Microbiol 5(9):1088–1095. https://doi.org/10.1038/s41564-020-0735-8
3. Selyutina A, Persaud M, Lee K, KewalRamani V, Diaz-Griffero F (2020) Nuclear import of the HIV-1 core precedes reverse transcription and uncoating. Cell Rep 32(13):108201. https://doi.org/10.1016/j.celrep.2020.108201
4. Diaz-Griffero F, Vandegraaff N, Li Y, McGee-Estrada K, Stremlau M, Welikala S, Si Z, Engelman A, Sodroski J (2006) Requirements for capsid-binding and an effector function in TRIMCyp-mediated restriction of HIV-1. Virology 351(2):404–419
5. Stremlau M, Perron M, Lee M, Li Y, Song B, Javanbakht H, Diaz-Griffero F, Anderson DJ, Sundquist WI, Sodroski J (2006) Specific recognition and accelerated uncoating of retroviral capsids by the TRIM5alpha restriction factor. Proc Natl Acad Sci USA 103(14): 5514–5519
6. Fricke T, Valle-Casuso JC, White TE, Brandariz-Nunez A, Bosche WJ, Reszka N, Gorelick R, Diaz-Griffero F (2013) The ability of TNPO3-depleted cells to inhibit HIV-1 infection requires CPSF6. Retrovirology 10(1):46. https://doi.org/10.1186/1742-4690-10-46
7. Fricke T, White TE, Schulte B, de Souza Aranha Vieira DA, Dharan A, Campbell EM, Brandariz-Nunez A, Diaz-Griffero F (2014) MxB binds to the HIV-1 core and prevents the uncoating process of HIV-1. Retrovirology 11(1):68. https://doi.org/10.1186/PREACCEPT-6453674081373986
8. De Iaco A, Santoni F, Vannier A, Guipponi M, Antonarakis S, Luban J (2013) TNPO3 protects HIV-1 replication from CPSF6-mediated capsid stabilization in the host cell cytoplasm. Retrovirology 10:20. https://doi.org/10.1186/1742-4690-10-20
9. Shi J, Zhou J, Shah VB, Aiken C, Whitby K (2011) Small-molecule inhibition of human immunodeficiency virus type 1 infection by virus capsid destabilization. J Virol 85(1): 542–549. https://doi.org/10.1128/JVI.01406-10
10. Lamorte L, Titolo S, Lemke CT, Goudreau N, Mercier JF, Wardrop E, Shah VB, von Schwedler UK, Langelier C, Banik SS, Aiken C, Sundquist WI, Mason SW (2013) Discovery of novel small-molecule HIV-1 replication inhibitors that stabilize capsid complexes. Antimicrob Agents Chemother 57(10): 4622–4631. https://doi.org/10.1128/AAC.00985-13
11. Fricke T, Buffone C, Opp S, Valle-Casuso J, Diaz-Griffero F (2014) BI-2 destabilizes HIV-1 cores during infection and prevents binding of CPSF6 to the HIV-1 capsid. Retrovirology 11:120. https://doi.org/10.1186/s12977-014-0120-x

Acknowledgments

References

Chapter 12

Chromatin Immunoprecipitation of Retroviral Genomes with Antibodies Recognizing Modified Histones and Specific Viral Proteins

Franziska K. Geis and Stephen P. Goff

Abstract

Mammalian cells have developed and optimized defense mechanisms to prevent or hamper viral infection. The early transcriptional silencing of incoming viral DNAs is one such antiviral strategy and seems to be of fundamental importance, since most cell types silence unintegrated retroviral DNAs. In this chapter, a method for chromatin immunoprecipitation of unintegrated DNA is described. This technique allows investigators to examine histone and co-factor interactions with unintegrated viral DNAs as well as to analyze histone modifications in general or in a kinetic fashion at various time points during viral infection.

Key words Unintegrated retroviral DNAs, Core histones, Linker histones, Histone variants, Histone modifications, Transcriptional silencing

1 Introduction

A milestone of the retroviral life cycle is the reverse transcription of the viral RNA genome to viral DNA [1]. Early in the life cycle, prior to integration, the viral DNA exists either as a linear form, or after circularization of viral DNA, as circular forms with either one copy of the long terminal repeats (1-LTR circles) [2] or two tandem copies of the repeats (2-LTR circles) [3]. Unintegrated viral DNAs, both the linear and the circular forms, are not able to replicate and gradually vanish during the course of subsequent cell divisions. Core and linker histones are rapidly deposited onto HIV-1 unintegrated DNAs and are marked with posttranslational histone modifications characteristic of silencing (high levels of H3K9 trimethylation, low levels of H3 acetylation) [4]. It has long been known and has been explored by many labs during the past years that transcription of unintegrated retroviral DNAs is heavily silenced in most cell types [5, 6]. In permissive cells, transcriptional silencing of unintegrated viral DNA is relieved after successful

Vinayaka R. Prasad and Ganjam V. Kalpana (eds.), *HIV Protocols*, Methods in Molecular Biology, vol. 2807,
https://doi.org/10.1007/978-1-0716-3862-0_12,

integration [5, 6]. Our work and that of other groups provided evidence that silencing of extrachromosomal DNAs is achieved not only through distinctive sets of regulatory proteins for different retroviruses but it also differs between cell types, suggesting that the mechanisms of silencing of extrachromosomal DNAs are more varied and multifactorial than expected [4, 7–11]. In this chapter, we describe Chromatin Immunoprecipitation (ChIP) as a technique to study the state of unintegrated viral DNAs. The approach involves analyses of histone compositions and histone modifications as well as the involvement of silencing co-factors of unintegrated viral chromatin. ChIP can be employed successfully to characterize silent viral chromatin.

2 Materials

2.1 Cells and Media

1. Human HeLa cell line (American Type Culture Collection, CCL-2).
2. Human Lenti-X 293T cell line (Clontech, 632180).
3. Dulbecco's Modified Eagle Media (DMEM) supplemented with 10% heat-inactivated fetal bovine serum (FBS), 100 U/mL penicillin, and 100 μg/mL streptomycin.
4. 1× Phosphate-Buffered Saline (PBS).
5. Trypsin-EDTA (0.25% Trypsin, 0.05 mM Ethylenediaminetetraacetic Acid (EDTA)).

2.2 Plasmids and Virus Constructs

1. HIV-1-based pNL4-3.ZsGreen.R$^-$.E$^-$ [12].
2. Integrase-deficient mutant pNL4-3.ZsGreen.R$^-$.E$^-$ [8].
3. VSV-G envelope encoded by pMD2.G (a gift from Didier Trono, Addgene plasmid, 12259).

2.3 Virus Production

1. 2 M $CaCl_2$.
2. 2xHEPES-Buffered Saline (pH 7.08).
3. DMEM was supplemented with 10 mM HEPES during particle production.
4. 0.45 μm pore-size-filters.
5. 5 U/mL RQ1 RNase-Free DNAse (DNase I).
6. 10 mg/mL Polybrene Transfection Reagent.

2.4 Chromatin Immunoprecipitation

1. 6% Formaldehyde.
2. 5 mg/mL Glycogen.
3. 10 mg/mL Yeast RNA.
4. 10 mg/mL Bovine Serum Albumin (purified).

Table 1
ChIP buffer preparation

	Nuclear lysis	ChIP dilution	Low salt wash	High salt wash	LiCl wash	TE (pH 8)	Elution
10% SDS	50 mL	500 uL	5 mL	5 mL	–	–	10 mL
Triton X 100	–	5.5 mL	5 mL	5 mL	–	–	–
0.5 M EDTA	10 mL	1.2 mL	2 mL	2 mL	1 mL	1 mL	–
1 M Tris-HCl (pH 8)	25 mL	8.4 mL	10 mL	10 mL	5 mL	5 mL	–
5 M NaCl	–	16.7 mL	15 mL	50 mL	–	–	4 mL
LiCl	–	–	–	–	5.3 g	–	–
NP40 (IGEPAL)	–	–	–	–	5 mL	–	–
Millipore H_2O	Up to 500 mL	Up to 500 mL	Up to 500 mL	Up to 500 mL	Up to 500 mL	Up to 500 mL	Up to 100 mL with TE, not H_2O

5. Protein A und G Dynabeads™.
6. 800 U/mL Proteinase K (molecular biology grade).
7. 1 mg/mL RNase A (affinity purified).
8. Protein-specific antibodies (preferably ChIP-grade).
9. QIAquick Gel Extraction Kit (QIAGEN).
10. ChIP Buffers (vacuum-filtered) (*see* Table 1).

2.5 Quantitative PCR

1. Fast Start Universal SYBR Green Master (ROX).
2. Sequence-specific primers (forward and reverse).

3 Methods

3.1 Virus Production and Titration

1. Seed 5 × 10^6 293T cells per 10-cm dish 1 day before transfection (*see* **Note 1**).
2. Calcium phosphate precipitation is the method of choice for retroviral particle production. Mix 10 μg retroviral plasmid DNA and 2 μg VSVg-expressing DNA in a total volume of 450 μL of 2xHEPES-buffered saline (pH 7.08). Add 50 μL of a 2 M $CaCl_2$ solution dropwise to the DNA while bubbling the solution with the pipet. After approximately 30 min (but not much longer), add the transfection mix to the 293T cell dish.
3. Supplement DMEM with 10 mM HEPES.
4. Remove Medium +Virus-Mix and replace with fresh medium 5 h after transfection.
5. Harvest viral supernatants 36 and 48 h after transfection and filter viral supernatants through 0.45 μm pore-size-filters.
6. Concentrate viral supernatants (100×) by ultracentrifugation (2 h, 25,000 rpm, 4 C).
7. Resuspend virus pellets in DMEM and store at −80 °C (*see* **Note 2**).
8. For titration, seed 7 × 10^4 HeLa cells per well (12-well plate format) 1 day before infection.
9. Apply serial dilutions of virus supernatants per well.
10. Remove Medium+Virus-Mix and replace with fresh medium 5 h after infection.
11. Harvest cells 48 h after infection with Trypsin-EDTA.
12. Add 1× PBS supplemented with 3% heat-inactivated FBS to the Cell-Trypsin-EDTA-Mix and transfer to FACS tube.
13. Carry out FACS analysis of the cells using, for example, the BD LSRII flow cytometer (BD Biosciences), and analyze using FlowJo software. Gate the data first for viable cells through

forward scatter height and side scatter height. Next, score for the ZsGreen-positive cells by gating from the previously defined viable population. Determine the mean fluorescence intensity (MFI) based on the gated ZsGreen-positive population.

14. Infectious Units/mL = (fluorescent-positive cells * total number of infected cells * 2/virus volume used for titration) × 1000) (*see* **Note 3**).

3.2 Virus Infection for Chromatin Immunoprecipitation

1. Seed 2 × 10^6 HeLa cells in a 10-cm dish 1 day before infection.
2. Pre-treat viral supernatant with 2 μL DNase I, supplemented with 10 mM $MgCl_2$ (10 μL of 1 M $MgCl_2$) in a volume of 1 mL media, for 1 h at 37 °C. DNase I is used to remove any residual plasmid DNA (*see* **Note 4**).
3. Mix 1 mL virus-DNase-mix with 9 mL and 8 μg/mL Polybrene Transfection Reagent and apply to HeLa cells.
4. Incubate for at least 4 h; then change medium.

3.3 Chromatin Immunoprecipitation

1. Harvest cells at time point of choice after virus application (e.g. 24 h after virus).
2. Rinse cells twice with 1× PBS (room temperature) and crosslink with 1% Formaldehyde for 10 min at 37 °C.
3. Add 0.125 M Glycine to Formaldehyde-Cell-Mix for 5 min at room temperature to quench crosslinker.
4. Wash three times with 10 mL 1× PBS. (Caveat: PBS contains Formaldehyde residuals and should be discarded appropriately.)
5. Lyse cells in 700 μL of Nuclear Lysis Buffer (50 mM Tris-HCL pH 8.0, 1% SDS, 10 mM EDTA) for 10 min at 4 °C on a shaker (*see* **Notes 5** and **6**).
6. Scrape and pipet cells to 15 mL Falcon tubes. Keep samples on ice as much as possible during procedure. (Optional: Scrape cells and shock-freeze cell pellet in liquid nitrogen and store at −80 °C. Thaw cells gently on ice to continue.)
7. Sonicate cell lysates. Perform sonication on ice.
8. The Branson 150 Digital Sonifier was used. Sonification was performed with the following settings: 30% power amplitude, sonication for 30 s pulse (four times) with 1 min between each sonication. Keep samples on ice between each sonication.
9. At the beginning and between samples: wash nozzle with H_2O, wipe, wash with 70% Ethanol, wipe. Transfer sample to 1.5 mL Eppendorf tube after sonication.
10. Leave samples on ice for 30 min after sonication.
11. Centrifuge at 13,000 rpm for 20 min at 4 °C.

12. Measure DNA concentration of supernatant (e.g. Nano Drop Spectrophotometer) and use 50 μg DNA/sample for ChIP. Keep 5 μg (10%) for Input-DNA calculation at −20 °C. (Optional: shock-freeze with liquid nitrogen and store at −80 °C. Measure DNA again after gently thawing samples on ice.)
13. Mix 50 μg DNA and 1 mL ChIP Dilution Buffer and 5 μg antibody in an Eppendorf tube. Use also an appropriate isotype antibody control. Rotate at 4 °C overnight.
14. Pre-treat beads for ChIP overnight. Per sample: 12.5 μL Protein A beads +12.5 μL Protein G beads (prepare Master Mix for all samples). Wash Master Mix twice with 1 mL ChIP Dilution Buffer (use magnetic separation rack).
15. Prepare Block Buffer Master Mix. Per sample: 40 μL Glycogen (5 mg/mL), 20 μL BSA (10 mg/mL), 20 μL tRNA (10 mg/mL), 920 μL ChIP Dilution Buffer.
16. Separate Master Mixes: 25 μL beads +1 mL Block Buffer in a 1.5 mL Eppendorf tube, rotate at 4 °C overnight.
17. The next day, wash beads twice with 1 mL ChIP Dilution Buffer (use magnetic separation rack).
18. Transfer DNA-Antibody-Beads-Mix in a clean Eppendorf tube and rotate at 4 °C for 2 h (*see* **Note 7**).
19. Wash twice with 1 mL Low Salt Wash (rotate at 4 °C for 5 min for each wash step) (use magnetic separation rack).
20. Wash twice with 1 mL High Salt Wash (rotate at 4 °C for 5 min for each wash step) (use magnetic separation rack).
21. Wash twice with 1 mL LiCl Wash (rotate at 4 °C for 5 min for each wash step) (use magnetic separation rack).
22. Spin down (no use of magnetic separation rack!) at 13,000 rpm for 1 min at room temperature and wash twice with 1 mL TE Wash (rotate at 4 °C for 5 min for each washing step).
23. For DNA elution use 200 μL Elution Buffer, supplemented with 15 μL Protease K and 10 μL RNase A (stock 1 mg/mL) and put Eppendorf tubes in a heat block at 65 °C, shake (1000 rpm) for 30 min.
24. Spin at 13,000 rpm for 1 min at room temperature.
25. Reverse Crosslink by heating at 65 °C. Pipet 200 μL supernatant in a clean Eppendorf tube. Put samples in a heat block at 65 °C (no shaking) overnight.
26. Reverse Crosslink the 10% Input DNA by adding 200 μL Elution Buffer, supplemented with 15 μL Protease K and 10 μL RNase A (stock 1 mg/mL), and put Input samples in a heat block at 65 °C (no shaking) overnight.

27. The next day, spin samples at 13,000 rpm for 1 min at room temperature.
28. Purify DNA by adding 600 μL QG Buffer and 200 μg Isopropanol (QIAquick Gel Extraction Kit, QIAGEN). Wash with PE-Buffer, elute 2× in 50 μL H_2O (total sample volume 100 μL). Store samples at −20 °C.

3.4 Quantitative PCR

1. Prepare Master Mix for SYBR Green-based qPCR. Per sample: 10 μL SYBR Green Mix, 1 μL Primer Mix forward + reverse (10 pmol/μL each), 4 μL H_2O (*see* **Note 8**).
2. Add 5 μL DNA to 15 μL Master Mix for qPCR analysis.
3. Run qPCR with the following settings: 50 °C for 2 min and at 95 °C for 10 min followed by 40 cycles of 15 s at 95 °C, 30 s at 60 °C, and 30 s at 72 °C (Applied Biosystems was used) (*see* **Note 9**).
4. Utilize Ct values to calculate $2^{\Delta Ct}$ relative to input DNA and express data in percentage relative to input DNA (*see* Fig. 1).

4 Notes

1. For virus production and the achievement of high titers, it is crucial that the packaging 293 cells are healthy in general. Furthermore, cell density should not be too high on the day of transfection (maximum 70% confluency).
2. In order to maintain high titers, prepare small aliquots of virus supernatants and store at −80 °C, avoid thaw and freeze cycles and keep virus supernatants on ice as often as possible during handling.
3. For the titer calculation, samples were only included with approximately 30% fluorescence-positive cells to exclude signals from multiple infected cells, which will influence titer outcome.
4. Virus constructs which are used for comparative studies (e.g. Integrase-Wild-Type Constructs versus. Integrase-Mutant Constructs) should be packaged simultaneously and equal amounts of supernatants should be applied to cells within an experiment for comparison. We typically utilized a multiplicity of infection of ≤0.5, which leads to approximately 30–50% fluorescent-positive cells.
5. ChIP Buffers should be vacuum-filtered and stored at 4 °C, except Elution Buffer and Nuclear Lysis Buffer which should be kept at room temperature.
6. During most of the ChIP protocol, it is crucial to keep cell lysates and samples after sonication at 4 °C as much as possible.

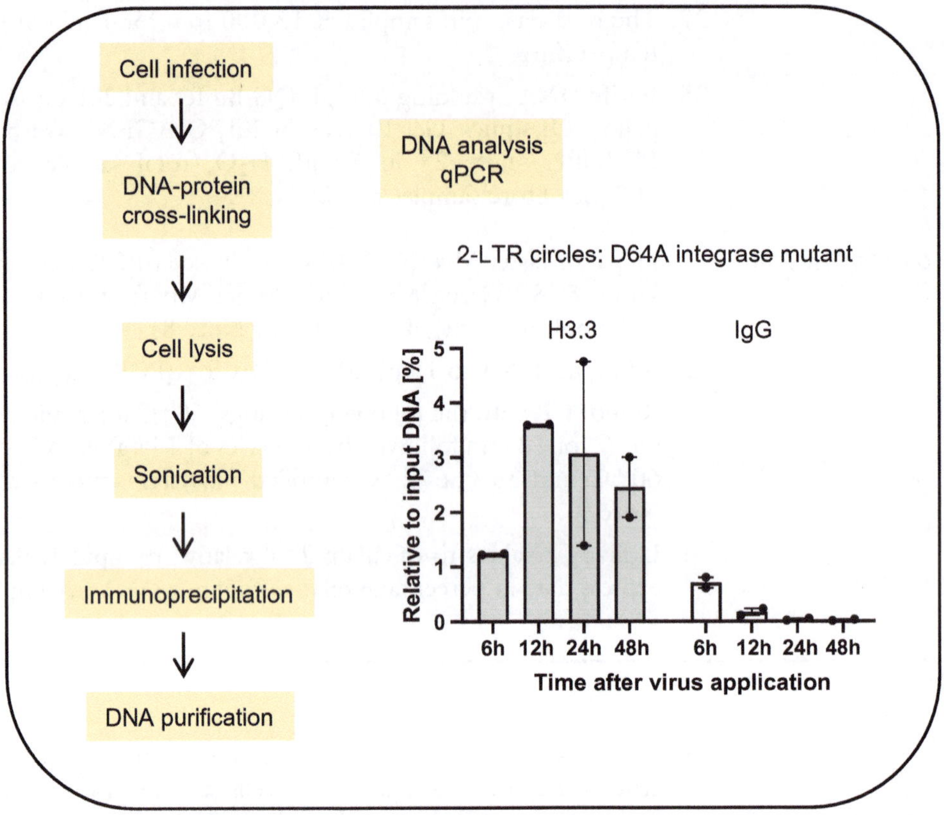

Fig. 1 Schematic summary overview of the different steps of the ChIP protocol and an example of a figure of ChIP results shown as a bar graph in percentage of immunoprecipitated DNA relative to input DNA

7. During the ChIP protocol, the washing steps toward the end (DNA-Antibody-Beads-Mix, Low Salt Wash, High Salt Wash or LiCl Wash) can be extended, if needed as long as all samples are treated comparably.
8. qPCR Primers which bind to the 2LTR region of the HIV-1 sequence were used in our laboratory to detect unintegrated HIV-1 virus. Primers specific for GAPDH was used as a transcriptional active gene control and primers binding to part of the beta globin sequence served as a transcriptional inactive gene control.
9. For quality control, we performed a melting curve for every qPCR run according to manufacturer's protocol.

The ChIP method described above rapidly gives information about the chromatin status and host proteins associated with incoming viral DNA. The limiting reagent for these studies is usually the availability of antibodies specific for a given modification or protein of interest, but more are becoming available daily. The findings will be crucial in unraveling the pathways responsible for silencing of incoming viral DNA. They also have the potential to improve many gene delivery approaches, which are still hampered by inefficient transcription.

Acknowledgments

This work was also supported in part by a grant from the NIH (R01 CA 030488) to SPG. F.K.G. was supported by the DFG (German Research Foundation) grant GE 3106/1-1.

References

1. Coffin JM, Hughes SH, Varmus HE (1997) The interactions of retroviruses and their hosts. In: Coffin JM, Hughes SH, Varmus HE (eds) Retroviruses. Cold Spring Harbor, New York
2. Yoshimura FK, Weinberg RA (1979) Restriction endonuclease cleavage of linear and closed circular murine leukemia viral DNAs: discovery of a smaller circular form. Cell 16(2):323–332. https://doi.org/10.1016/0092-8674(79)90009-6
3. Kilzer JM, Stracker T, Beitzel B, Meek K, Weitzman M, Bushman FD (2003) Roles of host cell factors in circularization of retroviral dna. Virology 314(1):460–467. https://doi.org/10.1016/s0042-6822(03)00455-0
4. Geis FK, Goff SP (2019) Unintegrated HIV-1 DNAs are loaded with core and linker histones and transcriptionally silenced. Proc Natl Acad Sci USA 116(47):23735–23742. https://doi.org/10.1073/pnas.1912638116
5. Sakai H, Kawamura M, Sakuragi J, Sakuragi S, Shibata R, Ishimoto A, Ono N, Ueda S, Adachi A (1993) Integration is essential for efficient gene expression of human immunodeficiency virus type 1. J Virol 67(3):1169–1174. https://doi.org/10.1128/JVI.67.3.1169-1174.1993
6. Schwartzberg P, Colicelli J, Goff SP (1984) Construction and analysis of deletion mutations in the pol gene of Moloney murine leukemia virus: a new viral function required for productive infection. Cell 37(3):1043–1052. https://doi.org/10.1016/0092-8674(84)90439-2
7. Geis FK, Sabo Y, Chen X, Li Y, Lu C, Goff SP (2022) CHAF1A/B mediate silencing of unintegrated HIV-1 DNAs early in infection. Proc Natl Acad Sci USA 119(4):e2116735119. https://doi.org/10.1073/pnas.2116735119
8. Zhu Y, Wang GZ, Cingoz O, Goff SP (2018) NP220 mediates silencing of unintegrated retroviral DNA. Nature 564(7735):278–282. https://doi.org/10.1038/s41586-018-0750-6
9. Wang GZ, Wang Y, Goff SP (2016) Histones are rapidly loaded onto unintegrated retroviral DNAs soon after nuclear entry. Cell Host Microbe 20(6):798–809. https://doi.org/10.1016/j.chom.2016.10.009
10. Dupont L, Bloor S, Williamson JC, Cuesta SM, Shah R, Teixeira-Silva A, Naamati A, Greenwood EJD, Sarafianos SG, Matheson NJ, Lehner PJ (2021) The SMC5/6 complex compacts and silences unintegrated HIV-1 DNA and is antagonized by Vpr. Cell Host Microbe. https://doi.org/10.1016/j.chom.2021.03.001
11. Machida S, Depierre D, Chen HC, Thenin-Houssier S, Petitjean G, Doyen CM, Takaku M, Cuvier O, Benkirane M (2020) Exploring histone loading on HIV DNA reveals a dynamic nucleosome positioning between unintegrated and integrated viral genome. Proc Natl Acad Sci USA 117(12):6822–6830. https://doi.org/10.1073/pnas.1913754117
12. Sabo Y, Walsh D, Barry DS, Tinaztepe S, de Los Santos K, Goff SP, Gundersen GG, Naghavi MH (2013) HIV-1 induces the formation of stable microtubules to enhance early infection. Cell Host Microbe 14(5):535–546. https://doi.org/10.1016/j.chom.2013.10.012

Part IV

HIV-1 RNA – Functions and Modification

Chapter 13

Measurement of HIV Rev-Rev Response Element Functional Activity

Patrick E. H. Jackson, Godfrey Dzhivhuho, Jing Huang, Marie-Louise Hammarskjold, and David Rekosh

Abstract

Retroviruses must overcome cellular restrictions to the nucleocytoplasmic export of viral mRNAs that retain introns in order to complete their replication cycle. HIV accomplishes this using a system comprised of a *trans*-acting viral protein, Rev, and a *cis*-acting RNA secondary structure in the viral genome, the Rev-Response Element (RRE). HIV primary isolates differ with respect to the sequence and functional activity of the Rev-RRE system. Here, we describe a high throughput assay system for analyzing Rev-RRE functional activity using packageable viral vectors.

Key words HIV, Lentivirus, Rev, Rev response element, RNA trafficking

1 Introduction

Retrovirus replication requires the export of viral mRNAs with retained introns (IR-mRNAs) from the host cell nucleus to the cytoplasm [1]. IR-mRNAs are often retained in the nucleus, and so-called "complex" retroviruses accomplish the nuclear export of their unspliced and incompletely spliced mRNAs via a system that includes both a *trans*-acting viral protein and a *cis*-acting RNA secondary structure [2]. In the case of HIV, the viral protein Rev is encoded by a completely spliced transcript that is readily exported to the cytoplasm and translated. Rev is then imported to the nucleus where it binds to an RNA secondary structure found on viral IR-mRNAs, the Rev-Response Element (RRE) [3–5]. RRE-bound Rev then recruits cellular factors to form a complex capable of exporting the mRNA to the cytoplasm [6, 7].

HIV primary isolates exhibit sequence variation throughout the genome including within *rev* and the RRE. These differences give rise to significant variations in Rev-RRE functional activity between viral variants [8, 9]. Variations in activity can affect the

Vinayaka R. Prasad and Ganjam V. Kalpana (eds.), *HIV Protocols*, Methods in Molecular Biology, vol. 2807, https://doi.org/10.1007/978-1-0716-3862-0_13,

relative expression of viral proteins and alter viral replication kinetics, and this may have implications for pathogenesis [10–13].

Previous studies of Rev-RRE activity have used systems based on transient transfection with reporter constructs in which activity differences are measured by protein expression [8, 14, 15] or lentiviral vector packaging [9]. In order to improve throughput and to determine Rev-RRE activity within a more physiological context, we developed an assay system based on packageable retroviral vector constructs [16]. Unlike with transient transfection assays, this new system permits measurement of Rev-RRE activity in lymphoid cell lines that are often hard to transfect, as well as primary cells. It is also possible to assess Rev-RRE activity in a context where single Rev and RRE sequences have been integrated into the host cell chromatin, more accurately reflecting the conditions during a natural HIV infection.

To perform this assay, two types of vector constructs are customized with *rev* and RRE sequences of interest. One vector, based on the murine stem cell virus (MSCV), carries an exchangeable Rev along with a fluorescent protein marker. The second vector, based on the HIV strain NL4-3, carries an exchangeable RRE along with two additional fluorescent markers. The assay vectors are packaged in 293T/17 cells and pseudotyped using VSV-G. Then, the titer of all vector stocks is determined using flow cytometry. Functional activity is assessed by co-transducing cells with both assay vectors and measuring the relative intensity of Rev-dependent to Rev-independent fluorescence.

2 Materials

2.1 Designing Assay Constructs (See Note 1)

1. Plasmid pMSCV-Rev-IRES-eBFP2 [16] (*see* **Note 2**).
2. Plasmid pNL4-3(gag-CHYSEL-eGFP)(Rev-)(Env-)(XmaI-RRE-XbaI)(Nef-)(mCherry+) [16] (*see* **Note 3**).

2.2 Packaging Assay Vectors

1. Equipment: laminar flow hood, cell culture incubator.
2. Plasmid DNA (100 ng/μL for each):
   - (a) Plasmid pMSCV-Rev-IRES-eBFP2 containing a Rev sequence of interest.
   - (b) Plasmid pNL4-3(gag-CHYSEL-eGFP)(Rev-)(Env-)(XmaI-RRE-XbaI)(Nef-)(mCherry+) containing an RRE sequence of interest.
   - (c) Plasmid pMSCV-IRES-mCherry (Addgene plasmid 52114).
   - (d) Plasmid pMSCV-IRES-GFP (Addgene plasmid 52107).
   - (e) Plasmid psPAX2 (Addgene plasmid 12260).

(f) Plasmid pMD2.G (Addgene plasmid 12259).
(g) Plasmid pHIT60-CMV-GagPol [17].

3. Cell line: 293T/17 cells.
4. 10 cm tissue culture plates.
5. 293T growth medium: Iscove's modified Dulbecco's medium supplemented with 10% v/v bovine calf serum and 50 ug/mL gentamicin.
6. Transfection medium: Iscove's modified Dulbecco's medium supplemented with 5% v/v bovine calf serum.
7. 1.5 mL tubes.
8. Opti-MEM reduced serum medium.
9. Polyethylenimine (1 mg/mL).
10. 15 mL conical tubes.

2.3 Determining Assay Vector Titer and Co-transduction of Target Cells

1. Equipment: laminar flow hood, cell culture incubator, flow cytometer with plate autosampler.
2. Software: FlowJo v 10.8.1 (FlowJo, LLC) or similar.
3. Cell line: SupT1 cells (*see* **Note 4**).
4. SupT1 growth medium: Roswell Park Memorial Institute 1640 medium supplemented with 10% v/v fetal bovine serum and 50 ug/mL gentamicin.
5. 15 cm tissue culture plates.
6. Automated cell counter.
7. Cell counting slides.
8. Trypan blue stain, 0.4%.
9. Transfection medium: Iscove's modified Dulbecco's medium supplemented with 5% v/v bovine calf serum.
10. 15 mL conical tubes.
11. 50 mL conical tubes.
12. Dulbecco's phosphate buffered saline (DPBS).
13. 96 well round bottom tissue culture treated plates with lids.
14. DEAE-dextran 0.4 mg/mL (*see* **Note 5**).

2.4 Data Analysis

1. Software: FlowJo v 10.8.1 or similar.

3 Methods

3.1 Designing Assay Constructs

1. Identify *rev* and RRE sequences to be tested in the functional assay (*see* **Note 6**). These regions can be bioinformatically clipped from genomic HIV sequences using the Los Alamos

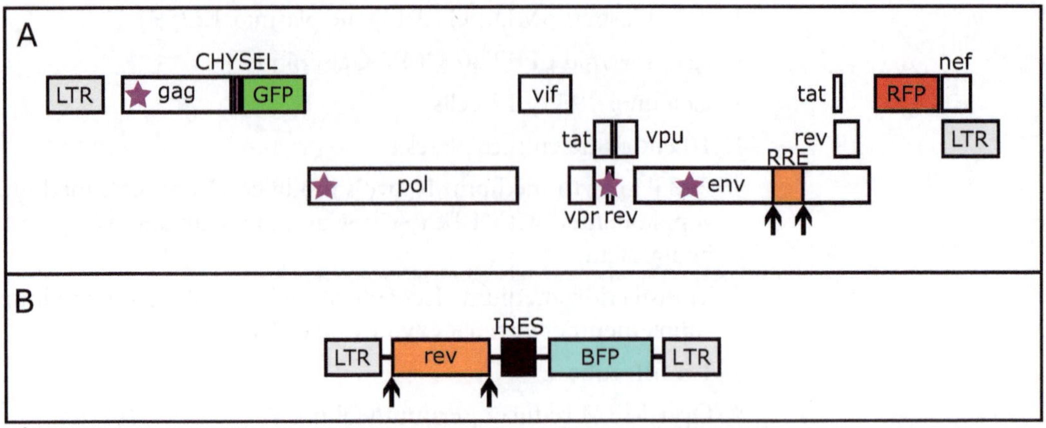

Fig. 1 Schematics of assay constructs. (**a**) pNL4-3(gag-CHYSEL-eGFP)(Rev-)(Env-)(XmaI-RRE-XbaI)(Nef-)(mCherry+) assay construct was created by modifying the HIV strain NL4-3 to include a CHYSEL-GFP (eGFP) cassette in the *gag* reading frame and an RFP (mCherry) gene in the *nef* reading frame. Additional modifications to render the construct replication incompetent are marked with stars. From left to right these are: deletion of the *gag* myristoylation site, deletion of the *gag-pol* frameshift, introduction of two mutations into the first exon of *rev*, and a frameshift mutation in *env*. The RRE in this construct is flanked by XbaI and XmaI restriction sites for easy exchange, indicated by black arrows. (**b**) An MSCV vector was modified to include a *rev* gene upstream of an IRES-BFP (eBFP2) cassette. The *rev* is flanked by EcoRI and XhoI restriction sites indicated by black arrows. *LTR* long tandem repeat, *CHYSEL cis*-acting hydrolase element, *GFP* green fluorescent protein, *RRE* Rev Response Element, *RFP* red fluorescent protein, *IRES* internal ribosome entry site, *BFP* blue fluorescent protein

HIV database Gene Cutter tool (https://www.hiv.lanl.gov/content/sequence/GENE_CUTTER/cutter.html). The *rev* sequences to be used must include both exons of *rev* with the intron deleted. The RRE sequences may be selected without reference to the overlying *env* reading frame (*see* **Note 7**).

2. After identification, the desired *rev* and RRE sequences are synthesized and cloned into the pMSCV or pNL4-3 plasmids (Fig. 1) (*see* **Note 8**). *Rev* sequences can be inserted into pMSCV-Rev-IRES-eBFP between the restriction sites EcoRI and XhoI. RRE sequences can be inserted into pNL4-3(gag-CHYSEL-eGFP)(Rev-)(Env-)(XmaI-RRE-XbaI)(Nef-)(mCherry+) between the restriction sites XmaI and XbaI.

3.2 Packaging Assay Vectors

Perform all steps in sterile fashion in a laminar flow hood.

1. For each assay of Rev-RRE functional activity, vector stocks of one or more versions of pMSCV-Rev-IRES-eBFP2 containing a Rev of interest and one or more versions of pNL4-3(gag-CHYSEL-eGFP)(Rev-)(Env-)(XmaI-RRE-XbaI)(Nef-)(mCherry+) containing an RRE of interest are required. Additionally, stocks of pMSCV-IRES-mCherry and pMSCV-IRES-GFP are needed for color compensation during FACS analysis.

Table 1
Packaging system parameters for pMSCV vectors

Plasmid name	Mass (ug)	Volume (uL)
pMSCV	15	150
pHIT60-CMV-GagPol	6.42	64.2
pMD2.G	2.54	25.4

Table 2
Packaging system parameters for pNL4-3 vectors

Plasmid name	Mass (ug)	Volume (uL)
pNL4-3	15	150
psPAX2	6.42	64.2
pMD2.G	2.54	25.4

2. For each assay vector to be packaged, plate 3.5×10^6 293T/17 cells in a 10 cm tissue culture dish with 10 mL of 293T growth medium. Incubate for 24 h at 37 °C in a 5% CO_2 cell culture incubator.
3. Remove the growth medium from the 293T/17 cells and replace it with 10 mL transfection medium. Return the plate to the incubator for 1 h prior to transfection.
4. To package vectors with the transducing constructs pMSCV-Rev-IRES-eBFP2, pMSCV-IRES-mCherry, or pMSCV-IRES-GFP, combine plasmid DNA in a 1.5 mL tube (Table 1).
5. To package vectors with the transducing construct pNL4-3 (gag--CHYSEL-eGFP)(Rev-)(Env-)(XmaI-RRE-XbaI)(Nef-)(mCherry+) combine plasmid DNA in a 1.5 mL tube (Table 2).
6. To each tube of plasmid DNA, add 239.6 μL Opti-MEM.
7. For each vector to be packaged, in a new 1.5 mL tube combine 143.76 μL of polyethylenamine (PEI) and 335.44 μL Opti-MEM. Vortex briefly to mix.
8. To each tube of plasmid DNA, add 479.2 μL of the PEI/Opti-MEM solution from the previous step. Incubate the DNA/PEI/Opti-MEM tubes, without mixing, at room temperature for 20 min.
9. Add the complete volume of one DNA/PEI/Opti-MEM tube to one plate of 293T/17 cells (*see* **Note 9**). Return cells to the incubator for 48 h.

10. Collect the medium from each plate of cells and transfer it to a 15 mL conical tube. Centrifuge the medium at 380 RCF at 4 °C for 5 min to pellet any cell debris. Transfer the supernatant from each tube to a new 15 mL conical tube, discarding the pellet (*see* **Note 10**).
11. Make 500 μL aliquots of clarified supernatant in fresh 1.5 mL tubes and freeze at −80 °C until needed.

3.3 Determining Assay Vector Titer

1. Split SupT1 cells by plating about 4×10^6 cells in a 15 cm tissue culture plate in 20 mL of SupT1 growth medium. Incubate cells for 24 h at 37 °C in a 5% CO_2 cell culture incubator.
2. Suspend any settled SupT1 cells by agitating the plate gently. Determine the concentration of viable SupT1 cells using an automatic cell counter with Trypan blue stain per the manufacturer's instructions (*see* **Note 11**).
3. To find the titer of up to 11 vector stocks, transfer a volume of SupT1 cell suspension containing 2.88×10^7 viable cells to a 50 mL conical tube. Pellet the cells by centrifuging at 380 RCF at room temperature for 5 min. Carefully pour off the supernatant and retain the cell pellet.
4. Gently resuspend the cell pellet in 5.76 mL DPBS warmed to 37 °C. Transfer 50 μL of the resulting cell suspension to each well of a 96 well plate, yielding 2.5×10^5 cells per well. Return the covered 96 well plate to the cell culture incubator until needed.
5. For each vector to be titered, thaw one 500 μL aliquot of the stock in a 37 °C water bath (*see* **Notes 12** and **13**). Vortex briefly to mix.
6. Make three 1:10 serial dilutions of each vector stock (i.e. 1:10, 1:100, and 1:1000) in 1.5 mL tubes with a total volume of 500 μL for each dilution (i.e. start by diluting 50 μL of the stock in 450 μL of transfection medium). Vortex each dilution to thoroughly mix prior to making the next dilution in the series.
7. To determine the titer of the vector stocks, add 200 μL of each undiluted stock to wells A and B (e.g. first stock to A1 and B1, second stock to A2 and B2, etc.) of the 96 well plate containing SupT1 cells. Then, add 200 μL of the 1:10 dilution of the stocks to wells C and D, 200 μL of the 1:100 dilution to wells D and F, and 200 μL of the 1:1000 dilution to wells G and H (Fig. 2).
8. Add 200 μL sterile transfection medium to the cells in each well of column 12 of the plate.
9. Add 5 μL DEAE-dextran solution to each well of the 96 well plate.

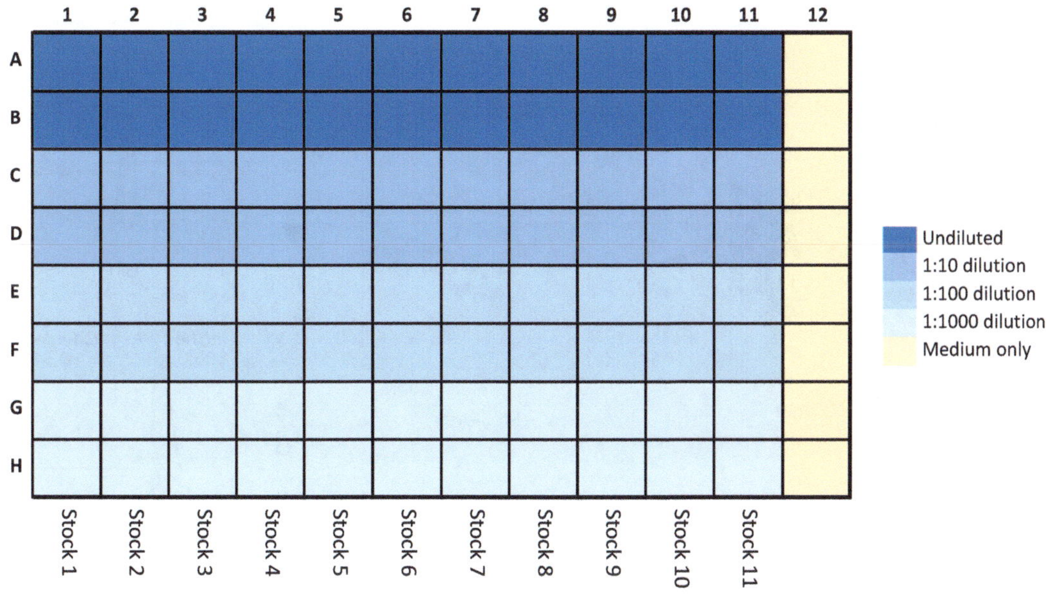

Fig. 2 Diagram of a 96 well plate set up for titer determination of vector stocks. In this example, the titer of eleven vector stocks can be determined using columns 1 through 11 of the plate. For each stock, 200 μL of undiluted medium is added to rows A and B, 200 μL of 1:10 diluted medium is added to rows C and D, etc. Column 12 is reserved for cells not transduced with a vector to be used for flow cytometry calibration and a negative control during data analysis

10. Centrifuge the covered plate at 380 RCF at room temperature for 1 h.
11. Carefully remove the supernatant from each well of the 96 well plate, leaving the cells on the bottom of the well undisturbed. Gently resuspend the cell pellet in each well in 250 μL of SupT1 growth medium warmed to 37 °C. Incubate the covered 96 well plate in the cell culture incubator for 72 h.
12. Remove the 96 well plate from the incubator and centrifuge at 380 RCF at 4 °C for 5 min to pellet the cells.
13. Carefully remove the supernatant from each well of the 96 well plate, leaving the cells undisturbed. Discard the supernatant. Resuspend the cell pellets in each well in 200 μL of cold DPBS.
14. Acquire the cells using the flow cytometer with plate autosampler. Collect data in the side scatter (SSC) and forward scatter (FSC) channels as well as channels corresponding to the mCherry and eBFP2 excitation/emission spectra (*see* **Note 14**). In an FSC-area vs SSC-area plot, use some untransduced cells from wells in column 12 to define an appropriate gate that identifies SupT1 cells and excludes debris and large cell aggregates. Adjust the gain for the channels corresponding to mCherry and eBFP2 such that all positive cells are within the dynamic range of detection and that the negative population

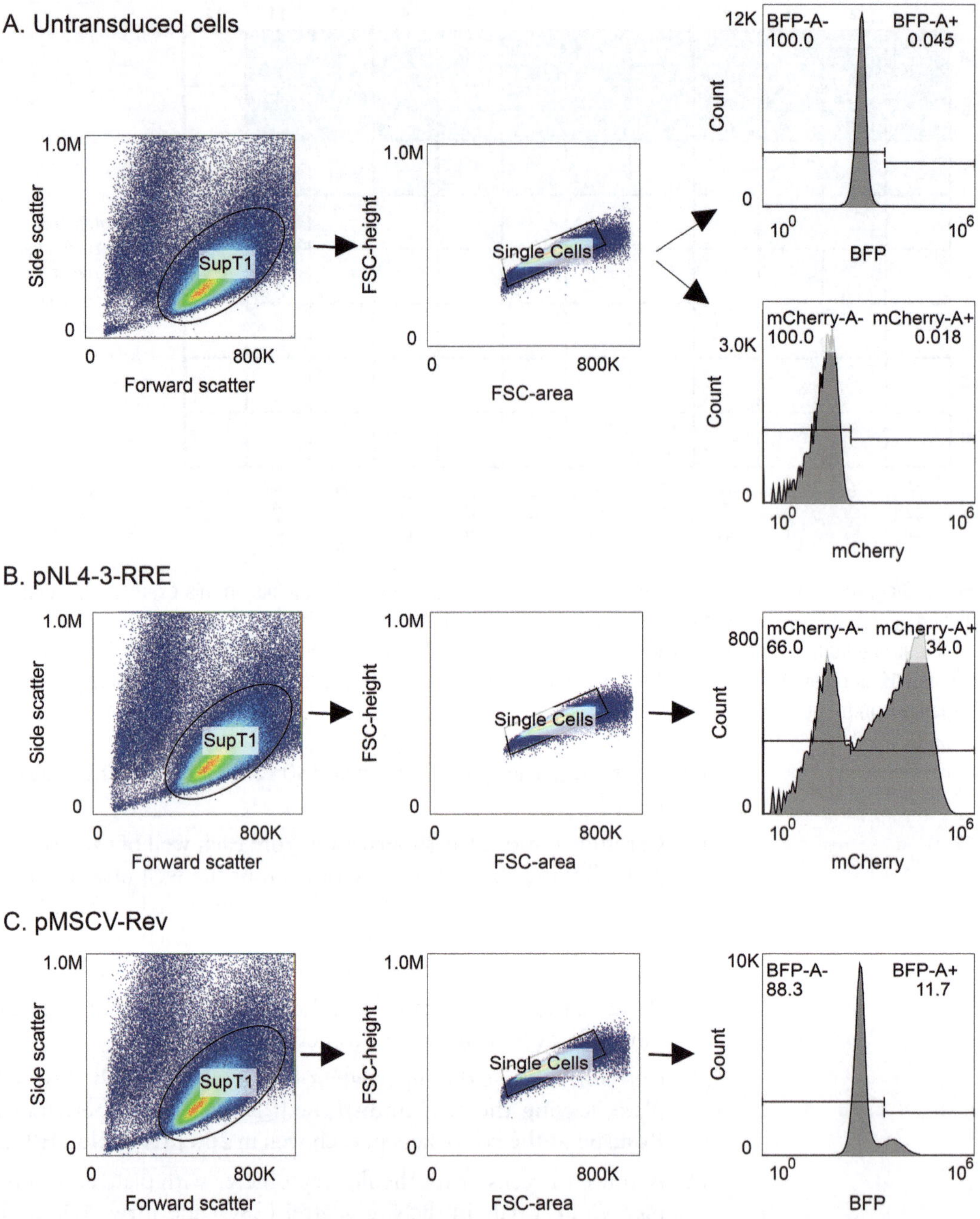

Fig. 3 Flow cytometry gating strategy for vector stock titer analysis. The gating strategy for determining vector stock titer is shown from left to right. First, SupT1 cells were distinguished from debris in a side scatter vs forward scatter plot. Next, the population of singlet cells was identified in a forward scatter height vs forward scatter area plot. Finally, the proportion of cells that were mCherry or BFP positive was determined. (**a**) Untransduced cells were used to define the mCherry negative and BFP negative gates. These gates were then applied to all of the sample files. (**b**) Cells were transduced with a pNL4-3-RRE vector. In this example well, 34% of singlet SupT1 cells were transduced. (**c**) Cells were transduced with a pMSCV-Rev vector. In this example, 11.7% of singlet cells in the well were transduced

can be easily distinguished (*see* **Note 15**). Acquire at least 50,000 events corresponding to healthy cells for each well. All the wells in the columns of the 96 well plate that include vector stocks should be sampled, as should at least one well of untransduced cells from column 12.

15. Analyze the resulting FCS files using FlowJo or similar software to determine the percentage of fluorescent cells in each well (Fig. 3). For pMSCV-Rev-IRES-eBFP2-type vectors, eBFP2 expression defines the successfully transduced population. For pNL4-3-type vectors, mCherry expression is used to determine the transduced population.
    (a) Using an untransduced well, define an FSC-A and SSC-A gate corresponding to healthy SupT1 cells. Then, create a daughter FSC-H and FSC-A gate corresponding to singleton cells.
    (b) From the untransduced singleton cells, define two daughter gates corresponding to the mCherry negative and eBFP2 negative populations using the appropriate parameters.
    (c) Apply these gates to all of the samples from **step 14** above.
    (d) For each transduced well, determine the proportion of singleton healthy cells that are mCherry positive (in the case of pNL4-3 constructs) or eBFP2 positive (in the case of pMSCV constructs).
16. For each well, calculate the titer in infectious units (IU) per mL of the original vector stock using the formula:

$$\text{Titer}\left(\frac{\text{IU}}{\text{mL}}\right)=\frac{\#\text{ cells transduced}\times-\ln\left(1-\frac{\%\text{ positive cells}}{100}\right)}{\text{stock volume (mL)}}$$

The number of cells transduced is the number added to each well in **step 4**. The stock volume in each well depends on the dilution of stock added. For wells with undiluted stock (i.e. rows 1 and 2), this is 0.2 mL, for the 1:10 dilution wells, this is 0.02 mL, etc. (*see* **Note 16**).

To calculate the titer of the pMSCV-Rev assay vector shown in Fig. 3c, assuming this represents a 1:100 dilution well:

$$\text{Titer}=\frac{2.5\times10^{5}\text{ cells}\times-\ln\left(1-\frac{11.7\%\text{ positive cells}}{100}\right)}{0.002\text{ mL stock}}$$
$$\approx 1.56\times10^{7}\text{ IU/mL}$$

The best estimate of the titer of the vector stock is found by averaging the calculated titer for two wells in a dilution range that yielded about 2–20% positive cells (*see* **Note 17**).

3.4 Co-transduction of Target Cells

1. To compare the functional activity of different Rev-RRE pairs, cells are simultaneously transduced with a Rev-containing pMSCV assay vector and an RRE-containing pNL4-3 assay vector. Assay runs are performed in a 96-well plate format. For each unique Rev-RRE pair, use three wells as technical replicates. Additionally, three single wells will be transduced with GFP, mCherry, or BFP single color vectors for color compensation. At least one well should contain cells that are not transduced (Fig. 4) (*see* **Note 18**).
2. Split SupT1 cells by plating about 4×10^6 cells in a 15 cm tissue culture plate in 20 mL of SupT1 growth medium. Incubate cells for 24 h at 37 °C in a 5% CO_2 cell culture incubator.
3. Suspend any settled SupT1 cells by agitating the plate gently. Determine the concentration of viable SupT1 cells using the automatic cell counter with Trypan blue stain per the manufacturer's instructions.
4. For one 96 well plate, transfer a volume of SupT1 cell suspension containing 2.88×10^7 viable cells to a 50 mL conical tube. Pellet the cells by centrifuging at 380 RCF at room temperature for 5 min. Carefully pour off the supernatant and retain the cell pellet.
5. Gently resuspend the cell pellet in 5.76 mL DPBS warmed to 37 °C. Transfer 50 μL of the resulting cell suspension to each well of a 96 well plate, yielding 2.5×10^5 cells per well. Return the covered 96 well plate to the cell culture incubator until needed.
6. Thaw aliquots of each Rev- and RRE-containing vector stock to be used in the assay run in a 37 °C water bath. Also thaw pMSCV-IRES-mCherry and pMSCV-IRES-GFP stocks for color compensation controls. In addition, thaw a pMSCV-Rev-IRES-eBFP2 stock to be used as a blue color compensation control.
7. Cells in Rev-RRE pair wells will be co-transduced with one pMSCV-Rev vector and one pNL4-3 vector at a multiplicity of infection (MOI) of 0.18 for each construct. Since each well contains 2.5×10^5 cells, a volume of vector stock sufficient to contain 4.5×10^4 infectious units is required for each well in which the Rev- or RRE-containing vector is included (*see* **Note 19**). Dilute each pNL4-3 vector to be used to a concentration of 4.5×10^5 IU/mL and make a total volume of 100 μL per well needed. Dilute the pMSCV-Rev vectors to a concentration of 9×10^5 IU/mL with a total volume of 50 μL per well needed. Use transfection medium as the diluent. Vectors used for color compensation controls do not require dilution.

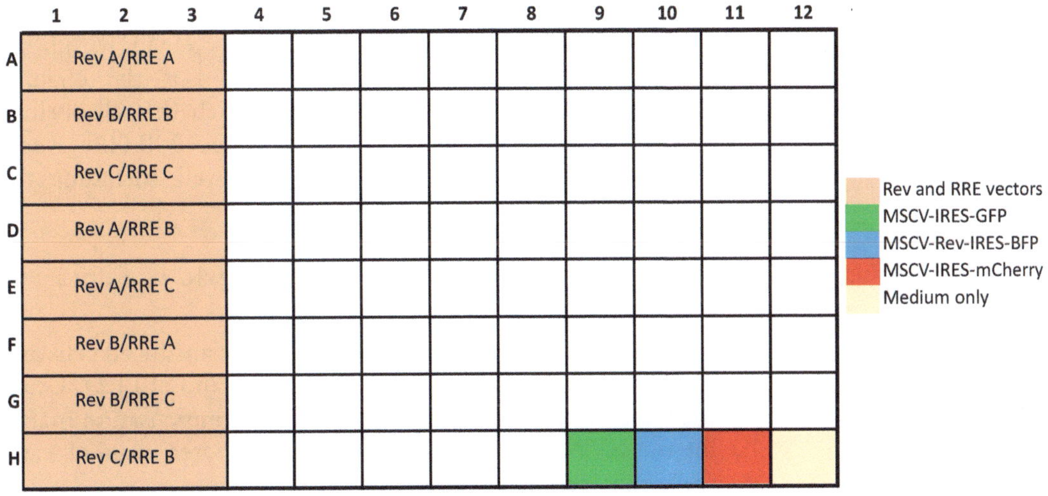

Fig. 4 Diagram of a 96 well plate set up for a Rev-RRE activity assay. In this example, the activity of eight unique Rev-RRE pairs was determined. Each Rev-RRE pair was tested in triplicate using the first three wells of each row. Single wells were transduced with GFP, BFP, and mCherry color compensation controls and a single well was reserved for untransduced cells for flow cytometer set up and as a negative control for data analysis

8. Add 100 μL of each pNL4-3 vector to the appropriate wells. Add 50 μL of each pMSCV-Rev vector to the appropriate wells. For color compensation, add 200 μL of undiluted pMSCV-IRES-mCherry, pMSCV-IRES-GFP, or pMSCV-Rev-IRES-eBFP2 stock to three single wells. Add 200 μL of sterile transfection medium to at least one well of untransduced cells.
9. Add 5 μL DEAE-dextran solution to each well of the 96 well plate.
10. Centrifuge the covered plate at 380 RCF at room temperature for 1 h.
11. Carefully remove the supernatant from each well of the plate, leaving the cells on the bottom of the well undisturbed. Resuspend the cell pellet in each well in 250 μL of SupT1 growth medium. Incubate the covered plate in the cell culture incubator for 72 h.
12. Centrifuge the covered plate at 380 RCF at 4 °C for 5 min to pellet cells.
13. Carefully remove the supernatant from each well, leaving the cells undisturbed. Resuspend the cell pellets in each well in 200 μL of cold DPBS.
14. Acquire cells in each well using the flow cytometer with plate sampler. Collect data in the SSC and FSC channels as well as channels corresponding to the mCherry, eBFP2, and GFP excitation/emission spectra. Use an SSC and FSC gate to identify a population of healthy cells. For the channels

corresponding to mCherry and eBFP2, use the same settings as for the titration, ensuring that positive cells remain within the dynamic range of the flow cytometer. For the channel corresponding to GFP, adjust the gain such that all positive cells remain in this range. Acquire about 200,000 events corresponding to single cells from each well, including the single color control and untransduced wells.

3.5 Data Analysis

1. Analyze the resulting FCS files using FlowJo or software of your choice.
2. Identify the wells corresponding to the untransduced cells and each single color control (eGFP, mCherry, and eBFP2). Using the Compensation tool, calculate a color compensation matrix for the GFP-area, mCherry-area, and BFP-area channels, and apply this compensation to all files.
3. Starting with an FCS file for an untransduced well, create a gate for healthy SupT1 cells using the FSC-A and SSC-A parameters. Then, make a daughter gate using the FSC-A and FSC-H parameters to eliminate doublets and cell aggregates. From the resulting population of singlet untransduced cells, define a quadrant gate in an mCherry versus BFP plot such that the untransduced cells fall into the fourth quadrant (mCherry negative and BFP negative) (Fig. 5). Create the calculated synthetic gates [Q1 AND Q2] corresponding to the mCherry positive population and [Q2 and Q3] corresponding to the BFP positive population. Apply all gates to all sample FCS files.
4. Calculate the percentage of transduced cells in each well, and calculate the arithmetic mean fluorescent intensity (MFI) of each fluorescent marker in the mCherry positive and BFP positive population (Fig. 6). Use the Table Editor tool to make the following calculations shown in Table 3, including a row for each Rev-RRE sample well.
5. For each Rev-RRE sample well, verify that the transduction was successful. The percentage of single cells transduced with any one assay vector should be less than about 30% to ensure that cells are not multiply transduced by the same vector construct (see columns 1 and 2 in the resulting data table) (*see* **Note 20**). The total number of cells transduced by both constructs should be at least 1000 for valid interpretation (see column 3 in the data table).

   In the example wells shown in Fig. 6, the mCherry positive proportion (i.e. the percent of singlet cells transduced with the pNL4-3 vector) was 7.4% for the high-activity pair and 7.5% for the low-activity pair. The BFP positive proportion (i.e. the percent transduced with the pMSCV vector) was 12.66% and 15.88%, respectively. The number of cells transduced by both

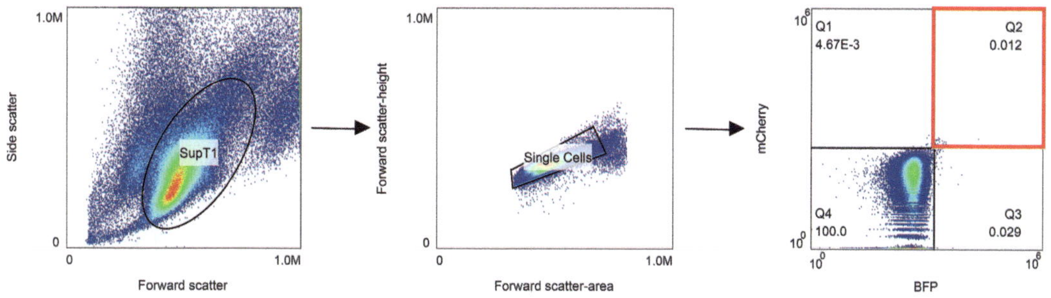

Fig. 5 Flow cytometry gating strategy for functional assay analysis. The gating strategy for functional assay analysis is shown from left to right. As with vector stock titer determination, SupT1 cells were identified in a side scatter vs forward scatter plot and then a single cell gate was defined in a forward scatter height vs forward scatter area plot. The singlet population was then used to define the mCherry negative and BFP negative population using the quadrant tool. Quadrant 2 in the right panel is highlighted as this gate contained the double-positive population used for Rev-RRE activity calculation in wells containing transduced cells

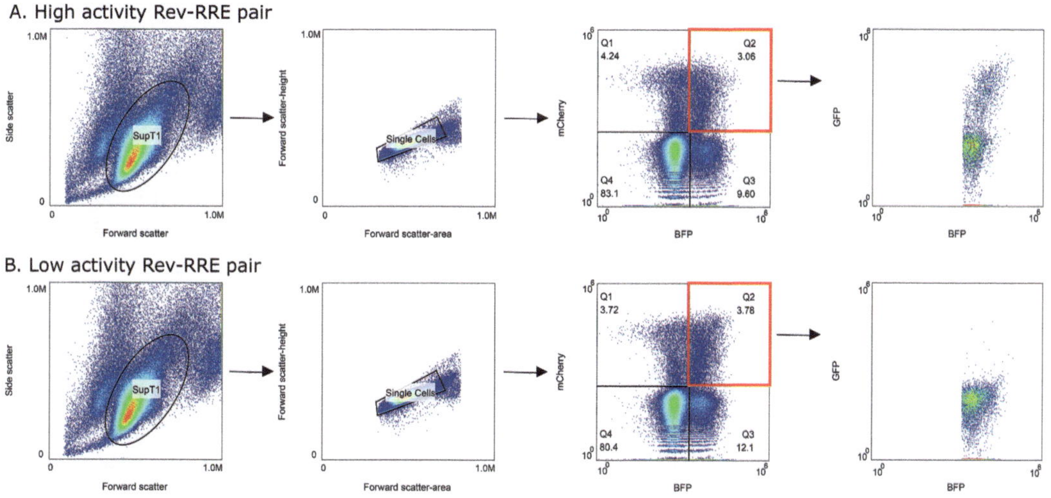

Fig. 6 Calculation of Rev-RRE functional activity. Application of the flow cytometry gating strategy shown in Fig. 5 to wells corresponding to high-activity (**a**) and low activity (**b**) Rev-RRE pairs. Only cells that are transduced by both the pNL4-3 and pMSCV assay vectors, and thus are mCherry and BFP positive, were used to determine relative Rev-RRE activity. The gate corresponding to this population is quadrant 2 in the third panel from left, and the GFP versus BFP distribution of the quadrant 2 population is displayed in the rightmost panel. Note that the high-activity pair displayed greater GFP intensity than the low-activity pair

constructs (i.e. the number of cells in quadrant 2 of the mCherry vs BFP plot) is not displayed in the figure, but was about 6000 for both Rev-RRE pairs. Both wells, therefore, were suitable for Rev-RRE activity calculation

6. For each sample well where transduction was successful, calculate the relative Rev-RRE activity as the ratio of GFP MFI (column 6 in the data table) to BFP MFI (column 4 in the data table) for the population of singlet cells transduced with

Table 3
FlowJo Table Editor setup

Column	Population	Statistic	Parameter	*Interpretation (not in Table Editor)*
1	SupT1/Single Cells/Q1&Q2	Frequency of parent		*Percentage of single cells transduced with pNL4-3 vector*
2	SupT1/Single Cells/Q2&Q3	Frequency of parent		*Percentage of single cells transduced with pMSCV vector*
3	SupT1/Single Cells/Q2	Count		*Count of single cells transduced with both assay vectors*
4	SupT1/Single Cells/Q2	Mean	Compensated BFP-A	*Arithmetic MFI of BFP signal in co-transduced cells*
5	SupT1/Single Cells/Q2	Mean	Compensated mCherry-A	*Arithmetic MFI of mCherry signal in co-transduced cells*
6	SupT1/Single Cells/Q2	Mean	Compensated GFP-A	*Arithmetic MFI of GFP signal in co-transduced cells*

both assay vectors. The three sample wells containing the same Rev-RRE combination should be considered technical replicates, and the calculated Rev-RRE activity value for these wells can be averaged to generate a single estimate of relative Rev-RRE activity (*see* **Note 21**). Represent relative Rev-RRE activity for each pair tested as a percentage of the reference Rev-RRE pair included in the plate.

In the example wells shown in Fig. 6, the high-activity Rev-RRE pair well had a GFP intensity of 2862 and a BFP intensity of 5607, yielding a relative activity value of 0.51. The low-activity Rev-RRE pair well had a GFP intensity of 110 and a BFP intensity of 5557, yielding a lower activity value of 0.02.

4 Notes

1. Plasmids used in this protocol are available from Addgene or from the authors. Reference numbers used by the authors are shown in Table 4.
2. The pMSCV-Rev-IRES-eBFP2 construct is derived from Addgene plasmid 52115, a gift of Dario Vignali. The plasmid was modified by cloning a *rev* sequence into the existing multiple cloning site upstream of the IRES between EcoRI and XhoI. The resulting construct constitutively expresses the customizable Rev and eBFP2 from a bicistronic transcript.
3. The pNL4-3(gag-CHYSEL-eGFP)(Rev-)(Env-)(XmaI-RRE-XbaI)(Nef-)(mCherry+) construct is derived from the HIV strain NL4-3 [18]. The HIV sequence was modified to render

Table 4
Plasmid reference numbers

Construct	Authors' plasmid reference number
pMSCV-NL4-3 Rev-IRES-eBFP2	5864
pNL4-3(gag-CHYSEL-eGFP)(Rev-)(Env-)(XmaI-NL4-3 RRE-XbaI)(Nef-)(mCherry+)	5936
pMSCV-IRES-mCherry	5301
pMSCV-IRES-GFP	5366
psPAX2	5691
pMD2.G	5693
pHIT60-CMV-GagPol	1854

it replication incompetent through deletion of the *gag* myristoylation site, two mutations within the first exon of *rev* to remove the start codon and add a premature stop codon, and a frameshift mutation in *env*. The ~350-nt RRE within *env* was flanked by an XmaI and XbaI site for easy exchange. *Gag* was truncated at amino acid 407 and a *cis*-acting hydrolase element (CHYSEL) and *eGFP* were inserted in frame. *Nef* was deleted and replaced with *mCherry*. The resulting construct expresses *mCherry* constitutively. Expression of *eGFP* occurs via an intron-containing transcript, similarly to *gag* expression in HIV infection, and thus is dependent on the Rev-RRE interaction.

4. A variety of lymphoid cell lines or primary cells can be used for the Rev-RRE functional assay. Vector stock titer should be determined with the same cell type used in the functional activity assay.
5. Make DEAE-dextran solution with DPBS. Sterilize the solution using a 0.2 uM syringe filter.
6. The assay system can be adapted to measure the functional activity of regulatory systems that are functionally homologous to the HIV Rev-RRE axis, such as the Human Endogenous Retrovirus (HERV)-K Rec-RcRE system [19, 20]. In this case, the *trans*-acting Rec sequence is used in place of Rev in the pMSCV construct, and the *cis*-acting RcRE RNA sequence is used in place of the RRE in the pNL4-3 construct. Similarly, the assay can be used to measure the activity of other *cis*-acting RNA sequences that accomplish nuclear export of intron-containing mRNAs in the absence of an additional *trans*-acting protein, such as the Mason-Pfizer Monkey Virus (MPMV) constitutive transport element (CTE) [21]. In this case, the

CTE sequence is substituted for the RRE in the pNL4-3 construct and the pMSCV construct is not used. Importantly, since RNA export is required for packaging of the RNA from the pNL4-3 construct, if the RRE is replaced with another element, be sure that an appropriate *trans*-acting protein or proteins are also expressed at appropriate levels in the packaging cells (e.g. Nxf1/Nxt1 in the case of the MPMV CTE [22]).

7. The minimum functional RRE is about 234-nt long [4], while the fully functional RRE is about 350-nt long [23]. Either may be used in this assay system, though the activity of the shorter RRE may be impacted by the abutting restriction enzyme site sequence from the vector backbone.
8. *Rev* and RRE sequences are each about 350-nt long and are readily synthesized by several companies including GenScript. When creating several custom constructs, it is often economical to also have cloning and DNA preparation performed commercially.
9. Dilute the plasmid DNA appropriately. The original concentration is not important to vector yield. As shown above, the combined plasmid DNA is diluted with an equal volume of Opti-MEM. A 3:1 mass ratio of PEI to DNA is used for the transfection, in a solution of PEI/Opti-MEM that is equal to the volume of the DNA/Opti-MEM. We try to keep the total volume of the transfection mixture to be added to the cells at 1 mL or less.
10. At this step, one could process the vector-containing supernatant by passing it through a 0.2 uM filter to further remove any debris. We have not felt this to be necessary as a routine practice, and filtering does appear to reduce vector titer slightly.
11. A hemocytometer can be used for manual cell counting rather than an automated device.
12. It is only necessary to determine the titer of vector stocks containing Rev or RRE sequences. The pMSCV-IRES-mCherry and pMSCV-IRES-GFP stocks are used only for color compensation and the precise titer is not required.
13. The titer of the vector stock is substantially reduced with each freeze-thaw cycle. The titer of the stock should be determined using an aliquot that was frozen at least overnight and then thawed. Using fresh, never frozen stock to determine titer would dramatically overestimate the titer of the frozen stocks used for the functional assay.
14. We use the flow cytometer channels for both titration and functional assay data collection shown in Table 5.

Table 5
Flow cytometer data collection channels

Fluorescent protein	Laser (nm)	Bandpass filter	Longpass filter
eBFP2	405	440/50	
eGFP	488	530/30	
mCherry	561	620/15	600

15. Transduced cells are very bright, and it is necessary to use a flow cytometer with sufficient dynamic range such that the brightest cell fluorescence intensity is quantifiable in the functional assay. The same detector settings should be used for the titration as are used for functional assay.
16. This formula is a derivation from the Poisson distribution and accounts for events where a cell is transduced by multiple vectors. When the fraction of positive cells in a well is less than about 20%, this approximates to a linear equation:

$$\text{Titer}\left(\frac{\text{IU}}{\text{mL}}\right) \sim \frac{\#\ \text{cells transduced} \times \frac{\%\ \text{positive cells}}{100}}{\text{stock volume (mL)}}$$

17. In our experience, pNL4-3 constructs yield a titer in the range of 10^5–10^7 IU/mL and pMSCV constructs yield a titer in the range of 10^6–10^8 IU/mL. These titers are sufficiently high for the functional assay described here without further concentration. Unexpectedly, low titers may be due to:
    (a) Unhealthy cells. Mycoplasma contamination, suboptimal growth conditions, and high passage number can decrease 293T/17 packaging efficiency. In this case, it may be necessary to obtain fresh cells and repeat the transfection.
    (b) Poor plasmid DNA. Plasmid DNA preparations that have become degraded will contain lower amounts of supercoiled DNA and less efficiently transfect cells. Assess the plasmid DNA on an agarose gel and prepare fresh DNA if needed.
    (c) Handling of the vector stock. The half-life of suspended vector in medium is around 6 h. Vector stocks should be collected, aliquoted, and frozen without delay in order to preserve titer.
    (d) If a very low-activity RRE sequence is used for the pNL4-3 construct, this may result in a lower titer. The RRE in the assay construct is required for packaging of the vector. In this case, vector can be concentrated by

ultracentrifugation. In our experience – assessing RREs obtained from HIV primary isolates – this is not required.

18. The functional assay gives consistent results of relative Rev-RRE activity, but the absolute activity value calculated in each run may vary depending on cell health and passage number. To aid in comparing results across assay runs, it is helpful to consistently include one Rev-RRE pair on a plate to be used as a reference value. The NL4-3 Rev-RRE pair works well for this purpose and other Rev-RRE pair activity values can be expressed as a percentage of NL4-3 Rev/NL4-3 RRE activity.
19. We suggest using an MOI of 0.18 as this minimizes the number of cells that receive two copies of any one assay vector while also ensuring a sufficient number of cells receive one copy of the Rev-containing and one copy of the RRE-containing vectors. The total volume of both vector dilutions together should not be greater than about 200 μL per well in order not to exceed the working volume of the well. If the titer of the vector stock is too low to achieve an MOI of 0.18 within this volume, alternative approaches include:
    (a) Use fewer cells in each well such that an MOI of 0.18 can be maintained with a lower number of infectious units (and thus a smaller volume of vector stock).
    (b) Use a lower MOI. In this case, it may be necessary to acquire a greater number of cells during flow cytometry in order to ensure a good sample of cells successfully transduced with both the Rev- and RRE-containing assay vectors.
    (c) Concentrate the vector stock by ultracentrifugation.
20. The assay may give an inaccurately high estimate of Rev-RRE functional activity if many cells in a well are transduced by more than one Rev- or RRE-containing assay vector. Based on the Poisson distribution, at an MOI of 0.18, 16.5% of cells should be transduced by the vector with only 1.4% of cells being transduced by multiple vector copies. The proportion of multiply transduced cells begins to exceed 5% when about 30% of the total population is positive. If the proportion of transduced cells in a well is too high, repeat the titer determination procedure for that vector stock.
21. Occasionally, a pipetting error will occur such that only one of the three replicate wells was transduced with too much vector stock or has too few co-transduced cells. In this case, the clearly erroneous well can be discarded and Rev-RRE activity calculated using the two valid wells for the Rev-RRE pair.

References

1. Rekosh D, Hammarskjold ML (2018) Intron retention in viruses and cellular genes: detention, border controls and passports. Wiley Interdiscip Rev RNA 9(3):e1470. https://doi.org/10.1002/wrna.1470
2. Frankel AD, Young JA (1998) HIV-1: fifteen proteins and an RNA. Annu Rev Biochem 67(1):1–25
3. Hadzopoulou-Cladaras M, Felber BK, Cladaras C, Athanassopoulos A, Tse A, Pavlakis GN (1989) The rev (trs/art) protein of human immunodeficiency virus type 1 affects viral mRNA and protein expression via a cis-acting sequence in the env region. J Virol 63(3):1265–1274
4. Malim MH, Hauber J, Le SY, Maizel JV, Cullen BR (1989) The HIV-1 rev trans-activator acts through a structured target sequence to activate nuclear export of unspliced viral mRNA. Nature 338(6212):254–257. https://doi.org/10.1038/338254a0
5. Hammarskjold ML, Heimer J, Hammarskjold B, Sangwan I, Albert L, Rekosh D (1989) Regulation of human immunodeficiency virus env expression by the rev gene product. J Virol 63(5):1959–1966
6. Fernandes JD, Booth DS, Frankel AD (2016) A structurally plastic ribonucleoprotein complex mediates post-transcriptional gene regulation in HIV-1. Wiley Interdiscip Rev RNA 7(4):470–486. https://doi.org/10.1002/wrna.1342
7. Neville M, Stutz F, Lee L, Davis LI, Rosbash M (1997) The importin-beta family member Crm1p bridges the interaction between Rev and the nuclear pore complex during nuclear export. Curr Biol 7(10):767–775. https://doi.org/10.1016/S0960-9822(06)00335-6
8. Sloan EA, Kearney MF, Gray LR, Anastos K, Daar ES, Margolick J, Maldarelli F, Hammarskjold ML, Rekosh D (2013) Limited nucleotide changes in the Rev response element (RRE) during HIV 1 infection alter over all Rev-RRE activity and Rev multimerization. J Virol 87(20):11173–11186. https://doi.org/10.1128/JVI.01392-13
9. Jackson PE, Tebit DM, Rekosh D, Hammarskjold ML (2016) Rev-RRE functional activity differs substantially among primary HIV-1 isolates. AIDS Res Hum Retrovir 32(9):923–934. https://doi.org/10.1089/AID.2016.0047
10. Sherpa C, Jackson P, Gray LR, Anastos K, Grice SFL, Hammarskjold M-L, Rekosh D (2018) Evolution of the HIV-1 RRE during natural infection reveals key nucleotide changes that correlate with altered structure and increased activity over time. bioRxiv:483511
11. Bobbitt KR, Addo MM, Altfeld M, Filzen T, Onafuwa AA, Walker BD, Collins KL (2003) Rev activity determines sensitivity of HIV-1-infected primary T cells to CTL killing. Immunity 18(2):289–299
12. Dzhivhuho G, Holsey J, Honeycutt E, O'Farrell H, Rekosh D, Hammarskjold M-L, Jackson PEH (2022) HIV-1 Rev-RRE functional activity in primary isolates is highly dependent on minimal context-dependent changes in Rev. Sci Rep 12(1):18416. https://doi.org/10.1038/s41598-022-21714-2
13. Jackson PE, Holsey J, Turse L, Hammarskjold ML, Rekosh D (2023) Rev–Rev Response Element Activity Selection Bias at the Human Immunodeficiency Virus Transmission Bottleneck. InOpen Forum Infectious Diseases 2023 Oct (Vol. 10, No. 10, p. ofad486). US: Oxford University Press. https://doi.org/10.1093/ofid/ofad486
14. Daelemans D, Afonina E, Nilsson J, Werner G, Kjems J, Clercq ED, Pavlakis GN, Vandamme AM (2002) A synthetic HIV-1 Rev inhibitor interfering with the CRM1-mediated nuclear export. Proc Natl Acad Sci USA 99(22):14440–14445. https://doi.org/10.1073/pnas.212285299
15. Hope TJ, Huang XJ, McDonald D, Parslow TG (1990) Steroid-receptor fusion of the human immunodeficiency virus type 1 Rev transactivator: mapping cryptic functions of the arginine-rich motif. Proc Natl Acad Sci USA 87(19):7787–7791
16. Jackson PE, Huang J, Sharma M, Rasmussen SK, Hammarskjold M-L, Rekosh D (2019) A novel retroviral vector system to analyze expression from mRNA with retained introns using fluorescent proteins and flow cytometry. Sci Rep 9(1):1–14
17. Soneoka Y, Cannon PM, Ramsdale EE, Griffiths JC, Romano G, Kingsman SM, Kingsman AJ (1995) A transient three-plasmid expression system for the production of high titer retroviral vectors. Nucleic Acids Res 23(4):628–633
18. Adachi A, Gendelman HE, Koenig S, Folks T, Willey R, Rabson A, Martin MA (1986) Production of acquired immunodeficiency syndrome-associated retrovirus in human and nonhuman cells transfected with an infectious molecular clone. J Virol 59(2):284–291
19. Magin C, Lower R, Lower J (1999) cORF and RcRE, the Rev/Rex and RRE/RxRE homologues of the human endogenous retrovirus family HTDV/HERV-K. J Virol 73(11):9496–9507

20. Magin-Lachmann C, Hahn S, Strobel H, Held U, Lower J, Lower R (2001) Rec (formerly Corf) function requires interaction with a complex, folded RNA structure within its responsive element rather than binding to a discrete specific binding site. J Virol 75(21): 10359–10371. https://doi.org/10.1128/JVI.75.21.10359-10371.2001
21. Bray M, Prasad S, Dubay JW, Hunter E, Jeang KT, Rekosh D, Hammarskjold ML (1994) A small element from the Mason-Pfizer monkey virus genome makes human immunodeficiency virus type 1 expression and replication Rev-independent. Proc Natl Acad Sci USA 91(4):1256–1260
22. Li Y, Bor Y-c, Fitzgerald MP, Lee KS, Rekosh D, Hammarskjold M-L (2016) An NXF1 mRNA with a retained intron is expressed in hippocampal and neocortical neurons and is translated into a protein that functions as an Nxf1 cofactor. Mol Biol Cell 27(24):3903–3912
23. Bai Y, Tambe A, Zhou K, Doudna JA (2014) RNA-guided assembly of rev-RRE nuclear export complexes. elife 3:e03656. https://doi.org/10.7554/eLife.03656

Chapter 14

Method for the Enrichment of N^6-Methyladenosine-Modified Cellular and HIV-1 RNA

Tarun Mishra, Stacia Phillips, and Li Wu

Abstract

N^6-methyladenosine (m^6A) modification of RNA is an important area in studying viral replication, cellular responses, and host immunity. HIV-1 RNA contains multiple m^6A modifications that regulate viral replication and gene expression. HIV-1 infection of $CD4^+$ T-cells or HIV-1 envelope protein treatment upregulates m^6A levels of cellular RNA. Changes in the m^6A modification of cellular transcripts in response to HIV-1 infection provide new insights into the mechanisms of posttranscriptional gene regulation in the host cell. To better investigate the functions of m^6A modification in HIV-1 infection and innate immune responses, it is helpful to standardize basic protocols. Here, we describe a method for the selective enrichment of m^6A-modified RNA from HIV-1-infected primary $CD4^+$ T-cells based on immunoprecipitation. The enriched RNA with m^6A modifications can be used in a variety of downstream applications to determine the methylation status of viral or cellular RNA at resolution from transcript level down to single nucleotide.

Key words HIV-1, Infection, RNA, Epitranscriptomic modification, N^6-methyladenosine (m^6A), Primary $CD4^+$ T-cells

1 Introduction

N^6-methyladenosine (m^6A) modification is the most abundant internal RNA modification found on eukaryotic RNA. m^6A modification is a reversible process mediated by cellular methyltransferases (termed writers) and demethylases (termed erasers). The m^6A modification is recognized by a set of proteins called readers. m^6A modification regulates important aspects of RNA fate in cells, such as localization, translation, splicing, and stability [1–3].

The profiling of HIV-1 genomic RNA has confirmed the presence of m^6A modification on the 5′ untranslated region (UTR) and 3′ UTR [4]. The levels of m^6A modification present on HIV-1 genomic RNA can be altered by controlling writer or eraser expression in HIV-1 producer cells [5]. Also, it was found that activators of methyltransferase complex enhances HIV-1 particles production

Vinayaka R. Prasad and Ganjam V. Kalpana (eds.), *HIV Protocols*, Methods in Molecular Biology, vol. 2807,
https://doi.org/10.1007/978-1-0716-3862-0_14,

[6] and knockdown of m^{6}A eraser proteins increases HIV-1 replication [7]. The overexpression of reader proteins (YTHDF1-3) in target cells reduces HIV-1 infection and restricts full-length Gag synthesis [4, 8]. The additional m^{6}A reader YTHDC1 protein present in the nucleus binds to m^{6}A modified HIV-1 RNA and regulates alternative splicing of HIV-1 RNA, hence facilitating HIV-1 gene expression [9]. Furthermore, infection of CD4$^+$ T cells by HIV-1 or treatment of CD4$^+$ T cells with the HIV-1 envelope protein gp120 enhances m^{6}A modification of cellular RNA [10], suggesting cellular responses to HIV-1 infection or gp120. Furthermore, we showed that the presence of m^{6}A modification on HIV-1 RNA prevents type-I interferon expression in macrophages, thereby suppressing innate immune responses [11]. Efficient enrichment of m^{6}A-modified RNA and accurate measurement of m^{6}A are critical for understanding dynamic functional changes of m^{6}A modification in cellular RNA and during HIV-1 infection. Although the method of m^{6}A RNA precipitation described here can enrich m^{6}A-modified RNA, it does not quantify the fraction of m^{6}A-modified transcripts present in the total cellular RNA.

Here, we describe a method for the enrichment of m^{6}A-modified cellular and HIV-1 RNA from HIV-1-infected primary CD4$^+$ T cells using an m^{6}A-specific antibody (Fig. 1). The protocol includes five major steps: (1) RNA isolation from HIV-1-infected or -uninfected CD4$^+$ T cells, (2) RNA fragmentation, (3) Immunoprecipitation, (4) Elution and concentration, and (5) m^{6}A dot blot. Using these basic methods, researchers will be able to enrich m^{6}A-modified RNA, which can be used for further analyses, including mass spectrometry, next generation sequencing, qRT-PCR validation [12], or other downstream applications of interest.

2 Materials

2.1 CD4$^+$ T-Cells Preparation and Culture

1. Whole blood from healthy blood donors for isolation of peripheral blood mononuclear cells (PBMCs).
2. Histopaque-1077.
3. Phosphate buffered saline without calcium or magnesium (Gibco).
4. Trypan Blue stain (Gibco).
5. RPMI 1640 complete medium (RPMI-1640) (ATCC), 10% fetal bovine serum (FBS), 1% penicillin-streptomycin.
6. Freezing medium: 50% RPMI, 40% FBS, and 10% DMSO.
7. 50 U/μL Interleukin 2 (IL-2).

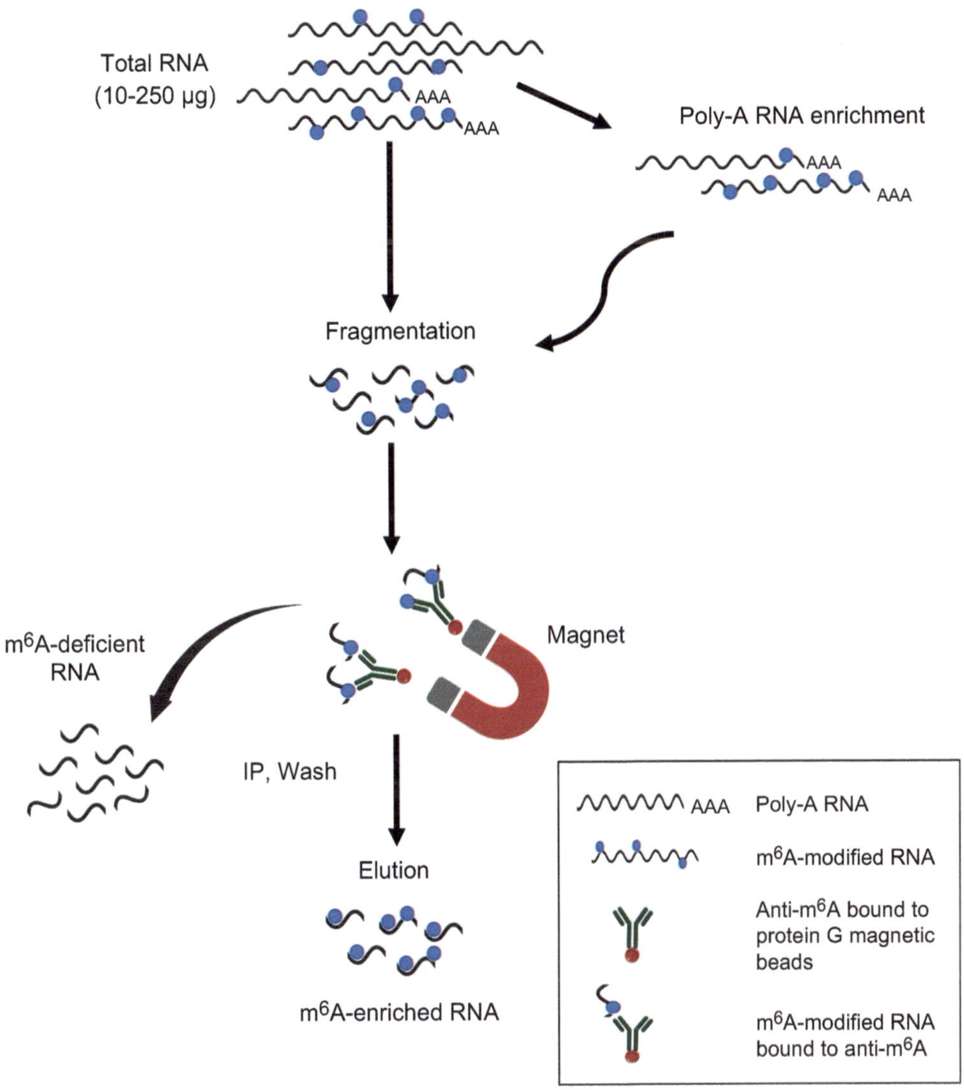

Fig. 1 Schematic representation of m^6A-modified RNA immunoprecipitation (meRIP) from total RNA or poly-A enriched RNA. The process incudes poly-A RNA enrichment, fragmentation, and immunoprecipitation. (This figure was created with BioRender.com)

8. EasySep™ Human $CD4^+$ T cell enrichment kit (STEMCELL technology).
9. Dynabeads Human T-activator CD3/CD28 for T Cell Expansion and Activation (Gibco).

2.2 RNA Isolation

1. TRIzol Reagent (ThermoFisher Scientific).
2. Chloroform (Sigma-Aldrich).
3. 100% Isopropanol (Sigma-Aldrich).
4. 70% ethanol.

5. Glycogen (Sigma-Aldrich).
6. Nuclease-free water.

2.3 mRNA Enrichment

1. Total RNA.
2. Binding Buffer: 20 mM Tris-HCl, pH 7.5, 1 M LiCl, 2 mM EDTA.
3. Wash Buffer B: 10 mM Tris-HCl, pH 7.5, 0.15 M LiCl, 1 mM EDTA.
4. Dynabeads Oligo$(dT)_{25}$ (ThermoFisher Scientific).
5. 1.5 mL tube magnetic rack.
6. Heating block.
7. Nanodrop.

2.4 RNA Fragmentation

1. RNA fragmentation 10× buffer: 100 mM Tris-HCL, pH 7.0, 100 mM $ZnCl_2$ in nuclease-free water.
2. 0.5 M EDTA.
3. 1 M $ZnCl_2$ (Sigma-Aldrich).
4. 3 M NaOAc, pH 5.2 (Sigma-Aldrich).
5. Nanodrop or Qubit.

2.5 m^{6}A RNA Immunoprecipitation (MeRIP)

1. Reaction buffer: 150 mM NaCl, 10 mM Tris-HCl, pH 7.5, 0.1% NP-40 in nuclease-free water. Always prepare fresh before use and keep on ice.
2. Protein G magnetic beads (NEB).
3. m^{6}A antibody (Synaptic Systems).
4. RNase Inhibitor (NEB).
5. Mouse IgG (Proteintech).
6. Magnetic rack (1.5 mL tubes).
7. Total RNA (fragmented or unfragmented).
8. RNA Clean and Concentrator™-5 (ZYMO Research).
9. Eppendorf RNA/DNA Low-bind microcentrifuge tubes (Sigma-Aldrich # Z666548).
10. RNase-free pipette tips.

2.6 m^{6}A Dot Blot and Quantification (Optional)

1. Dot blot apparatus (Bio-Rad).
2. UV crosslinker.
3. Nylon membrane positively charged (Sigma-Aldrich).
4. Tween-20 (RIP).
5. Anti-m^{6}A antibody (Synaptic Systems).
6. Goat anti-mouse IgG-HRP conjugate (Promega).

7. SuperSignal™ Femto Reagent (ThermoFisher Scientific).
8. Blocking Buffer: 5% Milk in Tris buffer saline (TBS).
9. Wash Buffer: TBS-T, 0.1% Tween-20 in TBS.
10. 20X SSC buffer (Sigma-Aldrich).
11. 1 mM EDTA pH 8.0.
12. LI-COR Odyssey Imager.
13. Methylene blue (Molecular Research Center).
14. ImageJ software (National Institutes of Health).

3 Methods

All procedures involving infectious HIV-1 should be performed in adherence to institutional biosafety protocols. All experiments using lentiviral vectors or replication-competent HIV-1 should be performed by trained lab personnel and with the appropriate Institutional Biosafety Committee approvals in place. Use best practices when handling RNA to avoid its degradation.

3.1 PBMC Preparation

1. Add 15 mL Histopaque to a 50-mL conical tube.
2. Dilute 15 mL of blood with equal volume of PBS containing 2% FBS.
3. Using a serological pipette, slowly layer the diluted blood on top of the Histopaque (Fig. 2a).
4. Centrifuge the tubes at 1000× *g* for 30 min at room temperature with deceleration set to zero (*see* **Note 1**).
5. Carefully observe the separation and look for PBMC layer between plasma layer and Histopaque layer (Fig. 2b). Using pasture pipette, collect the white PBMC layer without disturbing the plasma or Histopaque layer.
6. Transfer the PBMCs to a new 50-mL conical tube and bring to 50 mL with PBS (*see* **Note 2**).
7. Centrifuge at 300× *g* for 10 min at room temperature.
8. Carefully remove the supernatant and resuspend the pellet in 50 mL PBS.
9. Centrifuge at 300× *g* for 10 min at room temperature.
10. Remove the supernatant and resuspend the cells in 10 mL RPMI-1640 complete medium.
11. Count the cells and determine cell viability using Trypan blue stain.
12. Adjust the volume of medium to 2×10^6 cells/mL.

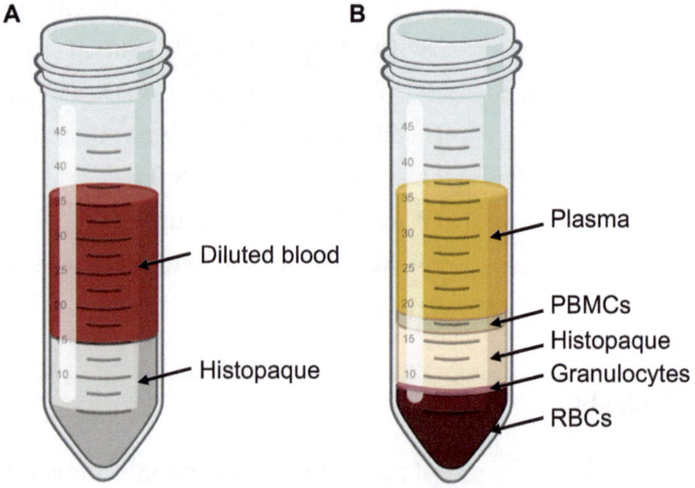

Fig. 2 Separation of PBMC from the whole blood using the Histopaque gradient centrifugation. (**a**) Layering of diluted blood on top of Histopaque. (**b**) Layers of blood components after centrifugation. RBCs, red blood cells. (This figure was created with Biorender.com)

13. PBMCs may be used immediately or cryopreserved in 50% RPMI, 40% FBS, and 10% DMSO at a concentration of 5×10^6 cells/mL for later use.

3.2 Enrichment and HIV-1 Infection of CD4⁺ T Cells

1. Enrich CD4⁺ cells using EasySep™ Human CD4⁺ T cell enrichment kit following the manufacturer's protocol. The purity of the enriched population can be assessed by CD4 staining and flow cytometry (*see* **Note 3**).
2. Count the enriched cells and resuspend in complete RPMI-1640 medium to 1×10^6 cells/mL and seed 1 mL per well in 24-well plate.
3. Activate primary CD4⁺ T cells using 25 μL of αCD3/CD28 beads per 1×10^6 cells for 3 days in the presence of 30 U/mL IL-2.
4. Transfer cells to 1.5 mL microcentrifuge tube and place in magnetic rack to remove beads.
5. Count the cells and resuspend in complete RPMI-1640 with 30 U/mL IL-2 to 1×10^6 cells/mL.
6. Infect CD4⁺ T cells with 40 ng of p24 virus per 1×10^6 cells by spinoculation at 460× *g* for 2 h in 5 mL round bottom tubes.
7. After 2 h, remove the virus particle-containing supernatant and resuspend the CD4⁺ T cells in 1 mL complete RPMI medium with 30 U/mL IL-2. Transfer infected cells to one well of 24-well plate (*see* **Note 4**).

8. Collect the infected cells 96 h post-infection by centrifugation at 300× *g* for 5 min at room temperature.
9. Completely remove supernatant and immediately add 1 mL TRIzol reagent, pipetting up and down several times to lyse cells.
10. Add 200 μL chloroform. Mix by vigorous shaking for 15 s and incubate for 3 min at room temperature.
11. Centrifuge for 15 min at 12,000× *g* at 4 °C.
12. Transfer the aqueous phase to a new tube. Avoid aspirating liquid from interphase to prevent DNA contamination (*see* **Note 5**).
13. Add 1 μL of 20 mg/mL glycogen, mix.
14. Add 500 μL of 100% isopropanol. Mix by inversion.
15. Incubate for 10 min at room temperature.
16. Centrifuge for 10 min at 12,000× *g* at 4 °C.
17. Carefully discard the supernatant without disturbing the RNA pellet.
18. Wash RNA pellet with 1 mL 75% ethanol (*see* **Note 6**).
19. Centrifuge at 12,000× *g* for 5 min at 4 °C. Carefully remove the supernatant without losing pellet as pellet is loosely bound to tube surface.
20. Pulse spin and remove all the residual ethanol. Let the pellet air dry for 5 min (*see* **Note 7**).
21. Rehydrate pellet in 50 μL of nuclease-free water.
22. Quantify RNA using Nanodrop and store at −80 °C.

3.3 Poly-A RNA Enrichment Using Oligo(dT) Beads (Optional)

1. Vortex oligo(dT) Dynabeads for 30 s and transfer required amount of beads to a 1.5 mL tube (25 μL beads per 10 μg total RNA).
2. Add 1 mL binding buffer to wash.
3. Collect beads using magnet and discard supernatant.
4. Resuspend beads in 100 μL binding buffer per sample.
5. Adjust the RNA to 100 μL using water (concentration should be 750 ng/μL or less).
6. Add 100 μL binding buffer to RNA sample.
7. Incubate at 65 °C for 2 min and place immediately on ice.
8. Add 100 μL washed beads to the 200 μL RNA sample.
9. Rotate for 5 min at room temperature.
10. Collect beads using magnet and save supernatants, which contain non-coding RNA (ncRNA).
11. Wash beads with 200 μL washing buffer B.

12. Collect beads using magnet and discard supernatant.
13. Repeat **steps 11–12**.
14. Resuspend beads in 10 μL 10 mM Tris buffer, pH 7.5.
15. Incubate at 80 °C for 2 min.
16. Transfer tubes immediately to magnet to collect beads and immediately transfer supernatant to a 1.5-mL tube (*see* **Note 8**). Save beads on ice.
17. Repeat **steps 14–16** and combine with the first elution.
18. Determine RNA concentration by Nanodrop.

3.4 RNA Fragmentation and MeRIP

The success of the MeRIP process depends on the quality of the starting RNA samples. For this, we isolated RNA from HIV-1 infected $CD4^+$ T cells by using TRizol reagent. Based on the downstream application, mRNA enrichment can be avoided by directly proceeding to RNA fragmentation (*see* **Note 9**). Another important criterion for m^6A RNA enrichment is the length of the fragmented RNA used, as large fragments may reduce total yield.

3.4.1 RNA Fragmentation

1. Add 20 μg of RNA to a 200 μL tube and add nuclease-free water to 18 μL (*see* **Note 10**).
2. Add 2 μL of 10× fragmentation buffer into tube. Mix by pipetting and spin briefly.
3. Incubate tubes at 94 °C for 5 min with lid temperature at 100 °C (*see* **Note 11**).
4. Add 2 μL of 0.5 M EDTA to stop the fragmentation reaction. Mix and place on ice.
5. Transfer the RNA samples to 1.5 mL low bind tubes.
6. Assess RNA fragmentation by running 2 μL RNA (1 μg/sample) on an agarose gel (Fig. 3).
7. Add 10 μL of 3 M sodium acetate pH 5.2, glycogen (100 μg/mL final concentration) and 4 volumes of 100% ice-cold ethanol.
8. Mix well by pipetting.
9. Place the mixture at −80 °C overnight.
10. Centrifuge at 16,000× *g* for 30 min at 4 °C.
11. Carefully discard the supernatant without disturbing pellet.
12. Wash pellet with 75% ethanol by centrifugation at 16,000× *g* for 5 min at 4 °C.
13. Discard supernatant, pulse spin, and remove all residual ethanol.
14. Air dry the pellet for 5 min.

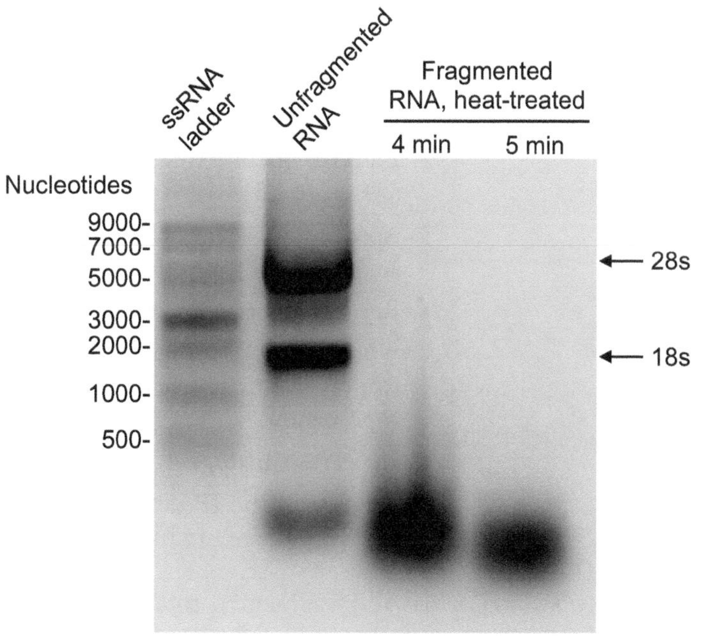

Fig. 3 Measurement of RNA fragmentation. Image of fragmented RNA (1 μg/sample) on 2% agarose gel stained with SYBR Safe. Four samples (from left to right): single-stranded RNA (ssRNA) ladder, unfragmented RNA control showing intact 28 s/18 s rRNA bands, fragmented RNA after 4 min and 5 min of heat treatment, respectively

15. Resuspend in 50 μL nuclease-free water.
16. Quantify RNA using Nanodrop or Qubit (*see* **Note 12**).

3.4.2 Magnetic Beads Preparation

1. Label 1.5 mL tubes (one for 10% input control, one for m^6A IP, and another for negative control IgG).
2. Pipette 25 μL of beads in fresh 1.5 mL tube for m^6A IP and negative control IgG sample. Apply magnetic field to collect beads and discard the supernatant.
3. Wash beads twice by resuspending the beads in 200 μL of reaction buffer and applying magnetic field to remove the supernatant.
4. Resuspend the beads in 250 μL of reaction buffer.
5. Keep 500 ng of untreated fragmented RNA as 10% input control.
6. Aliquot 10 μg of fragmented RNA in each of two 1.5-mL low RNA-binding tubes, one for IP sample and one for IgG control samples.

3.4.3 Attach Antibody to Protein G Beads

1. Add 2 μL anti-m^{6}A antibody for each 10 μg of RNA in IP sample and equal amount of Mouse IgG antibody in a negative control tube containing washed protein G beads (from **step 5** in the above section).
2. Incubate the mixture for 4 h to overnight at 4 °C with rotation (*see* **Note 13**).
3. Apply magnetic field to collect beads and discard the supernatant.
4. Resuspend beads in 200 μL reaction buffer, pipet up, and down 5 times to wash.
5. Apply magnetic field to collect beads and discard the supernatant.
6. Repeat wash **steps 4–5**.
7. Resuspend the IP sample tube and IgG control tube beads in 250 μL of reaction buffer.

3.4.4 Binding m^{6}A-Modified RNA to Antibody and Elution

1. Add fragmented total RNA or fragmented mRNA (10 μg) to IP sample and to negative control IgG tube containing resuspended beads (from **step** 7 in the above section) along with 100 units of RNase inhibitor (Prepare sample according to Table 1).
2. Incubate RNA and beads mixture for 3 h at 4 °C with orbital rotation.
3. Briefly spin the tubes to ensure no liquid is trapped in the lid.
4. Apply magnetic field and carefully remove supernatant.
5. Add 250 μL of reaction buffer to wash beads. Apply magnetic field and discard supernatant.
6. Repeat **steps 4–5** twice for a total of three washes.
7. Resuspend beads completely in 1 mL of Trizol. Pipette mix several times to separate RNA from beads.
8. Incubate at room temperature for 2 min.
9. Add 200 μL chloroform. Mix by vigorous shaking for 15 s and incubate for 3 min at room temperature.
10. Centrifuge for 15 min at 12,000× *g* at 4 °C.
11. Transfer the aqueous phase to a new tube. Avoid aspirating liquid from interphase to avoid DNA contamination.
12. Add 1 μL of 20 mg/mL glycogen, mix.
13. Add 500 μL of 100% isopropanol. Mix by inversion.
14. Incubate for 10 min at room temperature.
15. Centrifuge for 10 min at 12,000× *g* at 4 °C.
16. Carefully discard the supernatant without disturbing the RNA pellet.

Table 1
RNA sample preparation for immunoprecipitation

Components	Volume/amount
Primary antibody bound beads	250 μL
RNase inhibitor	100 U
RNA	20–100 μg
Nuclease-free water	To total 500 μL

17. Wash RNA pellet with 1 mL 75% ethanol (molecular grade).
18. Centrifuge at 12,000× *g* for 5 min at 4 °C. Carefully remove the supernatant without losing pellet as pellet is loosely bound to tube surface.
19. Pulse spin and remove all the residual ethanol. Let the pellet air dry for 5 min.
20. Rehydrate pellet in 50 μL of nuclease-free water.
21. Quantify RNA using Nanodrop or Qubit and store at −80 °C (*see* **Note 14**).

3.5 m^{6}A Dot Blot

Prepare RNA sample in 1 mM EDTA pH 8.0 (Follow Table 2 for sample preparation).

1. Add 66 μL 20× SSC buffer and 44 μL of 37% formaldehyde to each sample.
2. Vortex for 5 s and heat the samples at 65 °C for 30 min. Place immediately on ice.
3. Prepare nylon membrane by soaking it in 10× SSC buffer for 10 min.
4. Assemble dot-blot apparatus with nylon membrane and ensure no leakage by applying vacuum.
5. Add 200 μL of nuclease free water followed by 200 μL of 10× SSC buffer to the wells.
6. Load 100 μL sample to each well.
7. Apply gentle vacuum to pull sample through the wells and wash twice using 200 μL of 10× SSC buffer.
8. Stop the vacuum and take out the membrane and UV-crosslink sample three times.
9. Stain membrane with methylene blue for 2 min and rinse the membrane with nuclease free water to remove extra methylene blue.
10. Capture membrane image to quantify variation in loading volume.

Table 2
Preparation of RNA sample for m^6A dot blot

Components	Volume/amount
RNA	50–200 ng
1 mM EDTA	Variable to make 110 μL with RNA
20× SSC	66 μL
37% formaldehyde	44 μL
Total	220 μL

11. Block the membrane with 5% milk in TBST for 30 min at room temperature.
12. Add m^6A antibody (1:5000 diluted in TBST) and incubate for overnight at 4 °C or 1 h at room temperature.
13. Wash the membrane three times using 1X TBST.
14. Add HRP labelled secondary antibody (1:10,000) in 2.5% milk containing TBST and incubate for 1 h at room temperature.
15. Wash the membrane three times with TBST for 5 min.
16. Develop the membrane using SuperSignal Femto reagent.

4 Notes

1. Always perform the centrifugation with brakes off. Deceleration disrupts the formation of distinct layers.
2. Avoid collecting the Histopaque gradient while collecting the PBMC layer. Exposure to Histopaque for extended periods may be toxic to cells.
3. Generally, the purity of $CD4^+$ T cells is nearly 90–100%. Do not proceed for infection if purity is less than 90%.
4. Replace the culture medium of $CD4^+$ T cells 1 day after HIV-1 infection, as the cells are in growth phase.
5. Use 200 μL pipette to transfer aqueous phase. Avoid taking large volume at one time, this may lead to mixing of layers and leads to DNA contamination in RNA sample.
6. During ethanol wash, a small white pellet can be observed at the bottom of the tube. Always avoid touching the bottom of the tube as it may lead to loss of RNA pellet.
7. Do not overdry RNA pellet as it may cause difficulty in dissolving RNA.
8. Collect supernatant immediately as delay in supernatant collection may decrease RNA yield.

9. If ncRNA and long non-coding RNA (lncRNA) are not required for downstream application, we recommend proceeding for poly-A RNA enrichment.
10. The minimum amount of starting material for MeRIP can be 10 μg for total RNA. Based on the amount of RNA required for downstream application, the starting RNA amount should be decided.
11. Preheat the thermal cycler at 94 °C. Do not over incubate RNA samples for fragmentation more than 5 min. This will reduce the size of RNA fragments shorter than 100–200 nt.
12. The enriched RNA can be stored at −80 °C for 1 year. Usually more than 70% RNA is recovered after fragmentation.
13. DO not reduce the incubation time less than 2 h, it may reduce the enrichment efficiency.
14. Carefully use the enriched methylated RNA for quantification as the concentration might be very low.

Acknowledgments

We thank the Wu lab members for helpful discussions and suggestions. This work was supported by National Institutes of Health (NIH) grants AI159546, AI169659, and AI170070 to L.W. The content is solely the responsibility of the authors and does not necessarily represent the official views of the NIH.

References

1. Lou X, Wang JJ, Wei YQ, Sun JJ (2021) Emerging role of RNA modification N6-methyladenosine in immune evasion. Cell Death Dis 12(4):300. https://doi.org/10.1038/s41419-021-03585-z
2. Jiang X, Liu B, Nie Z, Duan L, Xiong Q, Jin Z, Yang C, Chen Y (2021) The role of m6A modification in the biological functions and diseases. Signal Transduct Target Ther 6(1):74. https://doi.org/10.1038/s41392-020-00450-x
3. Williams GD, Gokhale NS, Horner SM (2019) Regulation of viral infection by the RNA modification N6-methyladenosine. Annu Rev Virol 6(1):235–253. https://doi.org/10.1146/annurev-virology-092818-015559
4. Tirumuru N, Zhao BS, Lu W, Lu Z, He C, Wu L (2016) N(6)-methyladenosine of HIV-1 RNA regulates viral infection and HIV-1 Gag protein expression. elife 5. https://doi.org/10.7554/eLife.15528
5. Phillips S, Baek A, Kim S, Chen S, Wu L (2022) Protocol for the generation of HIV-1 genomic RNA with altered levels of N (6)-methyladenosine. STAR Protoc 3(3):101616. https://doi.org/10.1016/j.xpro.2022.101616
6. Selberg S, Zusinaite E, Herodes K, Seli N, Kankuri E, Merits A, Karelson M (2021) HIV replication is increased by RNA methylation METTL3/METTL14/WTAP complex activators. ACS Omega 6(24):15957–15963. https://doi.org/10.1021/acsomega.1c01626
7. Lichinchi G, Gao S, Saletore Y, Gonzalez GM, Bansal V, Wang Y, Mason CE, Rana TM (2016) Dynamics of the human and viral m(6)A RNA methylomes during HIV-1 infection of T cells. Nat Microbiol 1:16011. https://doi.org/10.1038/nmicrobiol.2016.11
8. Lu W, Tirumuru N, St Gelais C, Koneru PC, Liu C, Kvaratskhelia M, He C, Wu L (2018) N(6)-methyladenosine-binding proteins

suppress HIV-1 infectivity and viral production. J Biol Chem 293(34):12992–13005. https://doi.org/10.1074/jbc.RA118.004215

9. Tsai K, Bogerd HP, Kennedy EM, Emery A, Swanstrom R, Cullen BR (2021) Epitranscriptomic addition of m(6)A regulates HIV-1 RNA stability and alternative splicing. Genes Dev. https://doi.org/10.1101/gad.348508.121

10. Tirumuru N, Wu L (2019) HIV-1 envelope proteins up-regulate N (6)-methyladenosine levels of cellular RNA independently of viral replication. J Biol Chem 294(9):3249–3260. https://doi.org/10.1074/jbc.RA118.005608

11. Chen S, Kumar S, Espada CE, Tirumuru N, Cahill MP, Hu L, He C, Wu L (2021) N6-methyladenosine modification of HIV-1 RNA suppresses type-I interferon induction in differentiated monocytic cells and primary macrophages. PLoS Pathog 17(3):e1009421. https://doi.org/10.1371/journal.ppat.1009421

12. Phillips S, Mishra T, Khadka S, Bohan D, Espada CE, Maury W, Wu L (2023) Epitranscriptomic N(6)-methyladenosine profile of SARS-CoV-2-infected human lung epithelial cells. Microbiol Spectr 11(1):e0394322. https://doi.org/10.1128/spectrum.03943-22

Chapter 15

Direct Analysis of HIV mRNA m^{6}A Methylation by Nanopore Sequencing

Ethan Honeycutt, Fredrick Kizito, Jonathan Karn, and Thomas Sweet

Abstract

The post-transcriptional processing and chemical modification of HIV RNA are understudied aspects of HIV virology, primarily due to the limited ability to accurately map and quantify RNA modifications. Modification-specific antibodies or modification-sensitive endonucleases coupled with short-read RNA sequencing technologies have allowed for low-resolution or limited mapping of important regulatory modifications of HIV RNA such as N^6-methyladenosine (m^{6}A). However, a high-resolution map of where these sites occur on HIV transcripts is needed for detailed mechanistic understanding. This has recently become possible with new sequencing technologies. Here, we describe the direct RNA sequencing of HIV transcripts using an Oxford Nanopore Technologies sequencer and the use of this technique to map m^{6}A at near single nucleotide resolution. This technology also provides the ability to identify splice variants with long RNA reads and thus, can provide high-resolution RNA modification maps that distinguish between overlapping splice variants. The protocols outlined here for m^{6}A also provide a powerful paradigm for studying any other RNA modifications that can be detected on the nanopore platform.

Key words N^6-methyladenosine, Oxford nanopore sequencing, Direct RNA sequencing

1 Introduction

HIV has a complicated pattern of RNA expression, which includes the production of at least 50 overlapping unspliced, incompletely spliced, and completely spliced transcripts from the HIV provirus (Fig. 1a) [2, 3]. These three classes of transcripts are regulated by different transcriptional and nuclear export controls. The completely spliced transcripts are exported from the nucleus by the canonical mRNA export pathway, which utilizes NXF1 while the unspliced and incompletely spliced transcripts are exported by a separate pathway using the virally encoded Rev protein to recruit host cofactors CRM1 and Ran-GTP [4, 5]. The completely spliced transcripts are exported more rapidly from the nucleus, and the unspliced and incompletely spliced transcripts undergo nuclear retention, which delays their export to the cytoplasm (Fig. 1b, c).

Vinayaka R. Prasad and Ganjam V. Kalpana (eds.), *HIV Protocols*, Methods in Molecular Biology, vol. 2807, https://doi.org/10.1007/978-1-0716-3862-0_15,

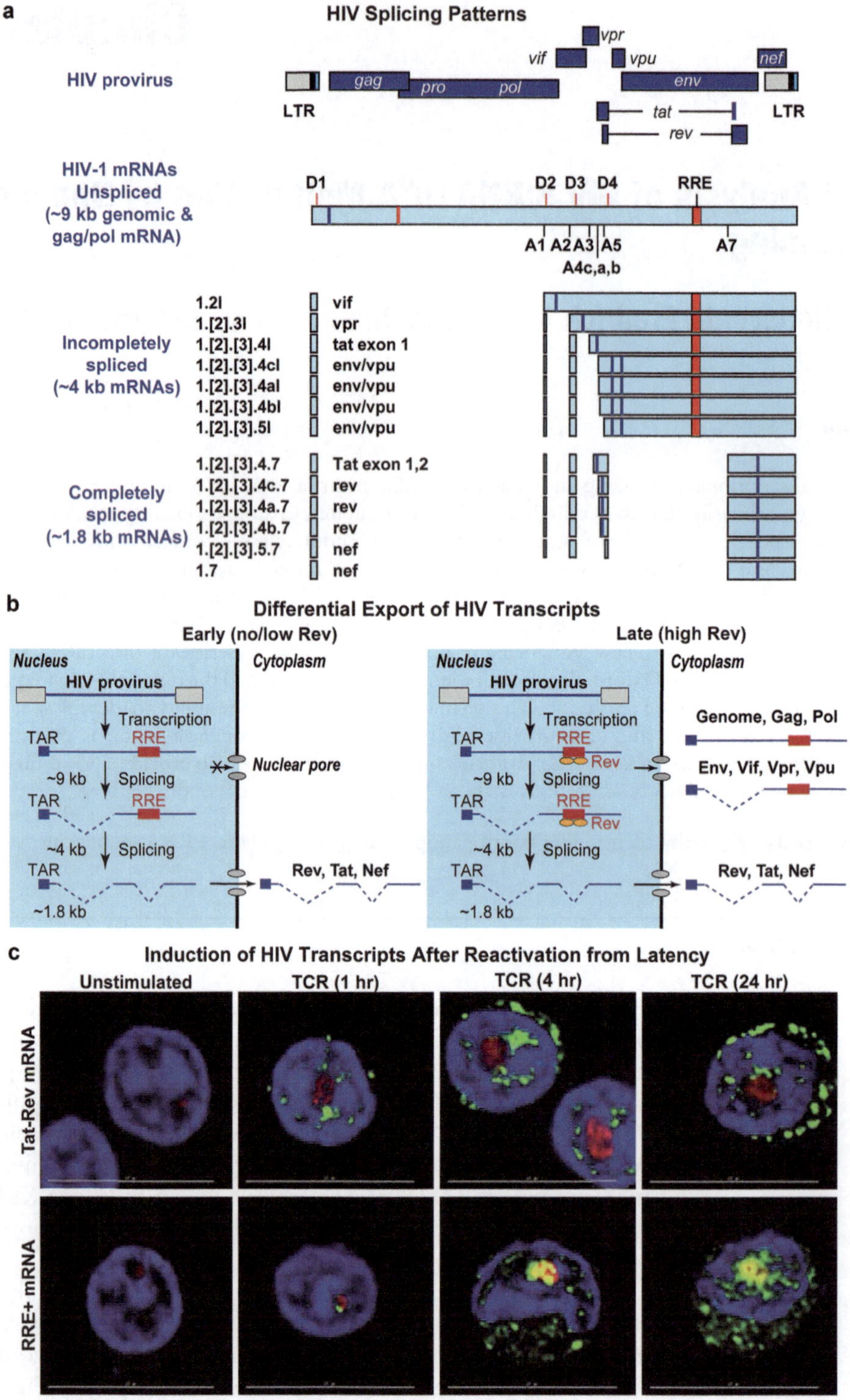

Fig. 1 Splicing and nuclear export of HIV RNA. (**a**) Schematic of the overlapping HIV transcripts and their splice patterns. (**b**) Schematic of the differential export of the major classes of HIV transcripts at early and late time points during infection. (**c**) Visualization of HIV transcription and nuclear export by in situ hybridization following activation of primary T cells latently infected with a *Δgagpol* HIV reporter virus through stimulation

How the different classes of HIV transcripts are post-transcriptionally regulated beyond alternative splicing and how RNA modifications may influence HIV RNA are currently of intense interest as these other layers of regulation are posited to play a major role in HIV gene expression [6–8].

Previous research has shown that N^6-methyladenosine (m^{6}A) modification of HIV transcripts regulates HIV RNA transcription, RNA stability, alternative splicing, and RNA-protein interactions, as well as translation [9–14], consistent with the influence of m^{6}A on cellular transcripts [15, 16]. m^{6}A modification of RNA is dynamically regulated by writer and eraser proteins. The writer complex that adds m^{6}A to RNA is composed of METTL3 catalytic subunit and other subunits that primarily direct the complex to mRNA and stabilize it. m^{6}A can also be removed from RNA in the nucleus by the erasers ALKBH5 and FTO. The presence of m^{6}A on an mRNA directs its interaction with readers, RNA-binding proteins that can influence mRNA stability, export, and translation in an m^{6}A-dependent manner [17].

Most of the original studies on the impact of m^{6}A methylation on HIV RNA biogenesis, trafficking, and metabolism have been performed in various cell lines that often do not recapitulate what occurs in primary T cells [18–20]. Additionally, the stage of HIV infection appears to influence the effect of m^{6}A on HIV replication. For example, prior to integration, m^{6}A may decrease the stability of HIV transcripts and inhibit viral replication as evidenced by m^{6}A reader protein overexpression in HeLa cells or Jurkat T cells [10, 13] or METTL3 knockdown in HeLa cells [9]. However, other studies report that, prior to HIV integration, overexpression of m^{6}A reader proteins in 293T cells and CEM-SS T cells enhances HIV gRNA and viral protein expression [11, 14]. One study suggests that at post-integration steps of HIV replication, the presence of m^{6}A on HIV transcripts may also differentially affect the expression levels of the different HIV genes at the transcript and protein level [9]. There are also conflicting results on the impact of m^{6}A on the export of HIV RNA from the nucleus with one report stating that m^{6}A in the RRE affects nuclear export [12] while others have found that m^{6}A in the RRE or other sites does not affect export [9, 21]. Although the modification sites of m^{6}A in primary T cells and Jurkat cells appear to be the same [13], the regulation and fate of these m^{6}A modified transcripts may differ substantially because primary T cells harboring HIV can be either activated effector cells or latently infected quiescent memory T cells. Therefore,

Fig. 1 (continued) of the T cell receptor following the QUECEL protocol [1]. Top panels: Tat-Rev mRNA (green), nucleolus (red), and nucleus (blue). Bottom panels: RRE+ mRNA (green), nucleolus (red), and nucleus (blue). Note the rapid export of the Tat-Rev mRNA from the nucleus and retention of the RRE+ transcripts in a peri-nucleolar region. Scale bars = 10 μm mechanistic studies of HIV RNA modification in primary T cells during productive infection and during the emergence of HIV from latency are needed.

Maps of where m^6A sites occur on HIV transcripts are key elements in developing a mechanistic understanding of how the transcripts are post-transcriptionally regulated and which regions of the transcripts are involved in these regulatory steps. Low-resolution maps of m^6A modifications on HIV transcripts have been generated by multiple groups using anti-m^6A antibody-dependent methods such as methylated RNA immunoprecipitation sequencing (MeRIP-seq) and m^6A-seq [22, 23], which couple immunoprecipitation with short-read RNA sequencing. These methods have identified an overall enrichment in m^6A on HIV transcripts in the 3′ UTR with other sites present in the 5′ UTR, Env, and Rev regions [11–13]. These observations align with what is known about m^6A distribution in cellular transcripts where it is often enriched in the 3′ UTR, near stop codons, and in long exons [24]. Unfortunately, these methods only provide a low-resolution map of m^6A sites with a resolution of about 100–200 nucleotides. This prevents the precise localization of single or multiple m^6A sites within this nucleotide window. For example, in the m^6A peak identified in the second exon of Rev [11–13], there are six possible consensus motifs in which m^6A typically occurs (D = A,G,U; R = A,G; H = A,C,U) or DRACH sites. Additionally, the m^6A antibodies used often cannot distinguish between m^6A modifications and other similar modifications like m^6Am, which could give rise to false-positive m^6A sites [25].

To overcome the limitations of these low-resolution techniques, single nucleotide resolution of m^6A sites can be obtained with alternative methods including m^6A individual-nucleotide resolution UV cross-linking and immunoprecipitation (miCLIP), site-specific cleavage, and radiolabeling followed by ligation-assisted extraction and thin layer chromatography (SCARLET), RNA digestion via m6A-sensitive RNase (MAZTER-seq), and m^6A-REF-seq. miCLIP uses a CLIP protocol to specifically pull down m^6A modified RNA and detects modification sites via reverse transcription truncations or mutations at sites of cross-linked anti-m^6A antibody. Yet, this technique prevents the sequencing of a transcript past the reverse transcription stall, limiting the depth of sequencing [26]. This is also a labor-intensive procedure that can take up to a week to perform and is still subject to m^6A antibody-dependent limitations, though it can map m^6A sites at a single nucleotide resolution. SCARLET combines m^6A site-specific cleavage, radiolabeling, and thin layer chromatography to map m^6A sites on RNA. Like miCLIP, SCARLET is labor-intensive and requires multiple enzymatic reactions, but it does allow for site-specific m^6A detection in low-abundance RNA species. However, this technique can only map one m^6A site per experiment, so multiple experiments must be performed in parallel to identify multiple m^6A sites, and it is not possible to study the co-occurrence of m^6A sites on the same RNA molecule [27]. MAZTER-seq and m^6A-REF-seq both utilize m^6A site-specific endonucleases and next-generation sequencing to map m^6A sites [28, 29]. These methods are limited by the fact that the endonuclease used to digest RNA can only recognize one motif in which m^6A occurs. Out of all possible DRACH motifs in which m^6A can occur, the endonuclease can only recognize the single ACA motif, which occurs in only about 16% of all m^6A sites [30]. These endonucleases have been known to cut at off-target sites, as well, which may generate false-positive results [25].

Due to the limitations of current m^6A mapping techniques, we have turned to using direct long-read RNA sequencing from Oxford Nanopore Technologies. This is currently the only technology available that can directly sequence native RNA without the need for conversion to cDNA and PCR amplification, two steps with significant risk to introduce bias to libraries. Unlike other sequencing technologies that rely first on the conversion of RNA to cDNA and then on fluorescent sequencing by synthesis, nanopore sequencing utilizes pores to directly sequence single-stranded RNA. The technology consists of an array of nanopores that span an artificial membrane. There is a constant electric current potential across this membrane, and this electric current is disturbed as RNA nucleotides pass through the pore. From this current perturbation, nucleotide identities can be determined. Additionally, as this technology sequences native RNA and uses electric currents to determine nucleotide identities, it can also detect differences in electric current between modified and unmodified nucleotides, allowing for identification of RNA modifications at single nucleotide resolution (Fig. 2a, b). This technology has already been used to map RNA modifications including m^6A on a variety of RNA species from plants to viruses [31–34]. While not all RNA modifications have been tested for mapping by nanopore, it has the potential to be able to identify any modifications that alter current measured including N^6,2′-O-dimethyladenosine (m^6Am), 5-methylcytosine (m^5C), 1-methyladenosine (m^1A), 7-methylguanosine (m^7G), N1-methylguanosine (m^1G), and pseudouridine (Ψ) [35]. The limiting step to mapping all RNA modifications with nanopore is the need for appropriate algorithms that can detect specific RNA modifications and differentiate them from modifications that may induce a similar change in electric current [36].

Due to the novelty of the nanopore sequencing technology, it is important to ensure that the sequencing data and the mapped RNA modifications are accurate. The first step upon RNA sequencing data alignment to a reference of interest is to visually check that the splicing patterns and sequenced transcripts appear as expected. RNA sequencing with nanopore is also known to have a 3′ bias [32, 34, 37]. If there are no reads at the 5′ UTR or the splicing is

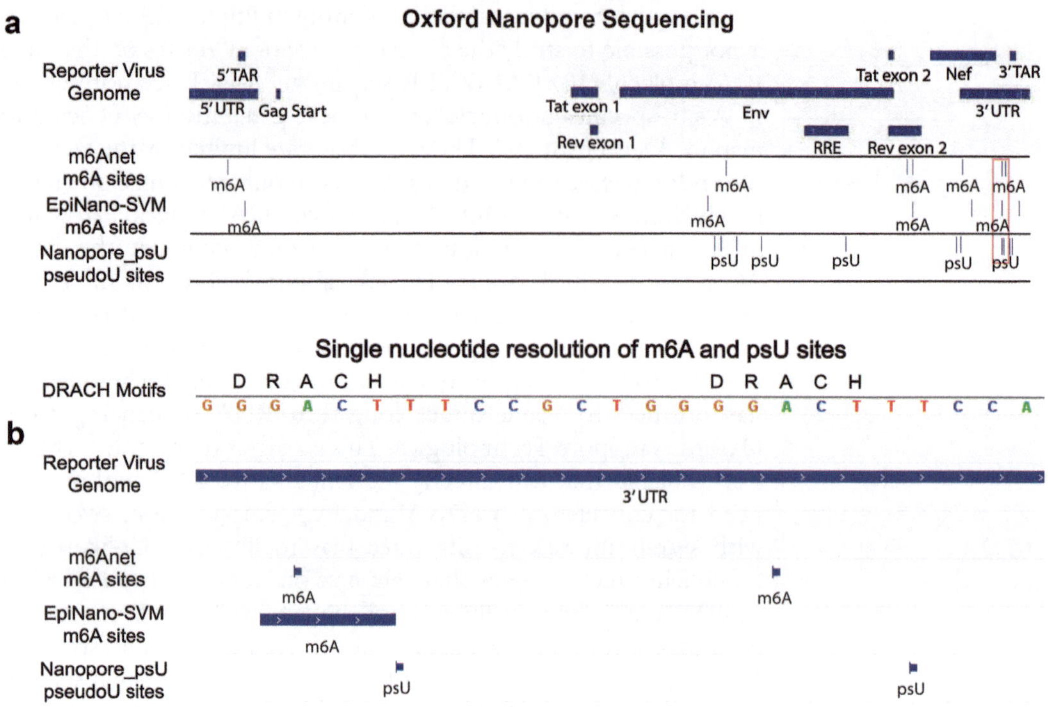

Fig. 2 m^{6}A and RNA modification map using Oxford Nanopore sequencing. (**a**) Maps of m^{6}A and pseudouridine (psU) sites on transcripts from a reactivated Δ*gagpol* HIV reporter virus infection of primary T cells. As in Fig. 1, the QUECEL protocol was used [1]. The plots are from direct long-read RNA sequencing. The m^{6}A sites were called with m6Anet and EpiNano-SVM, and the psU sites were called with Nanopore_psU. (**b**) A closeup view of the 3′UTR in the red box from (**a**) showing the single nucleotide resolution maps of m^{6}A and psU. Note the five-nucleotide stretch marked as m^{6}A represents the DRACH motif in which m^{6}A typically occurs

not correct, this may be due to a problem in the sample, library preparation, or alignment to the genome. When mapping m^{6}A sites, one should manually check that the mapped putative sites generally occur in the DRACH motif to ensure that the m^{6}A mapping programs are working correctly. It is also critical to use multiple m^{6}A or RNA modification mapping programs to analyze the sequencing data because the mapping programs that are compatible with Oxford Nanopore sequencing data often produce overlapping and non-overlapping regions of RNA modifications. To ascertain which sites are real, one should also use the appropriate controls for these experiments such as sequencing in vitro transcribed RNA that matches the transcript of interest or sequencing RNA from cells that have been depleted of the methyltransferase or modification enzyme of choice. These datasets can then be directly compared to datasets from modified RNA to more accurately map specific RNA modifications. Finally, it is necessary to confirm the mapped regions using orthogonal methods such as MeRIP-seq. High-confidence maps of m^{6}A can then be used to direct targeted functional studies using approaches such as site-directed mutagenesis [11].

2 Materials

2.1 RNA Extraction

1. TRIzol or column-based RNA extraction kit.
2. Chloroform.
3. Isopropanol.
4. 75% ethanol (freshly prepared).
5. Nuclease-free water.
6. Microcentrifuge.
7. Nanodrop Spectrophotometer or equivalent.

2.2 Poly(A) Selection

1. Poly(A) selection kit (New England Biolabs).
2. Magnetic rack.
3. Qubit fluorometer (Thermo Fisher) or TapeStation (Agilent).
4. RNA High-Sensitivity (HS) Qubit kit (Invitrogen) or High-Sensitivity RNA ScreenTape (Agilent).
5. Thermal cycler.

2.3 Library Preparation

1. T4 DNA ligase (2,000,000 units/mL) (New England Biolabs).
2. 5× Quick Ligation buffer.
3. Nuclease-free water.
4. SuperScript III (SSIII) reverse transcriptase (ThermoFisher).
5. 5× first-strand buffer (ThermoFisher).
6. 0.1 M DTT.
7. 10 mM dNTPs.
8. Agencourt RNAClean XP beads (Beckman Coulter).
9. Thermal cycler.
10. Direct RNA Sequencing kit (Oxford Nanopore Technologies).
11. 70% ethanol (freshly prepared).
12. Hula mixer.

2.4 Library Loading and Running

1. Flow Cell Priming Kit (Oxford Nanopore Technologies).
2. R9.4.1 flow cell.
3. MinION Mk1B or Mk1C.
4. Direct RNA Sequencing kit (Oxford Nanopore Technologies).
5. Nuclease-free water.

2.5 Computer and Software

1. A CPU with the following minimum specs: Running Windows—10 OR OSX—Mojave, Catalina OR Linux Ubuntu 20.04 or 18.04; 16 GB RAM or higher; Intel i7, i9, Xeon, or

better, with at least 4 cores/8 threads OR Ryzen 5, 7, or better, with at least 4 cores/8 threads; USB3 port; 1 TB internal solid-state hard drive; OR a GPU with the following minimum specs: NVIDIA GPU RTX 2060 SUPER or better, with at least 8 GB of GPU memory; 16 GB RAM or higher; USB3 port; 1 TB internal solid-state hard drive.

2. Guppy—proprietary program from Oxford Nanopore Technology (must have an ONT account to download the software).
3. Minimap2—available at https://github.com/lh3/minimap2
4. Samtools—available at https://github.com/samtools/samtools
5. Nanopolish—available at https://github.com/jts/nanopolish
6. hdf5plugin—available at https://github.com/silx-kit/hdf5plugin
7. Integrated Genome Viewer (IGV) https://software.broadinstitute.org/software/igv/
8. m6Anet—https://github.com/GoekeLab/m6anet
9. xPore—https://github.com/GoekeLab/xpore
10. Nanocompore—https://github.com/tleonardi/nanocompore
11. EpiNano—https://github.com/novoalab/EpiNano
12. DRUMMER—https://github.com/DepledgeLab/DRUMMER

3 Methods

3.1 RNA Extraction

1. Pellet cells of interest and aspirate supernatant. Add 1 mL of TRIzol reagent per 5 million cells. Mix well by pipetting and let sit at room temperature for 5 min (*see* **Note 1**).
2. Add 0.2 mL of chloroform per 1 mL of TRIzol reagent used and cap tube. Mix well by manual shaking. Incubate at room temperature for 2–3 min.
3. Centrifuge the sample for 15 min at 12,000× *g* at 4 °C.
4. Transfer the aqueous phase containing the RNA to a new tube careful not to disturb the interphase.
5. Add 0.5 mL of isopropanol to the aqueous phase per 1 mL of TRIzol used. Mix well by manual shaking. Incubate at room temperature for 10 min.
6. Centrifuge for 10 min at 12,000× *g* at 4 °C.
7. The RNA will form a white gel-like pellet at the bottom of tube. Use a pipette to remove the supernatant.

8. Wash the pellet in 1 mL of freshly prepared 75% ethanol per 1 mL of TRIzol used.
9. Vortex the sample briefly, then centrifuge for 5 min at 7500× *g* at 4 °C.
10. Discard the supernatant and repeat the wash step for a total of two washes.
11. Remove as much ethanol as possible and air dry the pellet for 5–10 min.
12. Resuspend the pellet in 20–50 μL of nuclease-free water.
13. Briefly vortex and spin down the sample to resuspend fully.
14. Use a Nanodrop Spectrophotometer or equivalent to measure RNA concentration.
15. (Optional) Treat the sample with DNase (*see* **Note 2**).

3.2 Poly(A) Selection (Optional) (See Notes 3–5)

1. Dilute 1–5 μg of total RNA with nuclease-free water to a final volume of 50 μL in a nuclease-free 0.2 mL PCR tube.
2. In a second nuclease-free 0.2 mL PCR tube aliquot 20 μL of well resuspended NEBNext Magnetic Oligo $d(T)_{25}$ Beads.
3. Wash the beads by adding 100 μL of RNA Binding Buffer to the beads. Pipette up and down six times to mix thoroughly.
4. Place the tube on the magnetic rack at room temperature for 2 min or until the solution is clear.
5. Remove the supernatant from the tube, careful not to disturb the beads.
6. Remove the tube from the rack and repeat the wash step for a total of two washes.
7. Resuspend the beads in 50 μL of RNA Binding Buffer and add the 50 μL of total RNA. Pipette to mix thoroughly.
8. Place the tube on the thermal cycler at 65 °C for 5 min, then hold at 4 °C.
9. Remove the tube from the thermal cycler once sample has reached 4 °C. Resuspend the beads by pipetting up and down six times.
10. Incubate the tube at room temperature for 5 min.
11. Resuspend the beads by pipetting up and down six times.
12. Incubate the tube at room temperature for 5 min.
13. Place the tube on the magnetic rack at room temperature for 2 min or until the solution is clear.
14. Remove the supernatant, careful not to disturb the beads.
15. Remove the tube from the rack.

16. Wash the beads by adding 200 μL of Wash Buffer. Pipette up and down six times to mix.
17. Place the tube on the magnetic rack at room temperature for 2 min or until the solution is clear.
18. Remove the supernatant, careful not to disturb the beads.
19. Repeat the wash step for a total of two washes.
20. Add 50 μL of Tris Buffer to the tube. Pipette up and down six times to resuspend the beads.
21. Place tube on the thermal cycler at 80 °C for 2 min, then hold at 25 °C.
22. Remove the tube from the thermal cycler once the sample has reached 25 °C.
23. Add 50 μL of RNA-Binding Buffer to the tube and pipette up and down six times to mix.
24. Incubate the tube at room temperature for 5 min.
25. Resuspend the beads by pipetting up and down six times.
26. Incubate the tube at room temperature for 5 min.
27. Place the tube on the magnetic rack for 2 min or until the solution is clear.
28. Remove the supernatant, careful not to disturb the beads.
29. Remove the tube from the magnetic rack.
30. Wash the beads once with 200 μL of Wash Buffer. Pipette up and down six times to resuspend the beads.
31. Place the tube on the magnetic rack at room temperature for 2 min or until the solution is clear.
32. Remove the supernatant, careful not to disturb the beads.
33. Spin down the tube and place on magnetic rack to remove all residual wash buffer.
34. Remove the tube from the magnetic rack.
35. Elute the mRNA from the beads by adding 10–11 μL of Tris Buffer. Mix by pipetting six times (*see* **Note 5**).
36. Place the tube on the thermal cycler at 80 °C for 2 min then hold at 25 °C.
37. Immediately place the tube on the magnetic rack for 2 min or until the solution is clear.
38. Collect the purified mRNA in a clean nuclease-free tube.
39. Measure poly(A) RNA concentration using 1 μL for RNA HS Qubit quantification.

3.3 Library Preparation

1. Before starting the library preparation, an R9.4.1 flow cell should be checked to ensure it has an acceptable number of pores available for sequencing.
2. In order to prepare a library for direct RNA sequencing using the SQK-RNA002 kit from Oxford Nanopore, 500 ng of total RNA or 50 ng of poly(A) RNA is recommended.
3. Use up to 9.5 μL of total or poly(A) RNA for the library preparation in a 0.2 mL PCR tube.
4. Add 3 μL of 5X Quick Ligation buffer, 1.0 μL RTA adapter from the ONT kit, 1.5 μL T4 DNA ligase, and nuclease-free water to a total volume of 15 μL. Mix by pipetting.
5. Let reaction incubate at room temperature for 10 min.
6. In another PCR tube, prepare the reverse transcription buffer. Add 9 μL nuclease-free water, 2 μL 10 mM dNTPs, 8 μL 5× first-strand buffer, and 4 μL 0.1 M DTT.
7. When ligation reaction is complete, add the 15 μL of ligation reaction to the reverse transcription buffer, then add 2 μL SSIII. Mix by pipetting.
8. Place the reverse transcription reaction in a thermal cycler at 50 °C for 50 min, then bring down to 4 °C (*see* **Note 6**).
9. Transfer the 40 μL reaction to a 1.5 mL Eppendorf tube. Resuspend the bottle of Agencourt RNAClean XP beads by vortexing until well mixed, then add 72 μL beads (1.8× volume:bead ratio) to the reverse transcription reaction. Pipette the mixture up and down to resuspend and incubate 5 min at room temperature on a hula mixer or on the bench top with periodic pipetting.
10. Place the Eppendorf tube on a magnet and let it sit for 5 min at room temperature or until the solution is clear. While waiting, prepare 200 μL fresh 70% ethanol.
11. Leave the tube on the magnet and pipette out the supernatant, careful not to disturb the beads. Add 150 μL 70% ethanol to the tube to wash the beads. Leaving the tube on the magnet, turn the tube 180° and allow the beads to pellet, then pipette off the ethanol. Spin down the tube and place on the magnet again, then pipette off the residual ethanol.
12. Remove the tube from the rack and resuspend in 20 μL of nuclease-free water. Let the resuspended mixture incubate at room temperature for 5 min, then pellet the beads on the magnet for 5 min or until the solution is clear. Pipette out 20 μL of the supernatant and place in a clean 1.5 mL Eppendorf tube.

13. To the eluted RNA, add 8 μL 5× quick ligation reaction buffer, 6 μL RMX adapter, 3 μL nuclease-free water, and 3 μL T4 DNA ligase. Mix by pipetting and incubate at room temperature for 10 min.
14. Add 16 μL of well-resuspended Agencourt RNAClean XP beads (0.4× volume:bead ratio) to the reaction and mix by pipetting. Incubate on a hula mixer for 5 min at room temperature or on the benchtop with periodic pipetting.
15. Spin down the tube and rack the beads on a magnet for 5 min or until the solution is clear. Pipette off the supernatant.
16. Remove the tube from the magnet and add 150 μL of Wash Buffer from the ONT kit. Agitate the tube by flicking to resuspend the beads. Rack the beads on the magnet, remove the supernatant, and repeat the wash step for a total of two washes.
17. Remove the supernatant and resuspend the beads in 20 μL of Elution buffer from the ONT kit. Resuspend the beads by flicking the tube, then spin down. Incubate at room temperature for 10 min (*see* **Note 7**).
18. Rack the beads on the magnet for 5 min or until the solution is clear. Remove the 20 μL eluate and place in a clean 1.5 mL Eppendorf tube.

3.4 Loading the Library onto the Flow Cell

1. Using a p1000 pipette set to 200 μL, place the tip into the priming port and draw up 20–30 μL of yellow storage buffer by spinning the pipette wheel to 220–230 μL. Do not remove more than 30 μL. This step removes any bubbles from the flow cell.
2. Add 30 μL of thawed flush tether (FLT) to one tube of thawed flush buffer (FB) from the ONT flow cell priming kit. Mix well by pipetting, then add 800 μL of this mix to the priming port making sure not to introduce any bubbles (*see* **Note 8**).
3. Allow the flow cell to sit for 5 min.
4. While waiting, prepare the library for loading. Add 17.5 μL of nuclease-free water to the 20 μL library from the previous section. Vortex the thawed running buffer (RRB) from the ONT direct RNA sequencing kit and then add 37.5 μL of RRB to the water and library mixture. Pipette up and down to mix.
5. Open the SpotOn port of the flow cell. Load 200 μL of the remaining flush buffer to the priming port. Liquid may bubble up from the SpotOn port, and this is expected.
6. As soon as possible after this step, load the 75 μL of library. Keeping both the priming port and SpotOn port open, add the

library to the SpotOn port in a dropwise fashion, allowing each drop to enter the flow cell before adding another.

7. Close both ports and attach the flow cell to the MinION device.

3.5 Setting Up a MinKNOW Run

1. Once the flow cell is attached to the MinION device, navigate to the sequencing menu and start a new sequencing run.
2. Choose the SQK-RNA002 kit for the sequencing run, then choose the options that are suitable for your experiment (*see* **Note 9**).
3. Once the run has started, monitor for at least 30 min to ensure your sample is being sequenced.

3.6 Basecalling and Mapping to a Reference Genome

The output of the sequencing run includes a summary file, which states basic information about the run including total run time, pore activity and health, and total number of reads. The sequencing data are stored in the raw fast5 files, which must be converted into fastq files for further analysis. Additionally, the fast5 files are vbz compressed by default. Therefore, the hdf5plugin is needed for reading of these compressed files by downstream analysis software. If this plugin does not work, the vbz compressed files can be converted to gzip compressed using the ont-fast5-api program. The data processing should be performed on a Linux or Mac as these operating systems are compatible with the required packages. All code can be executed through the command line.

1. Use ONT Guppy to basecall the fast5 files using the config files rna_r9.4.1_70bps_fast or rna_r9.4.1_70bps_hac depending on computer processing capacity. The fast option prioritizes speed over accuracy and is less computationally intense while the hac (high accuracy) option requires more time and computational power but is more accurate in assigning nucleotide identities.

   ```
   guppy_basecaller -i /path to fast5 files/ -s /path to output folder/ -c [config file of choice] --num_callers 2 --cpu_threads_per_caller 1
   ```

2. Merge all the passed fastq files into a single fastq file. In the passed fastq file directory, use the following code to merge the files into one large file.

   ```
   cat *.fastq > merged_fastq.fastq
   ```

   Locate this file and move it to a new folder for easier access.

3. Using the merged fastq file, align the reads to a reference genome using minimap2 with recommended settings for direct RNA sequencing.

```
minimap2 -ax splice -uf -k14 /path to fasta reference/ /path to merged fastq file/ > /path to output folder/file_name.sam
```

4. Convert the sam file into a sorted bam file by using this set of commands from the samtools package.

```
samtools view -bS /path to sam file/file_name.sam > /path to output folder/file_name.bam
samtools sort -o /path to output folder/file_name_sorted.bam /path to bam file/file_name.bam
samtools index /path to sorted bam file/file_name_sorted.bam
```

5. From here, the sorted and indexed bam files can be opened in IGV to visualize the mapped reads with the correct reference file.

3.7 Data Preprocessing with Nanopolish

Many of the RNA modification mapping programs use the eventalign output file from the nanopolish program. Thus, one common output file can be used as the input for multiple mapping programs. Nanopolish "resquiggles" the data to perform signal level analysis of Oxford Nanopore sequencing data. Through this process, it is able to detect base modifications. Although it was created for DNA sequencing data, it can also be used for RNA sequencing data.

1. Index the fast5 reads using the nanopolish index.

```
nanopolish_index -d /path to fast5 folder/ /path to merged fastq file/
```

2. Perform eventalign to align the fast5 files with the merged fastq file using the command below. It is important to look at the documentation of the mapping program of choice as different programs require different options to be included in this step.

```
nanopolish_eventalign --reads /path to merged fastq file/ --bam /path to sorted bam file/ --genome /path to fasta reference/ [optional flags] > /path to output folder/file_name_eventalign.[txt or tsv]
```

This eventalign text or tsv file is used as the input for many RNA modification mapping programs.

3.8 m^{6}A Mapping Programs

We have utilized m6Anet, EpiNano, and xPore successfully to map m^{6}A sites on transcripts (Fig. 2) [38–40]. There are also many other programs that can be used to map these sites although we have not utilized them ourselves.

3.9 RNA Modification Mapping Programs

Other programs can also be used to map other RNA modifications from Nanopore sequencing data such as Nanopore_psU or penguin (Fig. 2) [41, 42]. One can also compare modified transcripts to in vitro transcribed transcripts or transcripts from cells with knockouts/knockdowns in certain RNA modification adding enzymes to identify sites of modification using programs such as Nanocompore, xPore, and DRUMMER [35, 40, 43].

3.10 Conclusion

Although nanopore sequencing is currently the only technology that can map m^{6}A and RNA modifications at single nucleotide resolution on RNA transcripts recovered from cells, the technique does have its limitations, which must be considered when performing mapping experiments. The first limitation is that, when using a single flow cell for the MinION device, the maximum number of reads that can be generated is roughly 1.5 million, which is at least an order of magnitude lower than what can be generated by other RNA sequencing methods. This, in turn, decreases the read depth for each transcript, which must be taken into account during analysis. This limitation can be improved upon by using Oxford Nanopore Technology's larger sequencing devices such as the PromethION or by capturing specific RNAs for sequencing [44]. Another limitation to this technique is the 3′ bias exhibited by the nanopores. This means that the 3′ ends of transcripts are sequenced to a greater extent than the 5′ ends. This is likely due to the pores stopping sequencing toward the 5′ end of a transcript due to RNA secondary structures that block the pore, incomplete reverse transcription, or computational errors in recovering the 5′ ends of transcripts [32, 34, 37]. Additionally, because the direct RNA sequencing method from Oxford Nanopore is still new, the algorithms that call RNA nucleotide identities are error-prone and can call mutations where mutations are not actually present in the transcripts [45–47]. Along with basecalling algorithms, there is a need for RNA modification detection algorithms, as well. Nanopore sequencing has the potential to detect any RNA modification that induces a change in the electric current measured; however, there is currently a lack of algorithms that can detect these electric current changes and differentiate specific RNA modifications that may induce similar changes in electric current. Research is still ongoing to develop these algorithms so that nanopore sequencing can be used to map multiple RNA modifications at once at a single transcript resolution [36]. Finally, on a practical level, the input needed for direct RNA sequencing is relatively high especially if using primary samples with low RNA yields. This also poses a problem when trying to sequence HIV transcripts from samples with low infection rates. Despite these issues, direct RNA sequencing provides a unique, high-resolution snapshot of the critical modifications on specific HIV transcripts that may determine differential splicing, nuclear retention, and translation.

4 Notes

1. TRIzol extraction is preferred over kit extraction to prevent RNA fragmentation and preserve shorter or longer transcripts that may be lost using column-based extraction methods. However, a column-based kit can still be used depending on the size of the target RNA.
2. DNase treatment of extracted RNA is an optional step but is recommended if one has used a plasmid that contains a poly (A) tail as this may interfere with library preparation. DNase treatment is not required when using TRIzol as the DNA is partitioned in the interface. As long as the interface is not disturbed, there should be minimal DNA contamination.
3. Poly(A) selection can make the ligation step during library preparation more efficient. However, some RNA is lost during this selection process, so if one already has a low concentration of RNA to begin with, this step can be omitted. Alternatively, total RNA can be ribodepleted to remove the abundant rRNA species.
4. Multiple papers have used poly(A) RNA for their library preparations for Oxford Nanopore sequencing, yet they have used differing amounts. ONT recommends using 50 ng of poly (A) RNA for library preparation, but other studies have used up to almost 2 μg of poly(A) RNA [31–33]. However, we have had success in obtaining reads from samples with as little as 20 ng of poly(A) RNA. If using primary T cells, 50 ng of poly (A) RNA can be obtained from about 5 μg of total RNA.
5. If using a kit to poly(A) select RNA, it is important to elute in 10–11 μL of the kit's recommended buffer as only 9.5 μL can be used in library preparation. 1 μL of this should be used for RNA concentration quantification using an RNA HS Qubit kit.
6. We found that omitting the heat inactivation step of the SSIII protocol helped to preserve RNA integrity, which is necessary as ONT sequencing directly sequences the RNA strand. Additionally, this enzyme inactivation step is not necessary because the RNA is removed from the reverse transcription buffer immediately following the reverse transcription step.
7. The original protocol calls for elution in 21 uL of elution buffer so that 1 uL can be used for DNA HS Qubit quantification. However, we have omitted this step because we have found that the quantification after library preparation is not always accurate and that even concentrations below the range of detection can still yield useful sequencing data.
8. To help ensure no bubbles are introduced to the flow cell, pipette a drop of the flush buffer mixture on top of the priming

port hole and immediately place the pipette tip into the fluid bubble that is formed. This ensures that no air is introduced into the priming port when adding the flush buffer.

9. The following options can be chosen based on computing power and the research question being asked: if one's computer has enough computing power, live basecalling can be performed during the run. However, we do not recommend this as it can interfere with sequencing. The files should be saved as fast5 files on an internal solid-state hard drive with at least 1 TB of free storage. If interested in small RNAs, the bulk output file can be saved under the advanced options section. For this file, only the events-table and raw options need to be enabled. From this bulk file, data from short RNAs can be extracted [48]. Finally, the run can be set up to run for up to 72 h; however, our runs are typically done in 36–48 h.

References

1. Shukla M, Kizito F, Mbonye U, Nguyen K, Dobrowolski C, Karn J (2022) HIV reservoirs methods and protocols a reliable primary cell model for HIV latency: The QUECEL (Quiescent Effector Cell Latency) method. Springer US New York NY, pp 57–68
2. Nguyen Quang N, Goudey S, Ségéral E et al (2020) Dynamic nanopore long-read sequencing analysis of HIV-1 splicing events during the early steps of infection. Retrovirology 17:1–24. https://doi.org/10.1186/S12977-020-00533-1/FIGURES/8
3. Emery A, Zhou S, Pollom E, Swanstrom R (2017) Characterizing HIV-1 splicing by using next-generation sequencing. J Virol 91: 2515–2531. https://doi.org/10.1128/JVI.02515-16/ASSET/3185AA31-EF8E-4E28-B852-4ED36B5C54FA/ASSETS/GRAPHIC/ZJV9991824380010.JPEG
4. Pollard VW, Malim MH (2003) The HIV-1 REV protein. Ann Rev Microbiol 52:491–532. https://doi.org/10.1146/ANNUREV.MICRO.52.1.491
5. Truman CTS, Järvelin A, Davis I, Castello A (2020) HIV Rev-isited. Open Biol 10:200320. https://doi.org/10.1098/RSOB.200320
6. Karn J, Stoltzfus CM (2012) Transcriptional and posttranscriptional regulation of HIV-1 gene expression. Cold Spring Harb Perspect Med 2:a006916. https://doi.org/10.1101/CSHPERSPECT.A006916
7. Crespo R, Rao S, Mahmoudi T (2022) HibeR-NAtion: HIV-1 RNA metabolism and viral latency. Front Cell Infect Microbiol 12: 855092. https://doi.org/10.3389/FCIMB.2022.855092/BIBTEX
8. Riquelme-Barrios S, Pereira-Montecinos C, Valiente-Echeverría F, Soto-Rifo R (2018) Emerging roles of N6-methyladenosine on HIV-1 RNA metabolism and viral replication. Front Microbiol 9:576. https://doi.org/10.3389/FMICB.2018.00576
9. N'Da Konan S, Ségéral E, Bejjani F et al (2022) YTHDC1 regulates distinct post-integration steps of HIV-1 replication and is important for viral infectivity. Retrovirology 19:4. https://doi.org/10.1186/S12977-022-00589-1
10. Lu W, Tirumuru N, Gelais CS et al (2018) N6-Methyladenosine-binding proteins suppress HIV-1 infectivity and viral production. J Biol Chem 293:12992–13005. https://doi.org/10.1074/JBC.RA118.004215
11. Kennedy EM, Bogerd HP, Kornepati AVR et al (2016) Posttranscriptional m(6)A editing of HIV-1 mRNAs enhances viral gene expression. Cell Host Microbe 19:675–685. https://doi.org/10.1016/J.CHOM.2016.04.002
12. Lichinchi G, Gao S, Saletore Y et al (2016) Dynamics of the human and viral m6A RNA methylomes during HIV-1 infection of T cells. Nat Microbiol 1:16011. https://doi.org/10.1038/NMICROBIOL.2016.11
13. Tirumuru N, Zhao BS, Lu W et al (2016) N6-methyladenosine of HIV-1 RNA regulates viral infection and HIV-1 Gag protein expression. elife 5:e15528. https://doi.org/10.7554/ELIFE.15528

14. Tsai K, Bogerd HP, Kennedy EM et al (2021) Epitranscriptomic addition of m6A regulates HIV-1 RNA stability and alternative splicing. Genes Dev 35:992–1004. https://doi.org/10.1101/GAD.348508.121
15. Zhang Y, Geng X, Li Q et al (2020) m6A modification in RNA: biogenesis, functions and roles in gliomas. J Exp Clin Cancer Res 39(1):1–16. https://doi.org/10.1186/S13046-020-01706-8
16. Jiang X, Liu B, Nie Z et al (2021) The role of m6A modification in the biological functions and diseases. Signal Transduct Target Ther 6(1):74. https://doi.org/10.1038/s41392-020-00450-x
17. Zaccara S, Ries RJ, Jaffrey SR (2019) Reading, writing and erasing mRNA methylation. Nat Rev Mol Cell Biol 20(10):608–624. https://doi.org/10.1038/s41580-019-0168-5
18. Telwatte S, Morón-López S, Aran D et al (2019) Heterogeneity in HIV and cellular transcription profiles in cell line models of latent and productive infection: implications for HIV latency. Retrovirology 16:32. https://doi.org/10.1186/s12977-019-0494-x
19. Bartelt RR, Cruz-Orcutt N, Collins M, Houtman JCD (2009) Comparison of T cell receptor-induced proximal signaling and downstream functions in immortalized and primary T cells. PLoS One 4:e5430. https://doi.org/10.1371/journal.pone.0005430
20. Lin Z, Fillmore GC, Um TH et al (2003) Comparative microarray analysis of gene expression during activation of human peripheral blood T cells and leukemic Jurkat T cells. Lab Investig 83:765–776. https://doi.org/10.1097/01.LAB.0000073130.58435.E5
21. Chu CC, Liu B, Plangger R et al (2019) m6A minimally impacts the structure, dynamics, and Rev ARM binding properties of HIV-1 RRE stem IIB. PLoS One 14:e0224850. https://doi.org/10.1371/JOURNAL.PONE.0224850
22. Dominissini D, Moshitch-Moshkovitz S, Schwartz S et al (2012) Topology of the human and mouse m6A RNA methylomes revealed by m6A-seq. Nature 485:201–206. https://doi.org/10.1038/NATURE11112
23. Meyer KD, Saletore Y, Zumbo P et al (2012) Comprehensive analysis of mRNA methylation reveals enrichment in 3' UTRs and near stop codons. Cell 149:1635–1646. https://doi.org/10.1016/J.CELL.2012.05.003
24. Ke S, Pandya-Jones A, Saito Y et al (2017) m6A mRNA modifications are deposited in nascent pre-mRNA and are not required for splicing but do specify cytoplasmic turnover. Genes Dev 31(10):990–1006. https://doi.org/10.1101/gad.301036
25. Xiang ZH, Sheng ZX, Sui N (2020) Advances in the profiling of N6-methyladenosine (m6A) modifications. Biotechnol Adv 45:107656. https://doi.org/10.1016/J.BIOTECHADV.2020.107656
26. Linder B, Grozhik AV, Olarerin-George AO et al (2015) Single-nucleotide-resolution mapping of m6A and m6Am throughout the transcriptome. Nat Methods 12:767–772. https://doi.org/10.1038/NMETH.3453
27. Liu N, Pan T (2015) Probing RNA modification status at single-nucleotide resolution in total RNA. Methods Enzymol 560:149–159. https://doi.org/10.1016/BS.MIE.2015.03.005
28. Garcia-Campos MA, Edelheit S, Toth U et al (2019) Deciphering the "m6A code" via antibody-independent quantitative profiling. Cell 178:731–747.e16. https://doi.org/10.1016/J.CELL.2019.06.013
29. Zhang Z, Chen LQ, Zhao YL et al (2019) Single-base mapping of m6A by an antibody-independent method. Sci Adv 5:250–253. https://doi.org/10.1126/SCIADV.AAX0250/SUPPL_FILE/AAX0250_SM.PDF
30. Pandey RR, Pillai RS (2019) Counting the cuts: MAZTER-Seq quantifies m6A levels using a methylation-sensitive ribonuclease. Cell 178:515–517. https://doi.org/10.1016/J.CELL.2019.07.006
31. Price AM, Hayer KE, McIntyre ABR et al (2020) Direct RNA sequencing reveals m6A modifications on adenovirus RNA are necessary for efficient splicing. Nat Commun 11(1):1–17. https://doi.org/10.1038/s41467-020-19787-6
32. Parker MT, Knop K, Sherwood AV et al (2020) Nanopore direct RNA sequencing maps the complexity of arabidopsis mRNA processing and m6A modification. elife 9:e49658. https://doi.org/10.7554/ELIFE.49658
33. Campos JHC, Maricato JT, Braconi CT et al (2021) Direct RNA sequencing reveals SARS-CoV-2 m6A sites and possible differential DRACH motif methylation among variants. Viruses 13:2108. https://doi.org/10.3390/V13112108
34. Workman RE, Tang AD, Tang PS et al (2019) Nanopore native RNA sequencing of a human poly(A) transcriptome. Nat Methods 16(12):1297–1305. https://doi.org/10.1038/s41592-019-0617-2
35. Leger A, Amaral PP, Pandolfini L et al (2021) RNA modifications detection by comparative

nanopore direct RNA sequencing. Nat Commun 12(1):7198. https://doi.org/10.1038/s41467-021-27393-3

36. White LK, Hesselberth JR (2022) Modification mapping by nanopore sequencing. Front Genet 13:1037134
37. Depledge DP, Srinivas KP, Sadaoka T et al (2019) Direct RNA sequencing on nanopore arrays redefines the transcriptional complexity of a viral pathogen. Nat Commun 10:754. https://doi.org/10.1038/s41467-019-08734-9
38. Hendra C, Pratanwanich PN, Wan YK et al (2022) Detection of m6A from direct RNA sequencing using a multiple instance learning framework. Nat Methods 19(12):1590–1598. https://doi.org/10.1038/s41592-022-01666-1
39. Liu H, Begik O, Lucas MC et al (2019) Accurate detection of m6A RNA modifications in native RNA sequences. Nat Commun 10(1):1–9. https://doi.org/10.1038/s41467-019-11713-9
40. Pratanwanich PN, Yao F, Chen Y et al (2021) Identification of differential RNA modifications from nanopore direct RNA sequencing with xPore. Nat Biotechnol 39(11):1394–1402. https://doi.org/10.1038/s41587-021-00949-w
41. Huang S, Zhang W, Katanski CD et al (2021) Interferon inducible pseudouridine modification in human mRNA by quantitative nanopore profiling. Genome Biol 22:1–14. https://doi.org/10.1186/S13059-021-02557-Y/FIGURES/2
42. Hassan D, Acevedo D, Daulatabad SV et al (2022) Penguin: a tool for predicting pseudouridine sites in direct RNA nanopore sequencing data. Methods 203:478–487. https://doi.org/10.1016/J.YMETH.2022.02.005
43. Abebe JS, Price AM, Hayer KE et al (2022) DRUMMER—rapid detection of RNA modifications through comparative nanopore sequencing. Bioinformatics 38:3113–3115. https://doi.org/10.1093/BIOINFORMATICS/BTAC274
44. Tan CCS, Maurer-Stroh S, Wan Y et al (2019) A novel method for the capture-based purification of whole viral native RNA genomes. AMB Express 9:45. https://doi.org/10.1186/s13568-019-0772-y
45. Jain M, Abu-Shumays R, Olsen HE, Akeson M (2022) Advances in nanopore direct RNA sequencing state-of-the-art nanopore DRS. Nat Methods 19:1160–1164. https://doi.org/10.1038/s41592-022-01633-w
46. Rousseau-Gueutin M, Belser C, Silva C Da et al (2020) Long-read assembly of the *Brassica napus* reference genome Darmor-bzh. Gigascience 9:1–16. https://doi.org/10.1093/GIGASCIENCE/GIAA137
47. Grünberger F, Ferreira-Cerca S, Grohmann D (2022) Nanopore sequencing of RNA and cDNA molecules in *Escherichia coli*. RNA 28:400–417. https://doi.org/10.1261/RNA.078937.121
48. Lucas MC, Pryszcz LP, Medina R et al (2023) Quantitative analysis of tRNA abundance and modifications by nanopore RNA sequencing. Nat Biotechnol 42:72–86. https://doi.org/10.1038/s41587-023-01743-6

Chapter 16

Identifying Individual Pseudouridine (Ψ) Sites Across Transcripts from HIV-1 Infected Cells

Hannah L. Coutts and David G. Courtney

Abstract

The identification of RNA modifications at single nucleotide resolution has become an emerging area of interest within biology and specifically among virologists seeking to ascertain how this untapped area of RNA regulation may be altered or hijacked upon viral infection. Herein, we describe a straightforward biochemical approach modified from two original published Ψ mapping protocols, BID-seq and PRAISE, to specifically identify pseudouridine modifications on mRNA transcripts from an HIV-1 infected T cell line. This protocol could readily be adapted for other viral infected cell types and additionally for populations of purified virions from infected cells.

Key words Pseudouridine, RNA, Epitranscriptomics, Modifications, Bisulfite, HIV-1, Virus, Mapping

1 Introduction

The RNA modification landscape is becoming increasingly studied, as the role of these modifications in the post-transcriptional regulation of cellular mRNAs becomes more apparent. Indeed, the last few years have seen a wave of studies exploring the roles of RNA modifications including m^6A [1], m^5C [2], Ψ [3], N_m [4], and ac^4C [5] specifically on cellular mRNAs. This has started to spill over into the field of virology, where researchers have speculated over whether viral RNAs, which are being transcribed and trafficked within the same cellular context, may also be modified by the same modification machinery that associates so readily with cellular mRNA transcripts. Indeed, a wide range of both RNA and DNA viruses have now been shown to have modified transcripts in infected cells [6–16].

If there is an advantage to an RNA modification in a particular position over an unmodified base, it is entirely probable that the genome of rapidly evolving viruses such as HIV-1 will have

Vinayaka R. Prasad and Ganjam V. Kalpana (eds.), *HIV Protocols*, Methods in Molecular Biology, vol. 2807,
https://doi.org/10.1007/978-1-0716-3862-0_16,

explored this evolutionary niche and fixed mutations in place to allow this advantageous modification to occur [17]. This is possibly the case for N_m modifications on the HIV-1 genome, whose presence appears to assist in the evasion of innate immune recognition [18]. However, although we are still unsure of the exact role of a good number of these modified nucleotides, and with the caveat that some appear to be context-dependent within a transcript [19], the first step in this exploration would be to identify the exact nucleotide residues that are modified.

To this end, a number of groups have developed single-nucleotide resolution mapping techniques, such as MAZTER-seq for m^6A [20], BS-seq for m^5C [21], and RBS-seq for Ψ [22]. Recently, two groups published very similar methods utilizing bisulfite sequencing at a neutral pH to specifically map Ψ through inducing an RT deletion signature when a Ψ modification is present [23, 24]. Pseudouridine has previously been shown to be present on HIV-1 transcripts, through the utilization of an antibody-based method with a relatively large footprint [25]. However, as yet there is no peer-reviewed research examining what role these modifications could be playing in the HIV-1, or indeed any other viral, replication cycle. Herein, we describe an adapted protocol (Fig. 1) for the identification of Ψ residues on both viral and cellular transcripts of an HIV-1 infected T cell line at single nucleotide resolution.

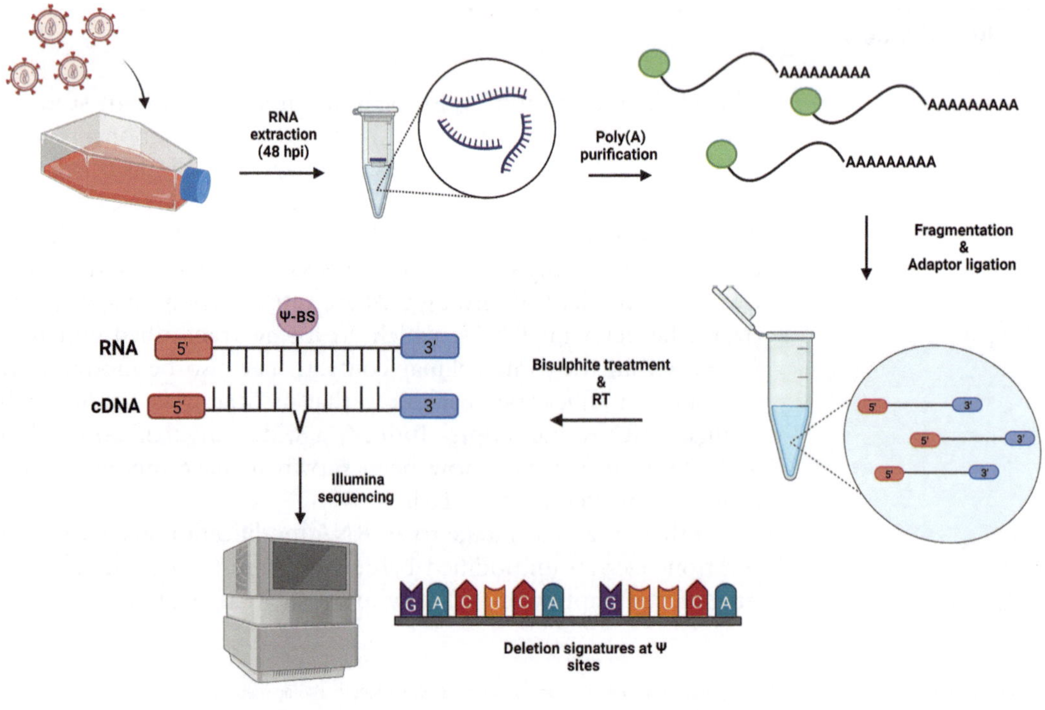

Fig. 1 Schematic of pseudouridine sequencing from HIV-1 infected cells

2 Materials

2.1 General Laboratory Equipment

1. Humidified CO_2 laboratory incubator.
2. Centrifuges, both refrigerated and non-refrigerated, capable of spinning 1.5 mL Eppendorf tubes and 50 mL falcon tubes.
3. BSL-2 tissue culture facility.
4. Thermal cycler.

2.2 Cell Culture and Media

1. CEM-SS cells.
2. 293T cells.
3. Roswell Park Memorial Institute (RPMI) medium, 10% heat-inactivated foetal bovine serum, 5% penicillin-streptomycin.
4. DMEM, 10% heat-inactivated foetal bovine serum, 5% penicillin-streptomycin.
5. Phosphate-buffered saline (PBS), pH 7.4.

2.3 Tissue Culture

1. 15 cm sterile tissue culture dishes or sterile T175 tissue culture flasks.
2. Sterile 1.5 mL Eppendorf tubes.
3. 15 mL falcon tubes.

2.4 Kits

1. Zymo Oligo Clean and Concentrator.
2. Zymo EZ RNA Methylation™ kit.
3. Invitrogen™ Poly(A)Purist™ MAG kit.
4. Invitrogen™ Fragmentation Reagents.
5. New England BioLabs® NEBNext® Small RNA Prep Set for Illumina® (Multiplex compatible).
6. Thermo Scientific™ Maxima H Minus First Strand cDNA synthesis kit.

2.5 Reagents

1. Magnetic microcentrifuge tube rack.
2. Illumina barcoded index primer(s).
3. 6% TBE gel.
4. TBE buffer: 20 g Tris, 10 g Boric acid, 8 mL 0.5 M EDTA, made up to 1 L.
5. Bisulfite buffer: 270 mg of sodium sulfite, 34 mg of sodium bisulfite in 900 μL of nuclease-free water.
6. T4 Polynucleotide kinase and the 10X Polynucleotide kinase (PNK) buffer provided with the enzyme.

7. 10 mM ATP.
8. 100% ethanol.
9. 70% ethanol.
10. Nuclease-free water.

3 Method

3.1 HIV-1 Infection and RNA Harvest

1. Culture CEM-SS cells in RPMI culture media and 293T cells in DMEM culture media at 5% CO_2 and 37 °C.
2. Seed two 15 cm dishes of 293T cells at a density of 2×10^6 cells/dish 24 h prior to transfection.
3. On the day of transfection, transfect 15 μg of HIV-1 molecular clone (pNL4-3) into 293T cells using PEI as a transfection reagent (*see* **Note 1**).
4. At 24 h post transfection, change the DMEM culture media for 15 mL RPMI culture media on the transfected 293T cells. Incubate for an additional 48 h.
5. Harvest conditioned media containing HIV-1 virions from 293T cells, and centrifuge at 4000× *g* for 10 min to pellet cell debris.
6. Pass the supernatant through a 0.45 μM filter.
7. Harvest CEM-SS cells and infect by resuspending 2×10^6 cells in 15 mL of HIV-1 containing filtered conditioned media from **step 6** in a T75 flask. Resuspend a further 2×10^6 cells in the remaining 15 mL of HIV-1-containing media. We generally do not calculate virus titer or MOI at this stage.
8. Incubate CEM-SS cells for 48 h in infectious media.
9. At 48 hpi, remove the cells and centrifuge at 4000× *g* for 5 min. Remove the spent media, and wash the cell pellet in with 5 mL of ice-cold PBS. Repeat this step for another wash. After removing the PBS, add 5 mL of Trizol™ to each pellet and resuspend the cell vigorously.
10. Combine each 5 mL of Trizol™ into a total of 10 mL in a falcon tube.
11. Store at −20 °C for later use or proceed with the extraction (*see* **Note 2**).

3.2 RNA Extraction

1. Remove the Trizol™ lysate from −20 °C and allow to thaw. Extract the RNA following the manufacturer's protocol, as follows.
2. Add 2 mL of chloroform to the RNA extraction. Mix well by shaking.

3. Centrifuge for 15 min at 12,000× *g*, 4 °C. The mixture will separate into an upper colorless aqueous phase, a white interphase, and a red lower phenol-chloroform phase.
4. Carefully transfer the RNA containing aqueous phase to a separate 15 mL falcon tube. It is important to transfer as much of the aqueous phase as possible without collecting the white interphase. This is best accomplished by tilting the pipette tip at a 45° angle during collection (*see* **Note 3**).
5. If the starting sample is small (<10^6 cells or <10 mg of tissue), add 50–10 μg of RNAse-free glycogen as a carrier to the aqueous phase. Add an equal volume of isopropanol to the RNA mixture (i.e. 5 mL of isopropanol for 5 mL of aqueous phase). Allow the sample to precipitate at −80 °C until the sample is completely frozen.
6. Transfer the precipitation from the freezer to the centrifuge and spin for 10 min at 12,000× *g*, 4 °C.
7. The RNA will form a white pellet at the bottom of the tube. Carefully discard the supernatant from the tube. Resuspend the pellet in 150 μL of 70% ethanol for every 500 uL of Trizol™. Briefly vortex to resuspend. Centrifuge for 5 min at 12,000× *g*, 4 °C. Wash twice in 70% ethanol.
8. After the second wash, discard the supernatant. Carefully remove the remaining ethanol with a micropipettor. Ensure the pellet is completely dry before resuspending in molecular-grade water, so that the final concentration is 600 ng/μL (*see* **Note 2**).

3.3 Poly(A) Purification

1. All solutions and reagents referred to in this section are supplied by the Invitrogen™ Poly(A)Purist™ MAG kit. Begin with 5–10 μg of total RNA.
2. Dilute the RNA to 600 ng/μL and mix 1:1 with the 2X Binding solution.
3. Gently resuspend the oligo dT beads (10 mg/mL) by inverting the tube and flicking. Do not vortex. Aliquot 2 μL of beads for every 10 μg of RNA. Allow the beads to sit on the magnetic rack for 2 min.
4. Wash the beads twice in 500 μL (per mg of RNA) of wash solution I. To wash the beads, gently add the solution to the tube, place the tube back on the rack, and then rotate the tube 180°. The beads will migrate to the magnet. Rotate 180° again.
5. Add the RNA to the beads. Invert to resuspend. Denature for 5 min between 65 and 75 °C.
6. After denaturing, rock the RNA for 30–60 min at room temperature. After incubation, allow the tube to incubate on the magnetic rack for 2 min.

7. Preheat the RNA Storage solution between 60 and 80 °C.
8. Wash the beads twice with wash solution I. The wash volume is the original RNA volume. Flick or invert the tube. Mix well. Gently spin down if any beads collect in the lid of the tube. Add the tube back to the magnetic rack and allow for the tube to site for 2 min for each wash.
9. Wash the beads twice with wash solution II. The wash volume is the original RNA volume. Flick or invert the tube. Mix well. Gently spin down if any beads collect in the lid of the tube.
10. Add 200 μL of the pre-warmed RNA Storage solution to the beads. Add with force to wash the beads off the wall of the tube. Flick or invert the tube. Mix well. Gently spin down if any beads collect in the lid of the tube. Allow the tube to sit on the magnetic rack for 2 min.
11. Transfer the 200 μL of elute to a new tube.
12. Repeat the elution process by adding another 200 μL of the RNA Storage solution to the beads. Add with force to wash the beads off the wall of the tube. Flick or invert the tube. Mix well. Gently spin down if any beads collect in the lid of the tube. Allow the tube to sit on the magnetic rack for 2 min.
13. Collect the eluate and add to the elution tube.
14. To the elution tube, add 40 μL of NH_4Ac, 1 μL of glycogen, and 1.1 mL of 100% ethanol. Allow to precipitate at −80 °C until the sample is frozen.
15. Remove sample and allow to thaw. Centrifuge the sample at 12,000× *g* for 30 min at 4 °C. Discard the supernatant.
16. Wash the pellet twice with 70% ethanol. For each wash, centrifuge at 12,000× *g* for 10 min at 4 °C.
17. Discard the ethanol. Carefully remove any remaining ethanol with a micropipettor.
18. Allow the pellet to air dry for 5 min.
19. Resuspend the pellet in 10–30 μL nuclease-free water. Store at −80 °C for later use or place on ice if immediately proceeding to the next step (*see* **Note 2**).

3.4 RNA Fragmentation and Bisulfite Treatment

1. Resuspend 250 ng of poly(A) RNA in 45 μL of nuclease-free water.
2. Fragment the RNA with 5 μL of 10X RNA fragmentation buffer at 94 °C for 4 min. Immediately add 5 μL of stop solution and place on ice.
3. Clean up the RNA with the Oligo Clean and Concentrator Kit.
4. Add 100 μL of oligo-binding buffer to a 50 μL sample.
5. Add 400 μL of ethanol (95–100%) and mix briefly by pipetting. Transfer the mix to the Zymo spin column in a collection tube.

6. Centrifuge at ≥10,000× *g* for 30 s. Discard the flow-through.
7. Add 750 μL of DNA wash buffer to the column. Centrifuge at ≥10,000× *g* for 30 s and discard the flow-through. Centrifuge at maximum speed for 1 min to collect any remaining wash buffer.
8. Transfer the column to a fresh microcentrifuge tube. Add 10 μL of nuclease-free water directly to the column matrix. Allow to water to incubate for 2 min. Then, centrifuge at ≥10,000× *g* for 30 s to elute the oligonucleotide.
9. At this stage, remove 1.5 μL of fragmented RNA and bring up to a total of 24 μL in nuclease-free water, label as "INPUT" and proceed to Subheading 3.5 (End Repair) (*see* **Note 2**).
10. Prepare *fresh* bisulfite buffer (270 mg of Na_2SO_3 and 34 mg of $NaHSO_4$ in 900 μL of nuclease-free water) (*see* **Note 4**).
11. Mix the 8.5 μL of remaining RNA with 45 μL of the freshly prepared bisulfite buffer and incubate to 70 °C in a thermal cycler for 4 h.
12. After incubation, add 100 μL of nuclease-free water to the bisulfite-treated RNA.
13. Add 200 μL of RNA-binding buffer from the Zymo EZ RNA Methylation kit.
14. Place a fresh Zymo spin column into a collection tube. Load the sample into the column. Add 400 μL of 100% ethanol to the column. Invert to mix.
15. Spin at >12,000× *g* for 30 s.
16. Discard the flow-through and wash with 200 μL of wash buffer. Spin at >12,000× *g* for 30 s.
17. Discard the flow-through. Add 200 μL of desulfonation butter to the column and allow to incubate at room temperature for 1 h.
18. After incubation, spin at >12,000× *g* for 30 s. Discard the flowthrough.
19. Wash twice with 400 μL RNA wash buffer. Spin at >12,000× *g* for 30 s for each wash. After the final wash, spin for 2 min.
20. Discard the remaining flow-through. Elute the RNA by adding 24 μL of nuclease-free water directly to the matrix of the column. Proceed to end repair (*see* **Note 2**).

3.5 End Repair

1. Proceed with "INPUT" and treated RNA samples as separate reactions throughout.
2. Perform 3′ end repair by adding 3 μL of 10X PNK buffer and 3 μL of T4 PNK. Incubate at 37 °C in a thermal cycler for 45 min.

3. Perform 5′ phosphorylation by adding a further 1.5 μL of T4 PNK and 1.5 μL of 10 mM ATP. Incubate at 37 °C in a thermal cycler for 45 min.
4. Clean up the RNA with the Zymo Clean and Concentrator kit and elute in 6 μL of nuclease-free water (*see* Subheading 3.4, **steps 4–8**).

3.6 NGS Prep

3.6.1 Ligation of the 3′ SR Adaptor (See Note 5)

1. Mix the following components in a sterile nuclease-free PCR tube. It is okay to premix reagents. Use immediately.

Component	Volume
Input RNA	6 μL
(green) 3′ SR adaptor for Illumina	1 μL
Nuclease-free water	Variable
Total volume	7 μL

2. Incubate the sample in a preheated thermal cycler for 2 min at 70 °C. Immediately transfer the tube to ice.
3. Add and mix the following reagents. It is okay to premix. Use immediately.

Component	Volume
(green) 3′ ligation reaction buffer	10 μL
(green) 3′ ligation enzyme mix	3 μL
Total volume	20 μL

4. Incubate for 1 h at 25 °C.

3.6.2 Hybridization of the Reverse Transcription Primer (See Note 6)

1. Add and mix the following regents to the ligation mixture from Subheading 3.6.1, **step 4**. Mix well. It is okay to premix the reagents.

Component	Volume
Nuclease-free water	4.5 μL
(pink) SR RT primer for Illumina	1 μL
Total volume	25.5 μL

2. Place in a thermal cycler with a heated lid set to >85 °C and run the following program:

   5 minutes at 75 °C

   15 minutes at 37 °C

   15 minutes at 25 °C

   Hold at 4 °C

3.6.3 Ligation of the 5′ SR Adaptor (See Note 7)

1. With 5 min remaining, resuspend the (yellow) 5′ SR adaptor in 120 μL of nuclease-free water.
2. Aliquot the (yellow) 5′ SR adaptor into a separate, nuclease-free PCR 200 μL tube, for the number of samples in the experiment plus an excess of 10%.
3. Incubate the adaptor in the thermal cycler at 70 °C for 2 min and then immediately place the tube on ice. Keep the tube on ice and use the denatured adaptor within 30 min of denaturation.
4. Add and mix the following components to the ligation mixture from Subheading 3.6.2, **step 2** and mix well. *Do not premix the reagents.*

Component	Volume
(yellow) 5′ SR adaptor for Illumina (denatured)	1 μL
(yellow) 5′ ligation reaction buffer	1 μL
(yellow) ligation enzyme mix	2.5 μL
Total volume	30 μL

5. Incubate for 1 h at 25 °C in a thermal cycler.
6. Clean up the RNA with the Zymo Clean and Concentrator kit and elute in 10 μL of nuclease-free water (*see* Subheading 3.4, **steps 4–9**).

3.7 Reverse Transcription

1. Set up the reverse transcription mixture by adding 1 μL of (pink) SR RT primer, 1 μL of dNTPs, 4 μL of RT buffer,1 μL of Maxima H RT enzyme, and 3 μL of nuclease-free water.
2. Perform the reverse transcription by running the following program:

   25 °C for 10 min

   50 °C for 30 min

   85 °C for 5 min

3. Store at −20 °C for later use or proceed to library PCR amplification (*see* **Note 2**).

3.8 Library PCR Amplification

1. Set up the following reaction in a sterile, 200 μL nuclease-free PCR tube.

   50 μL of 2X GoTaq Mastermix

   36 μL of nuclease-free water

   10 μL of cDNA

   2 μL of (blue) SR primer

   2 μL of barcode index primer

2. Run the following programme for 20 cycles (*see* **Note 2**).

   94 °C for 30 s for 1 cycle

   94 °C for 15 s
62 °C for 15 s
70 °C for 15 s — 20 cycles

   70 °C for 1 min for 1 cycle

   12 °C hold

3.9 Fragment Visualization and Overnight Digestion

1. Run 45 μL (not sure how much those TBE wells will hold) of the PCR product on a 6% TBE gel at 100 V for 1.5 h to visualize the fragments. Use a 50 bp ladder as a reference.
2. Image the gel to check that the fragment is the correct size (*see* **Note 8**).
3. Using a scalpel, *carefully* cut an X into the bottom of a 0.5 mL microcentrifuge tube. Carefully slice the 250 bp fragment from the gel and place the gel slice in the 0.5 mL tube. Place the 0.5 mL tube containing the gel slice in 1.5 mL microcentrifuge tube.
4. Centrifuge at >12,000× *g* for 5 min. It is important to ensure all the gel slice has gone into the 1.5 mL tube in order to increase the end concentration of the extracted RNA. Spin until all the gel has passed through to the 1.5 mL tube. Dispose of the 0.5 mL tube.
5. Add 500 μL of 0.3 M NaCl to the shredded gel slice. Place on a microcentrifuge shaker and allow to shake at 800 rpm overnight.
6. Centrifuge at >12,000× *g* for 10 min.

7. Using a thin-tipped gel loading tip, carefully pipette as much of the liquid into a fresh 2 mL tube. Dispose of the 1.5 mL microcentrifuge tube containing the gel once all the liquid is removed (*see* **Note 9**).
8. To the 2 mL tube, add 2.5× the volume of 100% ethanol and 1 μL of a glycogen co-precipitant. Mix well.
9. Allow the sample to precipitate at −80 °C until frozen. Once frozen, remove the sample and allow to thaw at room temperature.
10. Centrifuge at >12,000× *g* for 15 min, 4 °C.
11. Wash twice with 500 μL of 70% ethanol.
12. After the final wash, carefully remove any remaining ethanol and allow the pellet to air dry.
13. Resuspend the pellet in 12 μL of nuclease-free water. Quantify the RNA by Nanodrop spectrophotometer.
14. Store at −80 °C.

3.10 NGS Sequencing and Data Analysis

1. Sequence libraries on an Ilumina NGS sequencer, such as a NovaSeq 6000 or a NextSeq 2000.
2. Sequencing data can then be analyzed using standard trimming, alignment, and quantitative processes. Specifically, data analysis described by Dai et al. (2023), may be employed for any BID-seq dataset.
3. To be considered a valid Ψ candidate site, we utilize a number of exclusion criteria, which include: (1) Total coverage greater than 20 reads in both the INPUT and treated samples; (2) A deletion rate above 5% in treated samples; (3) A deletion rate below 1% in INPUT samples. Treated and input samples can be validated by aligning to ribosomal RNAs, such as 18 s, which have known and defined Ψ modifications.

Data analysis can be performed using any sequence aligner. Generally, we utilize software including AdapterRemoval to trim reads [26], Bowtie2 for read alignments [27], and Samtools for data processing and visualization [28]. We recommend aligning to the human genome first to remove any human reads, prior to alignment to your viral genome of choice. Either alignment file can be used for downstream data analysis described by Dai et al. (2023), depending on whether the user is interested in host or viral pseudouridine identification, respectively.

4 Notes

1. In reference to methods Subheading 3.1, **step 3**: avoid freeze-thawing the PEI transfection reagent too many times. To enhance transfection efficiency, thaw a fresh aliquot the day of.
2. There are several stopping points throughout this protocol. The stopping points are as follows: Subheadings 3.1, **step 11**, 3.2, **step 8**, 3.3, **step 19**, 3.4, **steps 9** and **20**, 3.7, **step 3**, and 3.8, **step 2**.
3. In reference to methods Subheading 3.2, **step 4**: it is important to avoid taking up the organic phase when taking off the aqueous phase, which is why we suggest holding the micropipette at a 45° angle. If some of the organic phase is accidentally taken up, *slowly* eject the liquid back into the tube and centrifuge the sample again for 5 min at 12,000× *g*, 4 ° C.
4. In reference to methods Subheading 3.4, **step 10**: the amounts of the sodium sulfite and sodium bisulfdite are quite small and can be difficult to precisely measure out without a well-calibrated scale. To get around this, a larger volume of the buffer can be made up. For example, 2.7 g of sodium sulfate and 0.34 g of sodium bisulfite in 9 mL of nuclease-free water would suffice.
5. In reference to methods Subheading 3.6.1: for total RNA inputs closer to 100 ng, dilute the (green) 3′ SR adaptor for Illumina 1:2 in nuclease-free water.
6. In reference to methods Subheading 3.6.2: for total RNA inputs closer to 100 ng, dilute the (pink) SR RT Primer for Illumina 1:2 in nuclease-free water.
7. In reference to methods Subheading 3.6.3: for total RNA inputs closer to 100 ng, additionally dilute the (yellow) 5′ SR adaptor for Illumina 1:2 in nuclease-free water.
8. In reference to methods Subheading 3.9, **step 2**: primer-dimer bands will appear between 100 and 150 bp. Fragments containing your NGS library should be visible around 250 bp.
9. In reference to methods Subheading 3.9, **step 7**: you may need to spin multiple times as the gel may come undone from the bottom of the microcentrifuge tube. You should expect to take off between 250 and 300 μL of liquid.

Acknowledgments

This research was funded in part by an ERC-STG grant, PTFLU 949506 D.G.C. This research received infrastructure support from the Wellcome-Wolfson Institute for Experimental Medicine at Queen's University Belfast.

References

1. Dominissini D, Moshitch-Moshkovitz S, Schwartz S et al (2012) Topology of the human and mouse m6A RNA methylomes revealed by m6A-seq. Nature 485:201–206
2. Edelheit S, Schwartz S, Mumbach MR et al (2013) Transcriptome-wide mapping of 5-methylcytidine RNA modifications in bacteria, archaea, and yeast reveals m5C within archaeal mRNAs. PLoS Genet 9(6):e1003602
3. Schwartz S, Bernstein DA, Mumbach MR et al (2014) Transcriptome-wide mapping reveals widespread dynamic-regulated pseudouridylation of ncRNA and mRNA. Cell 159:148–162
4. Marchand V, Blanloeil-Oillo F, Helm M et al (2016) Illumina-based RiboMethSeq approach for mapping of 2′-O-Me residues in RNA. Nucleic Acids Res 44:e135
5. Arango D, Sturgill D, Alhusaini N et al (2018) Acetylation of cytidine in mRNA promotes translation efficiency. Cell 175(7):1872–1886.e24
6. McIntyre W, Netzband R, Bonenfant G et al (2018) Positive-sense RNA viruses reveal the complexity and dynamics of the cellular and viral epitranscriptomes during infection. Nucleic Acids Res 46:5776–5791
7. Tsai K, Courtney DG, Cullen BR (2018) Addition of m6A to SV40 late mRNAs enhances viral structural gene expression and replication. PLoS Pathog 14:e1006919
8. Courtney DG, Chalem A, Bogerd HP et al (2019) Extensive epitranscriptomic methylation of A and C residues on murine leukemia virus transcripts enhances viral gene expression. MBio 10:e01209–e01219
9. Courtney DG, Kennedy EM, Dumm RE et al (2017) Epitranscriptomic enhancement of influenza A virus gene expression and replication. Cell Host Microbe 22:377–386
10. Courtney DG, Tsai K, Bogerd HP et al (2019) Epitranscriptomic addition of m5C to HIV-1 transcripts regulates viral gene expression. Cell Host Microbe 26:217–227
11. Gokhale NS, McIntyre ABR, McFadden MJ et al (2016) N6-methyladenosine in flaviviridae viral RNA genomes regulates infection. Cell Host Microbe 20:654–665
12. Price AM, Hayer KE, McIntyre ABR et al (2020) Direct RNA sequencing reveals m6A modifications on adenovirus RNA are necessary for efficient splicing. Nat Commun 11:1–17
13. Kennedy EM, Bogerd HP, Kornepati AVR et al (2016) Posttranscriptional m6A editing of HIV-1 mRNAs enhances viral gene expression. Cell Host Microbe 19:675–685
14. Kim D, Lee JY, Yang JS et al (2020) The architecture of SARS-CoV-2 transcriptome. Cell 181:1–29
15. Tsai K, Jaguva Vasudevan AA, Martinez Campos C et al (2020) Acetylation of cytidine residues boosts HIV-1 gene expression by increasing viral RNA stability. Cell Host Microbe 28(2):306–312.e6
16. Xue M, Zhao BS, Zhang Z et al (2019) Viral N 6-methyladenosine upregulates replication and pathogenesis of human respiratory syncytial virus. Nat Commun 10:1–18
17. Courtney DG (2021) Post-transcriptional regulation of viral RNA through epitranscriptional modification. Cells 10:1129
18. Ringeard M, Marchand V, Decroly E et al (2019) FTSJ3 is an RNA 2′-O-methyltransferase recruited by HIV to avoid innate immune sensing. Nature 565:500–504
19. Shi H, Wei J, He C (2019) Where, when, and how: context-dependent functions of rna methylation writers, readers, and erasers. Mol Cell 74(4):640–650
20. Garcia-Campos MA, Edelheit S, Toth U et al (2019) Deciphering the "m6A code" via antibody-independent quantitative profiling. Cell 178:731–747.e16
21. Schaefer M, Pollex T, Hanna K et al (2009) RNA cytosine methylation analysis by bisulfite sequencing. Nucleic Acids Res 37:e12
22. Khoddami V, Yerra A, Mosbruger TL et al (2019) Transcriptome-wide profiling of multiple RNA modifications simultaneously at single-base resolution. Proc Natl Acad Sci USA 116:6784–6789
23. Dai Q, Zhang LS, Sun HL et al (2023) Quantitative sequencing using BID-seq uncovers abundant pseudouridines in mammalian mRNA at base resolution. Nat Biotechnol 41:344–354
24. Zhang M, Jiang Z, Ma Y et al (2023) Quantitative profiling of pseudouridylation landscape in the human transcriptome. Nat Chem Biol 19(10):1–11

25. Campos CM, Tsai K, Courtney DG et al (2021) Mapping of pseudouridine residues on viral and cellular transcripts using a novel antibody-based technique. RNA 27(11): rna.078940.121
26. Dickson RJ, Gloor GB (2013) XORRO: rapid paired-end read overlapper. arxiv:13044620 9: 1–7
27. Li H, Handsaker B, Wysoker A et al (2009) The sequence alignment/map format and SAMtools. Bioinformatics 25:2078–2079
28. Langmead B, Salzberg SL (2012) Fast gapped-read alignment with Bowtie 2. Nat Methods 9: 357

Part V

Host-Virus Interactions

Chapter 17

Proximity Ligation Assay to Detect the Proximity Between Host Proteins and Viral Proteins of HIV-1

Sutanuka Chakraborty, Shobith Suresh, Hrimkar Buch, Arun Panchapakesan, and Udaykumar Ranga

Abstract

The study of host-pathogen interaction often requires interrogating the protein-protein interactions and examining post-translational modifications of the proteins. Traditional protein detection strategies are limited in their sensitivity, specificity, and multiplexing capabilities. The Proximity Ligation Assay (PLA), a versatile and powerful molecular technique, can overcome these limitations. PLA blends the specificity of antibodies, two antibodies detecting two different epitopes on the same or two different proteins, with the amplification efficiency of a polymerase to allow highly specific and sensitive detection of low-abundant proteins, protein-protein interactions, or protein modifications. In this protocol, we describe the application of PLA to detect the proximity between HIV-1 Tat with one of its cellular partners, p65, in an infected host cell. The protocol could be applied to any other context with slight modifications. Of note, PLA can only confirm the physical proximity between two epitopes or proteins; however, the proximity need not necessarily allude to the functional interaction between the two proteins.

Key words HIV-1, Tat, Proximity ligation assay, NF-κB, p65, Latency

1 Introduction

The detection and analysis of proteins and protein-protein interactions are crucial in biological research. Traditional methods for protein detection, such as Western blotting, ELISA, and Immunofluorescence, have been widely employed by various researchers over the years. However, these techniques are limited in sensitivity, specificity, and multiplexing capabilities. The Proximity Ligation Assay (PLA) is a powerful molecular technique that overcomes these limitations and provides a sensitive and specific approach to detect proteins or protein-protein interactions. PLA combines the specificity of antibodies with the amplification efficiency of a polymerase, thus surpassing the detection sensitivity of conventional immune-diagnostic assays by several orders of magnitude [1].

Vinayaka R. Prasad and Ganjam V. Kalpana (eds.), *HIV Protocols*, Methods in Molecular Biology, vol. 2807,
https://doi.org/10.1007/978-1-0716-3862-0_17,

The PLA technique employs two primary antibodies specific to the target molecule(s) of interest. These primary antibodies can be directed against two non-overlapping epitopes on the same protein, two epitopes on two different interacting proteins, or a post-translational modification and a different epitope on the same protein (Fig. 1). The technique can therefore visualize and quantify the spatial organization of proteins, protein-protein interactions, and protein modifications, respectively. Importantly, the primary antibodies must be raised in two different hosts (mouse, rabbit, goat, etc.). The two primary antibodies bound to the cognate epitopes on the protein(s) are interrogated using two secondary antibodies, each conjugated with oligonucleotides—one in the sense and the other in the anti-sense orientation. Subsequently, the two oligonucleotides linked to the two secondary antibodies will form a circular DNA molecule in the presence of the two connector oligonucleotides, following a ligation event. Importantly, the hybridization of the four oligonucleotides into an amplification DNA circle will be possible only when the two primary antibodies are in physical proximity (<40 nm), not when they are bound to distant epitopes [2]. Thus, the specificity of PLA is governed by the simultaneous binding of two independent antibodies to two cognate epitopes and by the need for the two epitopes to be in physical proximity.

The newly formed DNA circle will serve as the template for amplification (Fig. 1b). The 3′ ends of the antibody-conjugated oligonucleotides serve as primers to copy the circular DNA template by rolling circle amplification (RCA) [3]. RCA, a powerful isothermal amplification mediated by certain polymerases, such as phi29 polymerase, copies a circular DNA or RNA template several thousand times to create concatemer forms. The several thousand copies of concatemerized forms of the copied DNA are immobilized at the location of antibody binding, serving as a target for hybridization of a fluorescently labeled oligonucleotide probe (Fig. 2). Because several hundreds of fluorescent probe molecules are localized to the site of antibody binding, the fluorescent intensity of the signal is expected to be significantly high, detectable by a regular fluorescence microscope. Thus, the amplification power of the RCA, appended with the antibody sensitivity, makes PLA one of the most sensitive and specific immune techniques to detect proteins and protein-protein interactions in the cell, in addition to the versatility and simplicity of the technique.

The numerous advantages of PLA contribute to its widespread use in HIV-1 research and other fields of biology. First, PLA provides outstanding specificity, as it relies on the dual recognition of target molecules by specific antibodies and the proximity between the two molecules. This dual recognition ensures that only the desired targets are detected, minimizing the risk of false-positive results. Additionally, a combination of three antibodies has

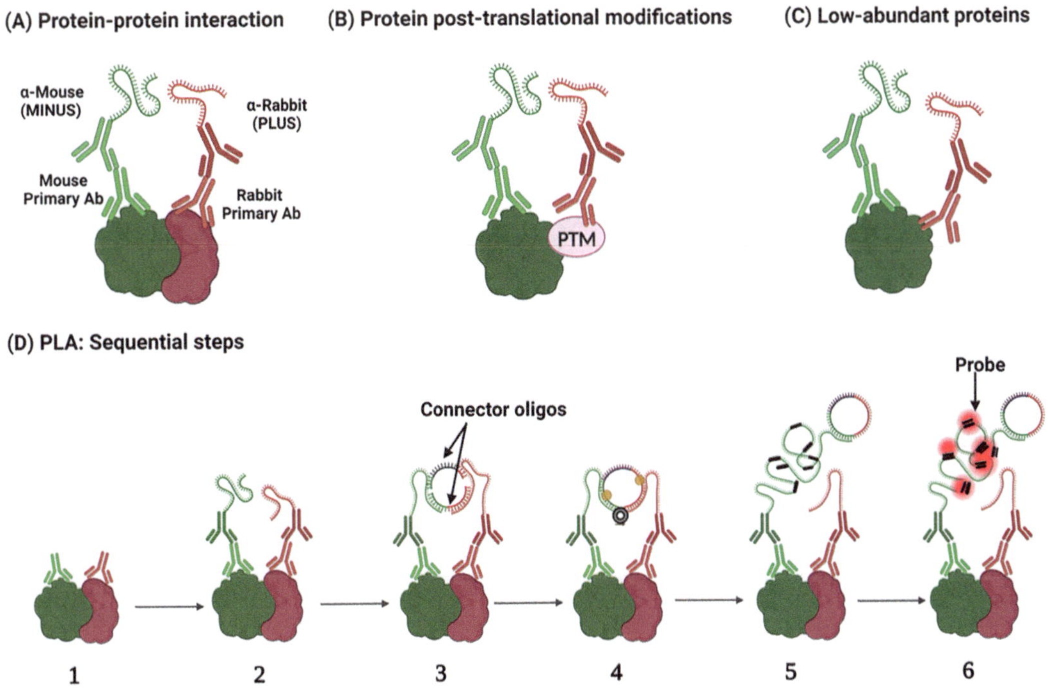

Fig. 1 General applications and sequential steps of the Proximity ligation assay. The PLA can identify (**a**) proximity between two proteins, (**b**) protein post-translation modifications, or (**c**) low-abundant proteins. (**d**) The sequential steps of a Proximity ligation assay. Step-1: The cells are permeabilized and incubated with two primary antibodies, which bind to the cognate epitopes. Step-2: The cells are incubated with the corresponding secondary antibodies, each conjugated with the corresponding single-stranded DNA oligonucleotide (MINUS or PLUS). Step-3: The cells are incubated with two additional connector oligonucleotides to form the open DNA circle for subsequent amplification. Step-4: Following the addition of the enzyme mix, the Ligase (yellow dots) will seal the gaps to create a covalently closed circle that will serve as a template for the polymerase (gray doughnut). Step-5: The closed circular DNA template undergoes rolling circle amplification to form multiple amplicon-concatemers that localize to the amplification site. Step-6: The cells are incubated with a single-stranded detector probe tagged with a fluorescent dye. Many copies of the probe hybridize with the amplified template developing bright fluorescent signal (the red specks) that can be detected using a fluorescence microscope or a flow cytometer. (Created with BioRender.com)

been reported to enhance the assay specificity further [4]. Further, multiple antibody pairs in a single PLA experiment can simultaneously detect multiple targets, providing valuable insights into complex molecular networks [5]. Second, PLA offers exceptional sensitivity, as it can detect individual protein-protein interactions and modifications within the complex mixture of proteins present in the cell. For example, PLA detected proteins present in a cell at a 1000-fold lower level than could a Western blot assay [6]. This high sensitivity enables the detection of low-abundance molecules to monitor subtle changes in molecular interactions that other techniques may fail to detect. Lastly, the most significant advantage of PLA is its versatility in accepting different sample types. The assay

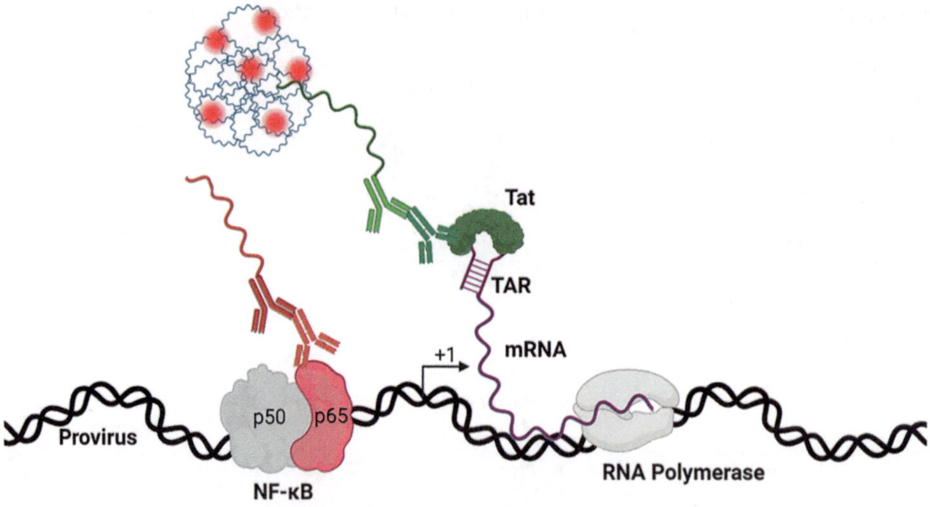

Fig. 2 PLA detects Tat-p65 proximity. p65 (pink) complexed with p50 is recruited to the NF-κB motif (only one motif has been shown for simplicity) on the viral DNA (black). Tat (green) is bound to the TAR element of the nascent viral transcript (purple). The PLA schematic is depicted in Fig. 1d. (Created with BioRender.com)

can be applied to various biological samples, including cultured cells, tissue sections, and body fluids. This flexibility permits the investigation of protein-protein interactions and protein modifications in the context of the native state of cells and tissues, providing a comprehensive understanding of biological processes.

In contrast to its numerous advantages, PLA also suffers from a few limitations. PLA requires a certain degree of optimization given the inherent nature of antibodies to cross react with non-specific proteins present in the cell. Choosing appropriate antibodies for the assay is crucial to minimise this problem. Commensurate with the enhanced detection sensitivity, PLA is also more expensive due to the quality of the reagents required. Further, PLA can work efficiently on an immobilized surface, such as adherent cells, whereas working with suspension cells is technically demanding [7]. Furthermore, PLA can only demonstrate the physical proximity between two proteins, not their functional interaction.

In this protocol, we present an important application of PLA—the proximity of Tat, a low-abundant HIV-1 protein, with p65, one of the cellular partners of Tat. We previously demonstrated two other applications of PLA that follow the same general protocol described here. In one of the applications, we confirmed the known interaction and colocalization of HIV-1 Gag with TSG101, a member of the cellular ESCRT pathway [8]. To this end, we exogenously expressed Gag and TSG101 by co-transfecting the expression vectors into host cells [9]. In a different application, we chromatinized the HIV-1 genome and allowed the virus to enter a state of transcriptional silence. Endogenous HIV-1 Tat is expected to be present in such cells at exceedingly low abundance. We demonstrated that PLA could detect Tat in an HIV-1 infected cell, particularly when the virus has entered latency, a condition that further limits the expression of Tat [10]. Of note, the suggested protocol may also be used for detecting post-translational modifications of HIV-1 proteins (exogenous or endogenous) without significant modifications.

2 Materials

2.1 Cell Culture, Transfection, Virus Production, Infection, and Sorting of Cells

1. Mammalian cells—HEK293, HEK293T, or HeLa cell lines.
2. Culture medium: Dulbecco's modified Eagle's medium (Sigma-Aldrich) supplemented with Fetal Bovine Serum to a final concentration of 10%.
3. Culture-ware: 12-well culture plate, 100 mm culture dish.
4. An HIV-based vector such as pcLGIT, containing an SV40 origin of replication, [11] co-expressing EGFP directly under the control of the LTR and the HIV-1C *tat* gene driven by an IRES element.
5. Third-generation lentiviral packaging vectors psPAX2 (NIH AIDS reagent program), pCMV-Rev (NIH AIDS reagent program), and pHEF-VSV-G (NIH AIDS reagent program), for generating pseudotyped viruses.
6. 2.5 M $CaCl_2$: Dissolve 36.8 g of $CaCl_2.2H_2O$ in ultrapure water and adjust the volume to 100 mL. Pass the solution through a 0.22 μm filter and store the solution in aliquots of 500 μL at −20 °C.
7. 1 M HEPES buffer (pH 7.1): Dissolve 23.83 g HEPES in 100 mL ultrapure water and adjust pH to 7.1 with 5 M NaOH. Pass through a 0.22 μm filter and store the buffer at 4 °C.
8. 2X HEPES buffered Saline (HBS): Dissolve 1.63 g NaCl and 21.294 mg Na_2HPO_4 in 10 mL ultrapure water and add 5 mL of 1 M HEPES pH 7.1. Adjust the volume to 100 mL, filter-sterilize through a 0.22 μm filter, and store the solution in aliquots of 500 μL at −20 °C.
9. DEAE-dextran: Dissolve 0.1 g of DEAE dextran in 10 mL of ultrapure water. Filter-sterilize and store in aliquots of 500 μL at −20 °C.
10. Sterile ultrapure water.

2.2 Immunostaining and PLA

1. Primary antibodies specific to Tat or p65: Rabbit polyclonal anti-Tat antibody (Abcam), and Mouse monoclonal anti-p65 antibody (Cell Signaling Technology).

2. Secondary antibodies conjugated to oligonucleotides: Anti-Mouse MINUS (Sigma-Aldrich) and Anti-Rabbit PLUS (Sigma-Aldrich).
3. DNA ligase (1 U/μL, Sigma-Aldrich).
4. DNA Polymerase (10 U/μL, Sigma Aldrich).
5. Permeabilization buffer: 0.1% Triton-X 100.
6. Fixing agent: 4% paraformaldehyde.
7. Duolink® Blocking Solution (Sigma-Aldrich).
8. Duolink® Antibody Diluent (Sigma-Aldrich).
9. Duolink® Wash buffers A and B (Sigma-Aldrich).
10. 5X Duolink® Ligation Buffer (Sigma-Aldrich).
11. 5X Duolink® Amplification Buffer (Sigma-Aldrich).
12. Duolink® Mounting Media with DAPI (Sigma-Aldrich).

2.3 Equipment

1. 8-well chambered glass slides.
2. Coverslips (18 mm circular, Blue Star).
3. Forceps, micropipettes, tips.
4. Hybridization chamber.
5. Flow cytometer (BD FACSAriaIII sorter, BD biosciences).
6. Confocal microscope with specific excitation lasers and emission filters.

3 Methods

The following protocol defines the physical proximity between the viral Tat protein recruited to the TAR element of the viral transcript and the host p65 protein bound to the NF-κB motifs in the LTR (*see* Fig. 2 for a schematic representation). The suggested experimental procedure is applicable to other proteins and other host cells, such as HEK293T or HeLa cells, with minor modifications.

3.1 Preparation and Titration of the HIV-1 Viral Stocks

1. Passage HEK293 or HEK293T cells in culture medium at a cell confluence of 50–60%. Incubate the cells at 37 °C in the presence of 5% CO_2 (*see* **Note 1**).
2. Prepare pseudotyped reporter LTR-GFP-IRES-Tat (LGIT) viruses in HEK293T cells in a 100 mm culture dish. To this end, prepare a plasmid DNA cocktail consisting of 10 μg of the reporter viral vector-LGIT, 5 μg psPAX2, 3.5 μg pHEF-VSVG, and 1.5 μg pCMV-Rev in 250 μL of sterile ultrapure water.
3. Substitute the spent DMEM from the culture dish an hour before transfection with 9 mL of fresh medium. Add 500 μL of 2X HBS and 250 μL of 2.5 M $CaCl_2$ solutions (*See* for recipes

in Materials) to 250 μL of plasmid DNA cocktail, mix well to homogeneity, and immediately apply the solution mix drop-wise to the cells.

4. Six hours post-transfection, replace the medium with complete DMEM. Harvest the culture supernatant, expected to contain the released virus, 48 h post-transfection, and pass the medium through a 0.22 μm filter. Store the viral stock in 1 mL aliquots in a deep freezer at −80 °C.
5. Estimate the relative infectious units (RIU) of the reporter viral stock by flow cytometry. To this end, infect HEK293 cells with a twofold dilution series of the viral stock ranging from 10 to 80-folds in a 96-well plate. After 48 h of incubation, estimate GFP expression using a flow cytometer and determine the percent fluorescent cells at each dilution. The viral stock dilution where approximately 10% of cells are GFP^{+ve} corresponds to a Multiplicity of Infection (MOI) value of 0.1. In other words, the viral stock at that dilution can infect 10% of the target cells under the given experimental conditions.

3.2 Infection and Sorting of Target Cells

1. Prepare a 100 mm culture dish or a 25 mL culture flask containing HEK293 cells at 50–60% confluency, where the total cell number could be approximately 5×10^6 cells. Dilute the viral stock to 0.1 MOI to make 5 mL of complete DMEM, supplemented with 25 μg/mL of DEAE-Dextran. DEAE-Dextran enhances viral infection of the target cells. After 6 h of infection, replace the DEAE-Dextran containing medium with complete DMEM and incubate the cells in a CO_2 incubator. Monitor the cells for GFP fluorescence using a fluorescence microscope.
2. Three days following infection, sort GFP^{+ve} cells using a flow sorter, such as BD FACSAriaIII. Cells falling within the median fluorescence range of ~5000–50,000 may be sorted (*see* **Notes 2–4**).
3. Suspend the sorted cells at a density of 5000 cells/mL in complete DMEM. Seed 200 μL of the cell suspension in each well of an 8-chambered glass slide. Each chamber of the slide would receive approximately 1000 sorted cells (*see* **Note 5**). Incubate the slide in a CO_2 incubator. The sorted cells may take up to 4 days to reach a 50–60% confluence. During incubation, approximately half of the cells may downregulate the expression of GFP, indicating latency establishment. Alternately, the cells may be directly sorted onto glass coverslips of 18 mm diameter immersed in the wells of a 12-well culture plate containing 500 μL of complete DMEM per well (Fig. 3b).

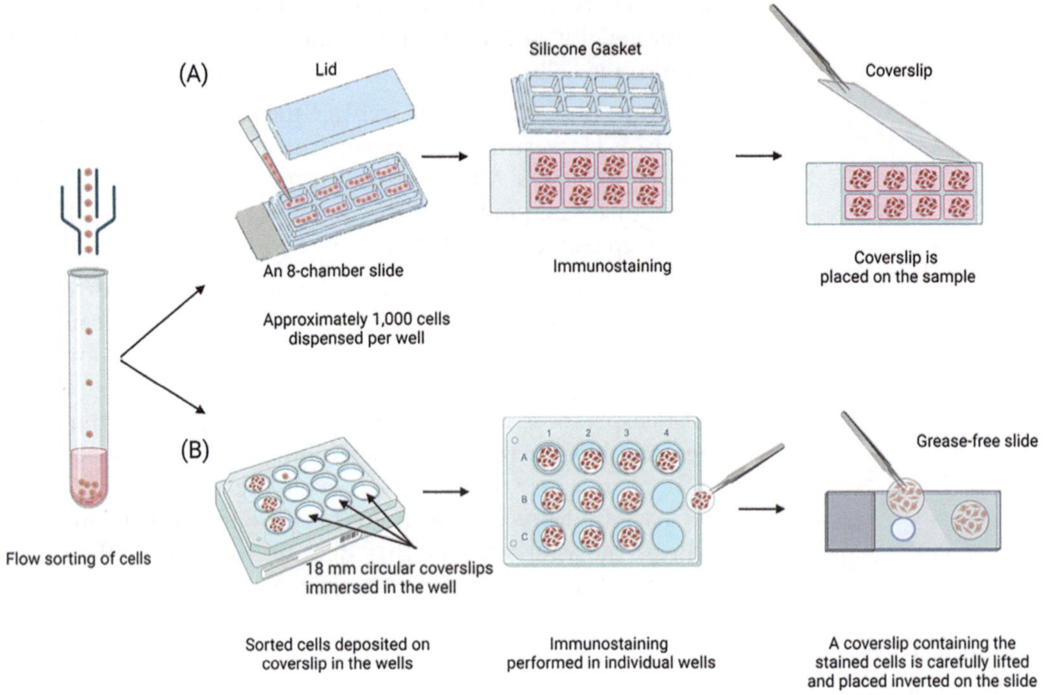

Fig. 3 Two ways of carrying out PLA assay. (**a**) 8-well-chamber slides. After fluorescent sorting, the cells are deposited in the chambers and stained as per the protocol. Before imaging, the silicone gasket is carefully peeled off, avoiding any cross-contamination between the adjacent samples. One drop of Mounting Medium with DAPI is added to each sample. Using a pair of flat-armed forceps, an 18 × 24 mm rectangular glass coverslip is placed on the slide to cover all the samples. Care must be taken not to trap air bubbles. Using a lint-free tissue paper, blot the excess Mounting Medium from the edges of the coverslip and seal the edges with nail polish. (**b**) A circular coverslip on which cells are grown. A circular glass coverslip of 18 mm diameter is suitable for a well in the 12-well culture plate. The coverslips are placed in each well, and the sorted cells suspended in 500 μL of complete DMEM are carefully deposited in each well. The plates are incubated until the cells reach 50–60% confluence. The spent medium is aspirated, cells are washed gently twice using 1X PBS, and processed for the PLA following the protocol described in the main text. A drop of the Mounting Medium with DAPI is placed on a clean, grease-free slide. The coverslip containing stained cells is lifted carefully from the well and placed on the Mounting Medium such that the cells are in direct contact with the Mounting Medium. Air bubbles must be avoided. Excess medium is removed from the edges and sealed with nail polish. (Created with BioRender.com)

3.3 Proximity Ligation Assay

The following immunostaining procedure is optimal for 8-chambered glass slides when the cells are 50–60% confluent.

1. Before the starting immunostaining, bring the stocks of Wash buffers A and B supplied in the kit to room temperature. Dilute the buffers to 1X solutions using ultrapure water. For long-term storage, keep the diluted wash buffers at 4 °C. Use freshly made wash buffers as far as possible since old wash buffers may contribute to non-specific signals.

2. On the day of immunostaining, carefully aspirate the culture medium from the chambers. Wash the cells twice with 200 μL of 1X PBS. Fix the cells with 4% paraformaldehyde for 20 min. Wash the cells three times using 200 μL of 1X PBS each time. Aspirate the buffer using a wide-bore tip to avoid cell loss. Perform all the steps of immunostaining at room temperature unless mentioned otherwise.
3. Permeabilize the fixed cells by adding 100 μL of 0.1% Triton-X-100 to cells dropwise carefully, not disturbing the monolayers. Incubate the cells for 10 min at room temperature. Aspirate the permeabilization solution and wash the cells by adding 1X PBS dropwise on the cells. Wash the cells thrice with PBS.
4. Vortex the vial containing the blocking reagent supplied with the kit to mix the contents. Add 200 μL of Blocking Solution to the cell layer in each chamber and incubate the samples for 1 h at 37 °C in a humidity chamber (*see* **Note 6**). Ensure that all the cells are sufficiently covered with the blocking reagent. Aspirate the blocking solution and tap the slide to remove excess Blocking Solution before adding the primary antibodies.
5. Prepare a mix of the rabbit polyclonal anti-Tat antibody and the mouse monoclonal anti-p65 antibody at 1:250 and 1:300 dilutions, respectively, using the PLA Blocking Solution (*see* **Note 7**). Deposit 80 μL of the antibody solution on the cell monolayer in each chamber. Cover the slide with the lid and incubate it in a humidity chamber at 37 °C for 1 h. The optimal dilution of the primary antibody must be determined empirically (*see* **Note 8**).
6. Remove the secondary antibody stocks from the freezer and bring them to room temperature. Note that one of the antibodies is conjugated to the PLUS oligonucleotide, whereas the other to the MINUS oligonucleotide. Mix the vials well and dilute the stocks fivefold using the Duolink® Antibody Diluent solution provided in the kit. For a single reaction of 80 μL, add 8 μL of each secondary antibody stock to 64 μL of the Duolink® Antibody Diluent. Prepare a sufficient volume of the antibody solution based on the number of total samples required.
7. Aspirate the primary antibodies and wash the samples twice with 200 μL of 1X Wash Buffer A for 5 min each time. Tap the slide to remove excess Wash Buffer (*see* **Note 9**).
8. Add 80 μL of the secondary antibody solution to each chamber and incubate the slide in a humidity chamber for 1 h at 37 °C.
9. Thaw the 5X Ligation Buffer to room temperature and mix the contents well (*see* **Note 10**). Dilute the buffer to 1X using ultrapure water; add 16 μL of the stock buffer to 64 μL of

water for a single reaction volume of 80 μL. Vortex the diluted buffer to make a homogeneous solution. Prepare a sufficient volume of the Ligation Buffer based on the total sample number. Add 2 μL of ligase (1 U/mL) per 80 μL of ligation buffer to prepare the Ligation Master mix.

10. Aspirate the secondary antibody solution from the samples and wash the cell monolayers twice with 200 μL of 1X Wash Buffer A for 5 min each time. Tap the slide to remove excess Wash Buffer.
11. Add 80 μL of the Ligation Master mix to each sample and incubate the slide in a humidity chamber at 37 °C for 30 min.
12. Thaw the 5X Amplification Buffer to room temperature in the dark. Dilute the 5X buffer to 1X using ultrapure water; add 16 μL of the buffer stock to 64 μL of water per a single reaction volume of 80 μL. Add 1 μL of polymerase per 80 μL of 1X Amplification Buffer to prepare the amplification master mix. Prepare the amplification master mix based on the number of samples. The Amplification Buffer also contains the fluorescent probe; therefore, the solution must be protected from light.
13. Aspirate the ligation solution from the wells and wash the samples twice with 200 μL of 1X Wash Buffer A for 5 min each time. Tap the slide to remove the excess wash buffer.
14. Add 80 μL of the amplification master mix to each chamber and incubate the slide in a humidity chamber at 37 °C for 100 min in the dark (*see* **Notes 11** and **12**).

3.4 Image Acquisition and Analysis

1. Aspirate the amplification master mix from the chambers. Wash the chambers twice with 200 μL of 1X Wash Buffer B for 10 min each time, protected from light.
2. Prepare 5 mL of 0.01X Wash Buffer B by diluting the buffer using ultrapure water. Wash the samples with 200 μL of 0.01X Wash Buffer B for 1 min in the dark. Aspirate excess wash buffer from each well.
3. Peel off the silicone gasket from the slide using a pair of forceps. Remove the gasket in one go and avoid cross-contamination between the adjacent samples (Fig. 3).
4. Add a drop of Mounting Medium with DAPI provided with the kit on each sample. Place a clean, grease-free 18 mm × 24 mm coverslip on the entire slide using a pair of flat forceps covering all the cell monolayers. Take care to avoid trapping air bubbles between the slide and the coverslip. Apply caution to avoid cross-contamination between adjacent monolayers. Soak excess Mounting Medium from the sides of the coverslip using sterile and lint-free tissue paper. Seal the edges of the coverslip with clear nail polish (*see* Fig. 3).

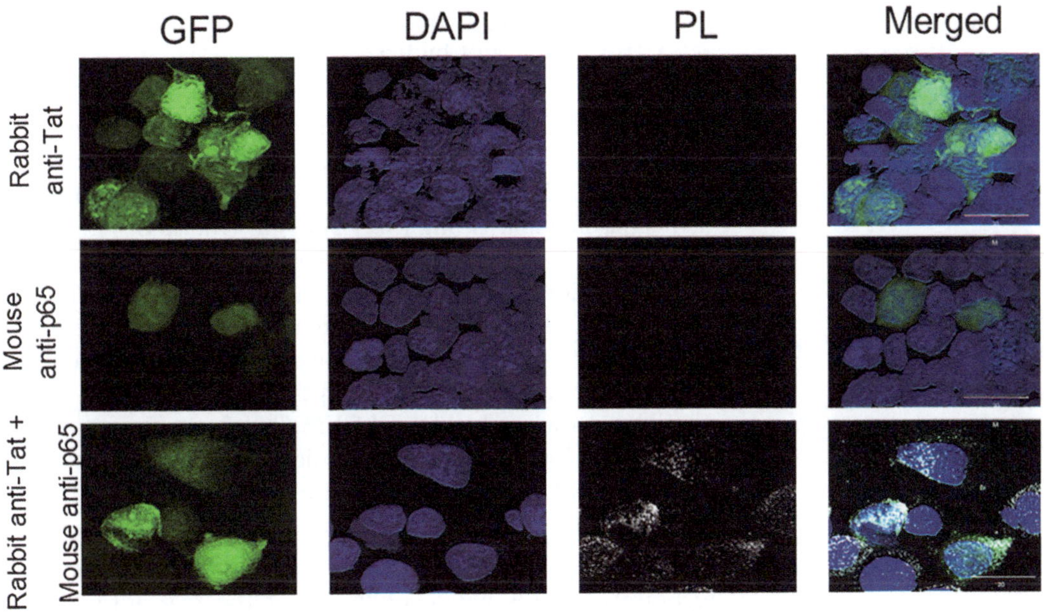

Fig. 4 Confocal images showing the PLA signals. Top Row: Rabbit anti-Tat single antibody control; Middle Row: Mouse anti-p65 single antibody control; Bottom Row: Staining with both antibodies. The GFP fluorescence indicates viral infection (Column 1), and the DAPI dye stains the nucleus (Column 2). The bright white dots in column 3 of the bottom row represent the saturating PLA signals indicative of Tat-p65 proximity, which are absent from the single antibody-stained wells

5. Analyze the image using a 20X or 100X objective in a confocal microscope. Use an appropriate filter combination to view the fluorescence signal (*see* **Note 13**). The PLA signals appear as distinct fluorescence specks of approximately 1 μm size within the cells (Fig. 4). Use appropriate technical and biological controls for the background correction (*see* **Notes 14–18**).
6. Use an appropriate image analysis software, such as ImageJ, to quantify PLA signals either as mean fluorescent intensity (MFI) of the fluorophore or as the total number of PLA spots per cell, or area within a subcellular organelle.

4 Notes

1. During viral infection, a cell confluence of 50–60% will ensure optimal protein expression and minimal cytopathic effects. In contrast, fewer cells at the beginning of the experiment may lead to significant cell loss due to downstream procedures, including multiple wash steps.
2. Integration of the provirus and its chromatinization are crucial to examine the viral and cellular factors and their interplay in the context of HIV-1 latency. Further, a single viral integration

per cell is desirable; therefore, the target cells must be infected at an MOI of 0.1, not higher.

3. Note that we used two different cell lines—HEK293T and HEK293 here, the former to make the viral stocks and the latter for experimentation. HEK293T is a daughter cell line derived from HEK293 cells by stable transfection of the cells with a plasmid containing the T antigen. Due to this modification, HEK293T cells are ideal for large-scale recombinant protein expression and recombinant virus production. HEK293 cells are more suited for HIV-1 latency evaluation as these cells represent a relatively low activation status given the absence of the T antigen, which is a transcription activator. Importantly, T-cell lines, such as Jurkat and CEM CCR5, are physiologically ideal targets for HIV-1 latency studies. However, standardizing PLA in suspension cell lines is more challenging.
4. Each PLA reaction will require approximately 0.2×10^6 GFP^{+ve} cells. Based on the total reactions required, including two technical controls where only one of the two antibodies is used for staining, calculate the total number of cells needed following sorting.
5. Chambered glass slides with removable silicone gasket chambers are best suited for immunofluorescence studies when multiple samples or biological replicates will be used. The gasket is biologically inert, prevents cross-contamination, and allows the processing of all samples simultaneously. Once immunostaining is completed, the silicone chamber is removed, the slide is sealed with a coverslip for imaging. Alternatively, the sorted cells may also be seeded on glass coverslips of 18 mm diameter when a large number of cells need to be imaged and analyzed.
6. If necessary, increase the duration of the incubation time with the blocking reagent to prevent high background signals.
7. Use the Blocking Buffer and the Duolink® Antibody Diluent provided in the kit to dilute the primary antibodies and the secondary antibodies, respectively.
8. The optimal dilution of the primary antibodies must be determined using the "single-recognition" PLA method. In a single-recognition PLA method, a serial dilution of the single primary antibody is used in combination with two secondary antibodies against the same primary antibody, each tagged with the PLUS and MINUS probes to generate a PLA signal.
9. Aspirate the wash buffer entirely before adding a new reagent. Residual wash buffers may dilute the reagents. Importantly, use a wide-bore tip while aspirating any solution or buffer to prevent dislodging or disturbing the cell monolayer.

10. Ligation buffer should be entirely thawed to homogeneity before use, else there could be nonuniform ligation across the sample.
11. Hold enzymes always on a freezer block (−20 °C) for optimal storage.
12. Prepare the Ligation and the Amplification Buffers during the intermittent washing steps. Add the master mixes to the samples immediately after the respective washing steps. Never allow the cells to go dry at any point.
13. If your experimental schematic already contains a fluorescence reporter protein, exercise caution while choosing the Duolink® PLA kit for the assay. The kits are available for a range of PLA fluorescence signals. For instance, if the cells express GFP, choose the red or the orange, not green, in situ detection reagents.
14. PLA spots are present in multiple focal planes across the entire depth of the cell. Perform Z-stack imaging to obtain a composite image of the cells.
15. It is normal to find a small number of PLA spots in the technical negative control monolayers. However, caution must be used when the spots are consistently seen in several or all fields of the "single-primary antibody" control cells, especially if the spot numbers are comparable to those of the test sample. This situation is indicative of a false-positive reaction. Troubleshooting will be required, including further titration of the primary antibody.
16. When abundant proteins are interrogated or the samples are overexposed under the microscope, the PLA spots might fuse to appear as bigger and merged spots. Follow the ligation and amplification incubation times strictly to prevent signal coalescence (**Steps 19** and **22**, Subheading 3.3).
17. Set the imaging parameters (filters, exposure time, gain, objectives, etc.) using the "single-primary antibody" controls. Use identical and uniform settings for all the test samples.
18. The slides may be stored at 4 °C for a week or at −20 °C for 6 months in the dark for future use.

References

1. Gustafsdottir SM, Schallmeiner E, Fredriksson S et al (2005) Proximity ligation assays for sensitive and specific protein analyses. Anal Biochem 345:2–9
2. Hnasko R (2015) ELISA: methods and protocols. ELISA Methods Protoc 1318:1–216
3. Söderberg O, Gullberg M, Jarvius M et al (2006) Direct observation of individual endogenous protein complexes in situ by proximity ligation. Nat Methods 3:995–1000
4. Schallmeiner E, Oksanen E, Ericsson O et al (2007) Sensitive protein detection via triple-binder proximity ligation assays. Nat Methods 4:135–137

5. Lundberg M, Thorsen SB, Assarsson E et al (2011) Multiplexed homogeneous proximity ligation assays for high-throughput protein biomarker research in serological material. Mol Cell Proteomics 10:1–10
6. Famulok M (2002) Bringing picomolar protein detection into proximity: combining target recognition by two aptamers, enzymatic ligation, minute amounts of proteins. Nat Biotechnol 20:448–449
7. Debaize L, Jakobczyk H, Rio AG et al (2017) Optimization of proximity ligation assay (PLA) for detection of protein interactions and fusion proteins in non-adherent cells: application to pre-B lymphocytes. Mol Cytogenet 10:1–13
8. Dussupt V, Javid MP, Abou-Jaoudé G et al (2009) The nucleocapsid region of HIV-1 gag cooperates with the PTAP and LYPX nL late domains to recruit the cellular machinery necessary for viral budding. PLoS Pathog 5: e1000339
9. Sharma S, Arunachalam PS, Menon M et al (2018) PTAP motif duplication in the p6 Gag protein confers a replication advantage on HIV-1 subtype C. J Biol Chem 293:11687–11708
10. Chakraborty S, Kabi M, Ranga U (2020) A stronger transcription regulatory circuit of HIV-1C drives the rapid establishment of latency with implications for the direct involvement of tat. J Virol 94:e00503-20. https://doi.org/10.1128/jvi.00503-20
11. Verma A, Rajagopalan P, Lotke R et al (2016) Functional incompatibility between the generic NF-κB motif and a subtype-specific Sp1III element drives the formation of the HIV-1 subtype C viral promoter. J Virol 90:7046–7065

Part VI

Studies on HIV Neuropathogenesis

Chapter 18

Modeling HIV-1 Infection in CNS via Infected Monocytes Using Immunocompetent Brain Organoids

Roberta S. dos Reis, Sathish Selvam, Marc C. E. Wagner, and Velpandi Ayyavoo

Abstract

The development of 3D-organoid models has revolutionized the way diseases are studied. Recently, our brain organoid model has been shown to recapitulate in in vitro the human brain cytoarchitecture originally encountered in HIV-1 neuropathogenesis, allowing downstream applications. Infected monocytes, macrophages, and microglia are critically important immune cells for infection and dissemination of HIV-1 throughout brain during acute and chronic phase of the disease. Once in the brain parenchyma, long-lived infected monocytes/macrophages along with resident microglia contribute to the establishment of CNS latency in people with HIV (PWH). Hence, it is important to better understand how HIV-1 enters and establishes infection and latency in CNS to further develop cure strategies. Here we detailed an accessible protocol to incorporate monocytes (infected and/or labeled) as a model of transmigration of peripheral monocytes into brain organoids that can be applied to characterize HIV-1 neuroinvasion and virus dissemination.

Key words 3D brain organoids, Neuroinflammation, HIV neuropathogenesis, Microglia, Monocytes, THP-1, Disease model

1 Introduction

Circulating CD4+T cells are the primary target of HIV-1 infection in the peripheral blood. However, several other circulating immune cells that bear the CD4 receptor and co-receptors necessary for viral entry can support productive and latent infection [1, 2]. Of particular interest are long-lived and highly migratory myeloid cells such as monocytes and macrophages [3]. Hence, infected monocytes and perivascular macrophages play an important role in dissemination of HIV-1 throughout the body, including the brain [4]. Although circulating monocytes and infected T cells may transmigrate into central nervous system (CNS) and contribute to viral spreading and persistence in the brain, studies of simian

Vinayaka R. Prasad and Ganjam V. Kalpana (eds.), *HIV Protocols*, Methods in Molecular Biology, vol. 2807,
https://doi.org/10.1007/978-1-0716-3862-0_18,

immunodeficiency virus (SIV)-infected rhesus macaques have identified the presence of SIV-infected perivascular monocytes/macrophages in the monkey brain as early as 10 days post infection [5, 6]. These results were supported by evidence from studies that detected viral DNA in perivascular macrophages, microglia, and astrocytes in the human brains [3–5, 7]. Collectively, these findings highlight the role of monocytes and macrophages in HIV neuroinvasion, seeding, and establishment of CNS infection.

Currently, there are no human-cell-based models that are physiologically relevant and reproducible to assess HIV-1 spreading from the periphery (via monocyte/macrophages) to other target cells (microglia and astrocytes) in the brain, and we have yet to elucidate the role that peripheral monocytes/macrophages and brain microglia individually play in the maintenance of HIV-latency in the brain. Three-dimensional culture systems such as brain organoids have the potential to mimic an in vivo tissue cytoarchitecture in a physiologically relevant manner in terms of structure, interaction, multi-cellular microenvironment, and function [8–13]. Therefore, an immunocompetent brain organoid offers an ideal platform to study neuroinflammatory diseases that will recapitulate HIV-1 infected human brain. Recently, we published a detailed protocol for generation of human-representative brain organoids containing HIV-infected microglia [14]. Here we expanded the previous protocol to generate a fully immunocompetent brain organoid to replicate the early steps of HIV neuroinvasion in vitro. We show the incorporation of a monocytic cell line (THP-1) or human primary CD14+ monocytes as a "proof of concept" for transmigration of HIV-infected monocytes into brain organoids containing CNS-resident microglia and astrocytes and neurons.

2 Materials

Use commercially available plasticware by Corning, Fisher brand, or Nunc for maintenance of 2D cultures. For 3D cultures, use ultralow attachment 12-well cell culture plates by Corning, Fisher brand, or Nunc. All prepared solutions, microcentrifuge tubes, and pipette tips used in cell culture were sterilized before use. Store the solutions at 4 °C unless indicated otherwise.

2.1 Cell Lines and Plasmids

1. pLenti-EGFP construct (Clontech) (*see* **Note 1**).
2. Human induced Pluripotent Stem Cell (hiPSC) (ATCC-BYS0113).
3. HEK293T cells (NIH HIV Reagent Program).
4. THP-1 monocytes (NIH HIV Reagent Program).
5. Adult Primary Human microglia (*see* **Note 2**).

2.2 Cell Culture Media and Supplements

1. HEK293T medium: For 500 mL of Dulbecco's modified Eagle's medium (DMEM) containing 4.5 g/L glucose, supplement with 5 mL of 200 mM L-glutamine solution (final concentration of 2 mM), 5 mL of 5000 U/mL of Penicillin-5000 μg/mL of Streptomycin solution (Final concentration of 50 U/mL and 50 μg/mL respectively), and 50 mL of heat-inactivated fetal calf serum (FCS) to a final concentration of 10%.
2. THP-1 medium: For 500 mL of medium Roswell Park Memorial Institute (RPMI)-1640, supplement with 5 mL of 200 mM L-glutamine solution (final concentration of 2 mM), 5 mL of 5000 U/mL of Penicillin- 5000 μg/mL of Streptomycin solution (Final concentration of 50 U/mL and 50 μg/mL respectively), 450 μL of 2-mercaptoethanol 55 mM to a final concentration of 0.05 mM, 5 mL of sodium pyruvate 100 mM to a final concentration of 1.0 mM, and 50 mL of FCS to a final concentration of 10%.
3. For Adult Primary Human microglia medium: Basal Microglia medium supplemented with 10 ng/mL of granulocyte-macrophage colony-stimulating factor (GM-CSF) and 100 μg/mL of Primocin with provided solutions (Microglia medium kit, ScienCell Research). Final supplement concentration of 1% microglia growth supplement, 50 U/mL of Penicillin and 50 μg/mL of Streptomycin, and 5% heat-inactivated FCS.
4. Human brain organoids maintenance medium Half of DMEM/F-12 with GlutaMAX medium and half of Neurobasal Medium, supplement with MEM/NEAA (100×), 2-mercaptoethanol (1000×), Insulin (1000×), N2 supplement (100×), and B27™ Supplement with Vitamin A (50×) (*see* **Note 3**).
5. Sterile phosphate-buffered saline (PBS) solution (1×), pH 7.4, stored at room temperature.
6. Hoechst 33342 solution.
7. 0.05% Trypsin–EDTA (1×) solution in sterile phosphate-buffered saline (PBS).
8. Poly-D-lysine coating solution. Prepare a solution of 50 μg/mL of PDL in sterile PBS.

2.3 Confocal Imaging of Neuroinvasion of Immunocompetent Brain Organoids

1. Sterile disposable rigid inoculation loops.
2. 4-well or 8-well Nunc™ LabTeK™ chambered cover glass
3. Glycerol (*see* **Note 4**).

3 Methods

A graphical overview of the methods is provided in Fig. 1. Carry out all the culture procedures aseptically and pre-warm the cell culture medium at 37 °C before use. All incubations should be performed in a humidified, temperature-controlled CO_2 incubator at 37 °C and 5% CO_2, unless specified otherwise. For maintenance of adult primary human microglia, it is recommended to follow the directions provided in the original manuscript [15]. The cells should be subcultured at least once before using them for experiments but no more than 10 passages. For THP-1 monocytes maintenance, it is recommended to follow the directions provided by ATCC. THP-1 cells were labeled with Celltrace™ far red dye (Invitrogen) per manufacturer's instructions. Normal donor iPSCs purchased from ATCC were maintained according to their instructions. iPSC-derived brain organoids were generated according to standard protocols [16]. Diligently follow the Biosafety Level disposal regulations when disposing viral and human materials.

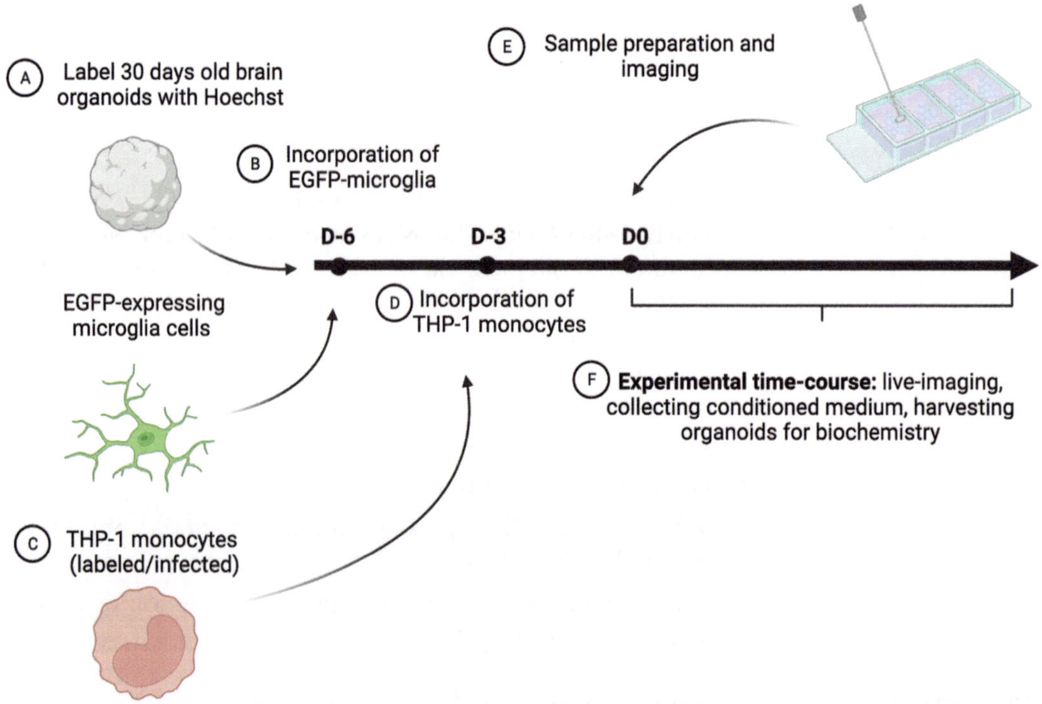

Fig. 1 Experimental timeline to develop the three-dimensional model of monocytes neuroinvasion by using human brain organoids. Schematic diagram of the key stages. (**a**) Mature organoids (30 days old) are labeled with Hoechst at day minus 6 (D-6) followed by (**b**) incorporation of EGFP-expressing primary human microglia. In order to model monocytes neuroinvasion, (**c**) THP-1 monocytes are labeled at D-6 and (**d**) incorporated into immunocompetent organoids at Day minus 3 (D-3). At day zero (D0), brain organoids are imaged in (**e**) to assess integration of both immune cells into organoids, which can be maintained in culture for time-course experiments in (**f**)

3.1 Generation of Microglia Expressing EGFP Reporter Gene

1. VSV-G-Env pseudotyped EGFP reporter lentiviral particles are produced by transient transfection of HEK293T cells using PolyJet™ transfection reagent (SignaGen) per manufacturer's instructions.
2. Human primary microglia are transduced with EGFP reporter lentiviral particles at a multiplicity of infection (MOI) of 1 by adding virus aliquot to the cells. Let the virus adsorption proceed for 16–18 h.
3. Wash the cells to remove unbound viruses and replace with microglia medium. Typically, human microglia express EGFP 3 days post transduction.

3.2 Generation of Immunocompetent Brain Organoid

1. Prior to labeling organoid nuclei, carefully remove most of the media of mature brain organoids (>30 days in culture). Wash the organoids once with 1 mL of pre-warmed 5% FBS-PBS.
2. Add 1 mL of pre-warmed 5% FBS-PBS containing 1 μL of 20 mM of Hoechst 33342 solution and incubate for 15 min at 37 °C to allow the organoids to incorporate the fluorescent dye.
3. In parallel, harvest EGFP expressing human primary microglia by adding 1 mL of trypsin solution for 5 min at room temperature. Add 5 mL of microglia medium and immediately centrifuge the cell suspension at 1000 RPM for 5 min.
4. Determine microglia cell density by a Coulter Counter or manually using a hemocytometer and separate 1×10^6 cells/organoid (*see* **Note 5**).
5. Spin down the EGFP-microglia at 1000 rpm for 10 min and wash twice with PBS to obtain fluorescent microglia to be incorporated into brain organoids.
6. Resuspend the cell pellet in 100 μL of brain organoid maintenance medium.
7. Wash off the excess of Hoechst 33342 solution by removing the dye solution and washing once with PBS and a second time with organoid media. Carefully remove most of the media of the organoids.
8. Pipette EGFP-microglia suspension on top of the organoids layer and allow cells to adhere to the organoids for 20–30 min.
9. Replenish the co-culture with 2 mL of brain organoids maintenance medium and incubate the co-culture overnight at 37 °C (*see* **Note 6**).

10. Set up a new ultralow attachment 12-well plate and add 2 mL of fresh maintenance medium per well/per device.
11. Carefully transfer the brain organoids with attached microglia to the new plate with the help of a sterile spatula or inoculation loop. Replace half of the medium every other day until the THP-1 neuroinvasion assay.

3.3 Incorporation of Monocytes to Model Neuroinvasion

1. Determine THP-1 cell density by a Coulter Counter or manually using a hemocytometer or harvest 1×10^5 monocytes/organoid in 1 mL of 1× PBS or plain RPMI.
2. Label THP-1 cells with 1 μL of CellTrace far-red dye and incubate for 20 min at room temperature.
3. Wash the cells in complete THP-1 medium to quench remaining dye and centrifuge the cell suspension at 1600 RPM for 5 min.
4. Resuspend the labeled THP-1 pellet in 100 μL of brain organoid differentiation medium.
5. Repeat **steps 9–11** of Subheading 3.2.
6. Carefully transfer the immunocompetent brain organoids with attached THP-1 cells or monocytes to the new plate with the help of a sterile spatula or inoculation loop. Replace half of the medium every other day to maintain the organoids for future experiments.
7. We have also incorporated CellTrace dye labeled CD14+ monocytes isolated from normal donor, and similar incorporation was observed (data not shown).

3.4 Imaging Whole Brain Organoids for Assessing Monocyte Neuroinvasion

1. Remove as much media as possible from organoid culture and gently wash two times with 1X PBS (*see* **Note** 7).
2. With the help of a sterile rigid inoculation loop, carefully scoop at least five brain organoids and transfer them to a 4- or 8-well glass bottom chamber slide.
3. Carefully remove any residual PBS from the chamber by using a P200 pipette. Do not disturb the organoids.
4. Pipette 200 μL of glycerol or enough volume to cover the organoids. Allow the organoids to incorporate the glycerol for 1 h to overnight at 4 °C before imaging (*see* **Note 8**).
5. When ready to perform the confocal imaging, centralize the organoids in the well with the help of a P200 tip.
6. On the microscope, localize the organoids in the well using the shortest wavelength (405 nm for Hoechst staining).
7. Adjust the z-plane up to the top of the organoid and select it in the imaging acquisition software as top. Adjust and select the bottom of the organoid.

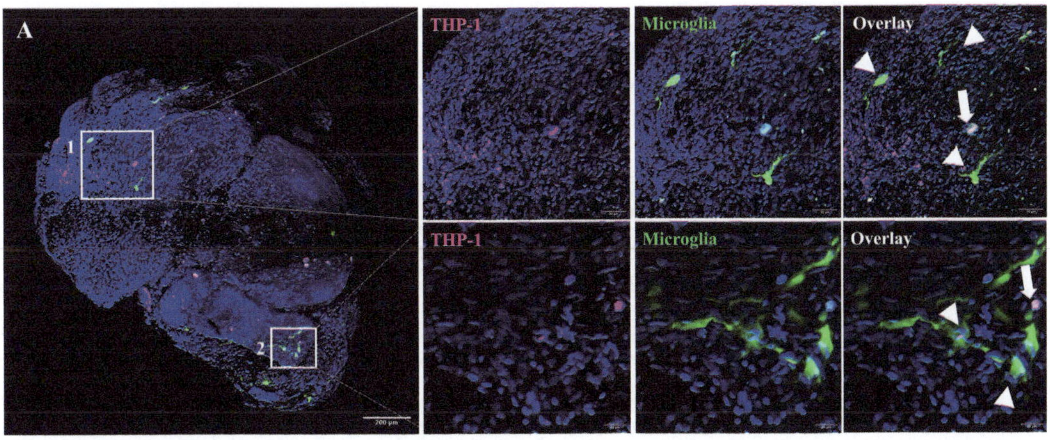

Fig. 2 Representative images of whole brain organoids with labeled THP-1 monocytes. (**a**) Representative confocal image of 30-days-old brain organoids (Hoechst+, blue) at D3 after incorporation of microglia (EGFP+, green) followed by neuroinvasion of labeled THP-1 (far-red) at 10× magnification scale. Zoomed images of inset 1 (upper right panels) and 2 (bottom right panels) showing labeled THP-1 (arrow), microglia (arrowhead), and the overlay of the three channels. A total of 130 z-stacks were collected using 3 μm stepwise in (**a**), and 260 images of using 1 μm stepwise in insets 1 and 2. Scale bar 200 μm for (**a**), 50 μm in upper panels, and 20 μm in bottom panels

8. Select the step size from 1 to 2.5 μm (*see* **Note 9**).
9. Create z-stacks.
10. Save the file once finished and perform th eimage analysis of using your choice of software (Fig. 2).

4 Notes

1. We used standard lentivirus packaging protocol by transfecting HEK293T cells using Virapower lentiviral packaging mix (InVitrogen) per manufacturer's recommendations, with minor modifications. In brief, 2.75 μg of pLenti-EGFP vector, 1.0 μg of pLP1 (gag/pol), 0.50 μg of pLP2, and 0.75 μg of pLP/VSV-G-Env were co-transfected into HEK293T cells with the help of 15 μL polyjet transfection reagent. Forty-eight and seventy-two hours after transfection, supernatants were harvested, filtered (0.45-μm filter, Millipore) and concentrated by centrifugation (20,000 RPM, 1 h at 4 °C). The lentivirus infectivity in supernatants can be determined by infecting U87MG CD4+ CCR5+ permissive cells (NIH HIV Reagent Program) with serial dilutions of the virus stocks and assessing the number of EGFP+ cells under fluorescent microscope [17].

2. The protocol for postmortem adult microglia cells was established by Guo et al. (2016), and these cells were kindly donated by Dr. Changiz Geula from the Laboratory for Cognitive and Molecular Morphometry at Northwestern University [15]. Alternatively, commercially available immortalized microglia cell lines, as HMC3 (ATCC CRL-3304), can be used. We find that HMC3 microglia attach and embed into the brain organoids and produce strong inflammatory response upon HIV-1 infection [18].
3. The brain organoids medium suggested in this chapter refers to the original brain organoids protocol described by Lancaster et al. (2014) [16]. Many published brain organoids protocols describe different formulations as "maintenance medium" including the protocol developed and recently described by our group [14]. Therefore, use the maintenance media according to the brain organoid protocol applied.
4. Glycerol must be autoclaved if performing time-course experiments, in order to maintain the sterile conditions of the organoids culture.
5. Alternatively, human primary microglia can be labeled with Carboxy Fluorescein-diacetate Succininidyl Ester (CFSE) (Sigma) per manufacturer's instructions before incorporating into brain organoids.
6. Microglia will spontaneously interact with and incorporate into mature brain organoids. To increase microglia attachment, place the culture plate on orbital shaker at 37 °C and incubate overnight at low speed (60 r.p.m. max). The orbital shaker must be cleaned by spraying with 70% ethanol before it is placed in the incubator.
7. Use cut P1000 tip with the tip cut off. This will prevent mechanical disruption of the organoids when removing the media.
8. We found that brain organoids bigger than 300 μm require longer incubation period for glycerol embedding. Glycerol-embedded organoids can be washed with PBS three times and returned to culture in brain organoid maintenance media for time-course experiments. When ready to image them again, repeat the glycerol embedding process.
9. To better capture the three-dimensional physiology in a high resolution, it is recommended to decrease the step-sizes to 1–2.5 μm and keep constant the z stacks throughout the samples. Glycerol provides higher refractive index than PBS or other commercial formulations while providing stability to the organoid during image capture due to its viscosity.

Acknowledgments

The THP-1 cell line was obtained through the NIH HIV Reagent Program, Division of AIDS, NIAID, NIH. We thank Dr. Gengiz Geula from the Laboratory for Cognitive and Molecular Morphometry at Northwestern University for donating the adult brain primary microglia cells for our study. This study was supported by NIH grant R56NS122567 to VA.

References

1. Woodham AW, Skeate JG, Sanna AM et al (2016) Human immunodeficiency virus immune cell receptors, coreceptors, and cofactors: implications for prevention and treatment. AIDS Patient Care STDs 30:291–306. https://doi.org/10.1089/apc.2016.0100
2. Preza GC, Tanner K, Elliott J et al (2014) Antigen-presenting cell candidates for HIV-1 transmission in human distal colonic mucosa defined by CD207 dendritic cells and CD209 macrophages. AIDS Res Hum Retrovir 30: 241–249. https://doi.org/10.1089/aid.2013.0145
3. Crowe S, Zhu T, Muller WA (2003) The contribution of monocyte infection and trafficking to viral persistence, and maintenance of the viral reservoir in HIV infection. J Leukoc Biol 74:635–641. https://doi.org/10.1189/jlb.0503204
4. Koppensteiner H, Brack-Werner R, Schindler M (2012) Macrophages and their relevance in Human Immunodeficiency Virus Type I infection. Retrovirology 9:82. https://doi.org/10.1186/1742-4690-9-82
5. Thompson KA, Varrone JJ, Jankovic-Karasoulos T et al (2009) Cell-specific temporal infection of the brain in a simian immunodeficiency virus model of human immunodeficiency virus encephalitis. J Neurovirol 15:300–311. https://doi.org/10.1080/13550280903030125
6. Williams KC, Corey S, Westmoreland SV et al (2001) Perivascular macrophages are the primary cell type productively infected by simian immunodeficiency virus in the brains of macaques: implications for the neuropathogenesis of AIDS. J Exp Med 193:905–915. https://doi.org/10.1084/jem.193.8.905
7. Veenhuis RT, Abreu CM, Costa PAG et al (2023) Monocyte-derived macrophages contain persistent latent HIV reservoirs. Nat Microbiol 8:833–844. https://doi.org/10.1038/s41564-023-01349-3
8. Lancaster MA, Huch M (2019) Disease modelling in human organoids. Dis Model Mech 12:dmm039347. https://doi.org/10.1242/dmm.039347
9. Benito-Kwiecinski S, Lancaster MA (2020) Brain organoids: human neurodevelopment in a dish. Cold Spring Harb Perspect Biol 12: a035709. https://doi.org/10.1101/cshperspect.a035709
10. Lancaster MA (2018) Brain organoids get vascularized. Nat Biotechnol 36:407–408. https://doi.org/10.1038/nbt.4133
11. Lancaster MA, Corsini NS, Wolfinger S et al (2017) Guided self-organization and cortical plate formation in human brain organoids. Nat Biotechnol 35:659–666. https://doi.org/10.1038/nbt.3906
12. Renner M, Lancaster MA, Bian S et al (2017) Self-organized developmental patterning and differentiation in cerebral organoids. EMBO J 36:1316–1329. https://doi.org/10.15252/embj.201694700
13. Luo C, Lancaster MA, Castanon R et al (2016) Cerebral organoids recapitulate epigenomic signatures of the human fetal brain. Cell Rep 17:3369–3384. https://doi.org/10.1016/j.celrep.2016.12.001
14. Dos Reis RS, Sant S, Ayyavoo V (2023) Three-dimensional human brain organoids to model HIV-1 neuropathogenesis. Methods Mol Biol 2610:167–178. https://doi.org/10.1007/978-1-0716-2895-9_14
15. Guo L, Rezvanian A, Kukreja L et al (2016) Postmortem adult human microglia proliferate in culture to high passage and maintain their response to amyloid-β. J Alzheimers Dis 54: 1157–1167. https://doi.org/10.3233/JAD-160394
16. Lancaster MA, Knoblich JA (2014) Generation of cerebral organoids from human pluripotent stem cells. Nat Protoc 9:2329–2340. https://doi.org/10.1038/nprot.2014.158

17. Montefiori DC (2009) Measuring HIV neutralization in a luciferase reporter gene assay. Methods Mol Biol 485:395–405. https://doi.org/10.1007/978-1-59745-170-3_26
18. Dos Reis RS, Sant S, Keeney H et al (2020) Modeling HIV-1 neuropathogenesis using three-dimensional human brain organoids (hBORGs) with HIV-1 infected microglia. Sci Rep 10:15209. https://doi.org/10.1038/s41598-020-72214-0

Chapter 19

An In Vitro Model of Blood-Brain Barrier for Studies on HIV Neuroinflammation and CNS Antibody Penetration

Yuekun Lang, Dina Mofed, Vasudev R. Rao, Samantha J. Welninski, Anjelica F. Reyes, Jeffrey R. Schneider, and Vinayaka R. Prasad

Abstract

The blood-brain barrier (BBB) is one of several barriers between the brain and the peripheral blood system to maintain homeostasis. Understanding the interactions between infectious agents such as human immunodeficiency virus type 1 (HIV-1), which are capable of traversing the BBB and causing neuroinflammation requires modeling an authentic BBB in vitro. Such an in vitro BBB model also helps develop means of targeting viruses that reside in the brain via natural immune effectors such as antibodies. The BBB consists of human brain microvascular endothelial cells (HBMECs), astrocytes, and pericytes. Here we report in vitro methods to establish a dual-cell BBB model consisting of primary HBMECs and primary astrocytes to measure the integrity of the BBB and antibody penetration of the BBB, as well as a method to establish a single cell BBB model to study the impact of HIV-1 infected medium on the integrity of such a BBB.

Key words In vitro blood-brain barrier, IgG CNS penetration, Astrocytes, Human brain microvascular endothelial cells

1 Introduction

Despite antiretroviral therapy, a significant proportion of people living with human immunodeficiency virus (HIV) experience Central Nervous System (CNS) complications including HIV-associated neurocognitive disorder (HAND) and mental health-related disorders. Approximately 30–50% of HIV-infected individuals in all parts of the world display some form of HAND [1], which include asymptomatic neurocognitive impairment (ANI), mild neurocognitive disorder (MND), and severe HIV-associated dementia (HAD) [2]. While ANI has no impact on daily life, the MND and HAD both impact daily function of the affected individual. HIV-1 enters the CNS within days after exposure to HIV-1 [3]. HIV-1 or HIV-infected monocytes, prior to entering the brain, encounter the blood-brain barrier (BBB) that separates the

Vinayaka R. Prasad and Ganjam V. Kalpana (eds.), *HIV Protocols*, Methods in Molecular Biology, vol. 2807, https://doi.org/10.1007/978-1-0716-3862-0_19,

peripheral blood from the CNS environment [4]. BBB consists of a layer of human brain microvascular endothelial cells (HBMECs) of the microcapillaries interspersed on the brain side with astrocytes. The astrocyte processes penetrate the spaces between HBMECs via their end feet, which form tight junctions in the intercellular surfaces, thus forming a selectively permeable barrier. Such barriers are further fortified by pericytes, resulting in increased selectivity of the BBB. Mononuclear phagocytes (MPs) have the ability to cross the BBB, which is exacerbated in HAND. Microglia are known to develop from primitive myeloid progenitors that arise before embryonic day 8 and reach CNS during development [5]. In addition, HIV-infected T lymphocytes are known to cross the BBB in an infected individual. HIV-infected monocytes, once in the CNS, establish a β-chemokine gradient with a key role played by HIV-1 Tat protein [6, 7] to increase the number of HIV-infected MPs (monocyte, macrophages), which leads to neuroinflammation as a result of several viral products, and proinflammatory cytokines and chemokines released by the HIV-infected MPs that together culminate in synaptodendritic injury [8].

There are several barriers between the brain and the peripheral blood system. One is choroid plexus that forms a barrier between the cerebrospinal fluid (CSF) and blood [9]. The other is the aforementioned BBB, which consists of human brain microvascular endothelial cells (HBMECs; also known as brain microvascular endothelial cells or BMECs) that are in direct contact with circulating blood and the astrocytes that are part of the brain. The characteristic feature of HBMECs is their ability to form a tightly knit monolayer due to intercellular tight junctions made up of tight junction proteins such as Claudin-5, selective permeability, and high trans-endothelial electrical resistance (TEER) [10, 11]. The astrocytes from the brain compartment send out their long processes to plug any intercellular spaces between HBMECs with their end feet to further enhance the integrity of the barrier. Finally, pericytes confer an additional layer to complete the process of making BBB as selectively permeable as possible [11].

In vitro studies of BBB have been performed using single-cell (HBMECs alone), dual-cell (HBMECs and astrocytes), and triple-cell (HBMECs, astrocytes, and pericytes) models using an insert containing a filter with 3 μm pores. In dual-cell models, the HBMECs are plated inside the top chamber where they are facing the "blood side" of BBB and the astrocytes are plated on the reverse side of the same filter facing the lower chamber or the "brain side." The integrity of the BBB layer that is formed can be tested by measuring TEER or by measuring the passage of hydrophilic tracer molecules from the upper chamber to the lower chamber or vice versa [12]. These dyes range in size from 180 to 4000 Daltons and include: FITC Dextran (4 kDa), Evans Blue (960 Da), Sucrose

(842 Da), Lucifer yellow (444 Da), Sodium Fluorescein (376 Da), and Mannitol (180 Da) [12].

Once generated, such in vitro BBB models can be used to study the impact of HIV-1 on BBB permeability, monocyte migration under the influence of a chemokine gradient across the BBB, the effect of proinflammatory molecules such as IL1b, TNFα, HIV-1 Tat, or gp120 on the integrity of the BBB, and whether a given immunoglobulin (Ig) can cross the BBB and with what efficiency.

2 Materials

2.1 Cells and Viruses

1. Human Brain Microvascular Endothelial cells (HBMEC). Also known as Brain Microvascular Endothelial Cells (BMVECs) (kind gift of Dr. Servio Ramirez, Temple University).
2. Primary astrocytes (kind gift of Dr. Herbert Lachman).
3. Elutriated primary human monocytes (Obtained from University of Nebraska Medical Center).
4. HIV-1_{ADA} high titer virus produced in 293T cells.

2.2 Media and Buffers

1. Astrocyte Medium (Sciencell catalog number 1801) Contains 500 mL of basal medium, 10 mL of fetal bovine serum (FBS), 5 mL of astrocyte growth supplement (Sciencell AGS, catalog number 1852), and 5 mL of penicillin/streptomycin solution.
2. Endothelial Cell Medium (Sciencell catalog number 1001) consists of 500 mL of basal medium, 25 mL of FBS, 5 mL of endothelial cell growth supplement (Sciencell ECGS, catalog number 1052), and 5 mL of antibiotic solution.
3. Type I Rat Tail Collagen solution: First prepare 0.02 N glacial acetic acid. Dilute glacial acetic acid 1:700 into sterile water by adding 14 μL of glacial acetic acid to 10 mL sterile water. To prepare 50 μg/mL collagen solution, dilute the stock using 0.02 N glacial acetic acid (*see* **Note 1**).
4. A solution of 3 μg/mL of bovine fibronectin.
5. A solution of 3 μg/mL poly-L-Lysine.
6. Phosphate-buffered saline.
7. Sodium fluorescein should be prepared in Phenol Red free medium at a final concentration of 10 μg/mL.
8. Sterile water.
9. Sterile trans-well inserts with 3 μm pores (Falcon catalog number 353096) for 24-Well plates.
10. BBB-penetrating and non-BBB-penetrating Immunoglobulin G (J. Schneider, unpublished) transfected in 293 cells and isolated via protein G+ agarose.

11. Sulfo-Cyanine5 N-hydroxysuccinimide ester.
12. 40 kDa molecular weight cutoff gel filtration column (desalting spin columns).
13. Sodium bicarbonate.
14. Monocyte–macrophage differentiation medium: DMEM supplemented with 10% human serum, PenStrep, L-glutamine, and 6.6 ng/mL MCSF (Sigma).

2.3 Equipment

- EVOM-2 volt-ohmmeter: For measuring trans-epithelial electrical resistance or trans-epithelilal voltage.

3 Methods

3.1 Establishing an In Vitro BBB Model

The in vitro BBB model is depicted in Fig. 1.

1. Coating the plastic surfaces with collagen type-1: All plastic surfaces where HBMECs or astrocytes are to be plated need to be coated with collagen type-1 (*see* **Note 2**). This includes T75 or T150 flasks, 6-well plates, and the cell culture inserts.
2. Add diluted type I rat tail collagen solution to the tissue culture vessel (e.g., T25 flask and T75 flask) to be coated and rock the flask back and forth to be sure the entire surface is covered. 1 mL is sufficient for a T25 flask and 3 mL for a T75 flask.
3. Place flasks on a level surface at room temperature for 60 min.
4. Aspirate off the collagen solution.

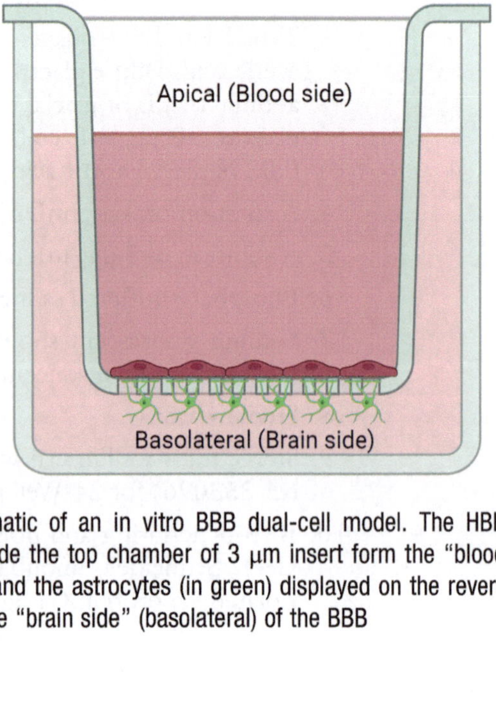

Fig. 1 Schematic of an in vitro BBB dual-cell model. The HBMECs (in red) displayed inside the top chamber of 3 μm insert form the "blood side" of the BBB (apical) and the astrocytes (in green) displayed on the reverse side of the insert form the "brain side" (basolateral) of the BBB

5. Rinse the flask with at least four volumes of sterile PBS.
6. Aspirate off the PBS solution.
7. Use immediately or leave flasks uncapped in the hood to dry.
8. Unused flasks should be capped and can be stored at 4 °C for up to 2 weeks.

3.1.1 Day 1: Seed Astrocytes on the "Brain Side" of the Trans-well Inserts

1. The lower chamber of the two-chamber system represents the brain compartment, and thus, in order to seed the astrocytes on the "brain side" of the inserts, the coated inserts should be flipped over. The low rim on the bottom end of the insert for a 24-well plate forms a reservoir for a small well and can hold 50 μL of the medium.
2. First, astrocytes should be trypsinized and counted (*see* **Note 3**). About 3×10^5 cells are suspended in 50 μL of astrocyte medium (for a 24-well format) (*see* **Note 4**).
3. Then 50 μL of cell suspension is applied over the "brain face" of each insert in the hood.
4. The trans-well inserts sit in the hood for 4–5 h with the application of media for every 5–15 min, to allow for the attachment of astrocytes (*see* **Note 5**).
5. Subsequently, 500 μL of astrocyte medium was added to the bottom chamber of the 24-well plates, and the trans-well inserts are flipped and placed in the wells in the normal orientation.

3.1.2 Day 3: Seed HBMECs on the "Blood Side" of the Trans-well Inserts

(a) Two days after seeding the astrocytes, HBMECs should be plated at a density of 3×10^4 cells/well on the "blood side" of each trans-well (in the top well in the trans-well insert placed upright). The entire setup (*see* Fig. 2) is incubated at 37 °C in a CO_2 incubator for 14 days.

3.1.3 Day 4 to Day 16: HBMEC-Astrocyte Coculture Maintenance

1. Change medium regularly. In the first week, change the medium every 3 days (twice in 1 week) and in the second week, change the medium on alternate days. It is best to change the medium in both chambers at the same time. Do not allow the medium to change to yellow color.

3.1.4 Day 17: Permeability Test with Sodium Fluorescein

1. This test will allow one to assess the changes in the BBB permeability to molecules near the size of 376 Daltons (Sodium Fluorescein).
2. Carefully rinse the inserts with PBS. Add 500 μL Phenol Red free medium in the 24-well permeability test plate. Each type of insert needs three wells to perform the experiment.
3. Place the BBB inserts in the first wells. At the same time, place an empty insert as well to serve as negative control.

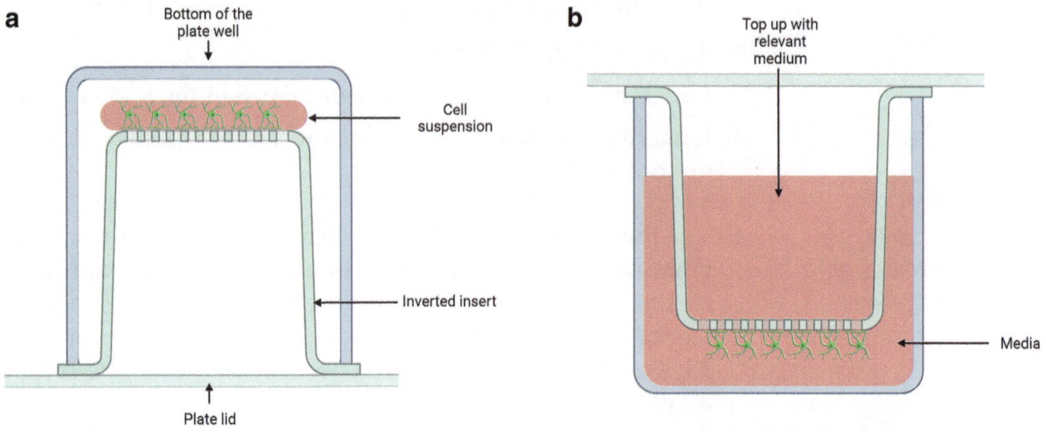

Fig. 2 Schematic of how the astrocytes are seeded on the "brain side" of the trans-well insert (**a**). Seed 3×10^5 astrocytes that are suspended in 50 μL of Astrocyte medium over the "brain side" of the inverted insert in the hood. Then cover the insert using a fresh, inverted bottom piece of the 24-well plate over the insert, and let the trans-well insert sit in the CO_2 incubator for 4–5 h with application of fresh media every 15 min. (**b**) After 4–5 h, first add 500 μL of Astrocyte medium to the bottom chambers of a new 24-well plate. Then, gently flip the trans-well insert and place it upright in the wells

4. Carefully add 200 μL of 10 μg/mL sodium fluorescein solution in the upper chamber of the inserts and incubate at 37 °C for 20 min.
5. Then move the inserts to the well #2 next to it containing 500 μL of clear Phenol Red free medium and incubate at 37 °C for another 20 min.
6. Repeat **step 5** above, move the inserts to the wells #3 containing 500 μL of clear Phenol Red free medium and incubate at 37 °C for another 20 min.
7. Carefully rinse the inserts with PBS and put them back to the original 24-well tissue culture plate.
8. Transfer 100 μL of the incubated Phenol Red free medium from the bottom chamber to a 96-well fluorescein reading plate from each of the three wells (for the BBB containing insert as well as for the blank insert).
9. Read the fluorescence signal with Perkin Elmer Wallac 1420 Victor2 Microplate Reader and plot with Microsoft Excel. Finally, calculate the Pe value as described in Subheading 3, **step 6** (should be below 10). The Pe values will start high and gradually decrease and reach values between 5 and 10 by day 14–18, at which time, they are ideal to use. P_e values below 3 indicate that the cells are overgrown and start to progressively become more and more impenetrable for any molecule.

3.1.5 Calculation of Endothelial Permeability Coefficient (Pe)

In order to calculate the endothelial permeability coefficient (Pe) [13–15], one must first measure the flux of the selected tracer across monolayers and across cell-free inserts as described above. Transport is expressed as microliter of tracer diffusing from luminal ("blood side") to abluminal ("brain side") compartments.

$$\text{Clearance}\ (\mu\text{L}) = [\text{concentration}_{\text{abluminal}}] \times [\text{volume}_{\text{abluminal}}] \times [\text{concentration}_{\text{luminal}}]^{-1}$$

The average volume cleared is plotted versus time, and permeability × surface area product value for endothelial monolayer (PS_e) is calculated by the following formula:

$$\text{PS}_{\text{endothelial}}{}^{-1} = \text{PS}_{\text{time}-\text{plotted}}{}^{-1} - \text{PS}_{\text{insert}}{}^{-1}$$

PS_e divided by the surface area generates the endothelial permeability coefficient [P_e (cm/s)].

3.2 Use of In Vitro BBB to Measure Antibody Penetration to Brain

An in vitro BBB setup can be used to detect and measure the efficiency of BBB penetration by an antibody. The method of setting up the BBB in trans-well inserts and measuring the Pe is the same as described above. Below, we describe a method to label IgGs with Cy5 flurophore, measuring the labeling efficiency and its use to determine the efficiency of penetration across BBB. We have used a pair of IgGs with and without the ability to penetrate in vitro BBB (J. Schneider et al. unpublished observations) as examples.

3.2.1 Labeling IgGs with Cy5 (adapted from Schneider et al. [16])

1. Dilute antibody of interest to 1 mg/mL in 100 μL and buffer exchange into PBS using 40 kDa Zeba column (following manufacturer's instruction).
2. Add 4 μg (2 μL of 2 μg/μL stock) of Sulfo-Cyanine5 NHS Ester to antibody and add 10 μL of 1 M sodium bicarbonate to raise the pH for the labeling reaction.
3. Store in the dark at room temperature for 1 h.
4. Following labeling reaction, remove free dye by passing the labeled antibody serially through two 40 kDa Zeba columns equilibrated with PBS. The eluate from the second column should be clean and should not contain any more free dye (*see* **Note 6**).
5. Using the nanodrop, obtain the A280 and A650 from the labeled antibody and calculate the fluorophores/protein (F/P) ratio using Beer-Lambert's Law equation:

$$\text{DOL} = \frac{M\text{flourophore}}{M\text{protein}} = \frac{A\lambda\max\ \varepsilon\text{protein}}{\varepsilon\lambda\max(A280 - A\lambda\max C280)}$$

Where the extinction coefficient for IgG is $\Sigma\text{protein} = 210{,}000\ \text{M}^{-1}\ \text{cm}^{-1}$, the 280 nm correction factor for Cy5 is $c_{280\text{nm}} = 0.05$, and the extinction coefficient for Cy5 is $\Sigma max = 250{,}000\ M^{-1}\ cm^{-1}$. After incorporation of these coefficients, the following equation will be used to calculate the labeling efficiency.

$$\text{IgG (F/P)} = \frac{A650(210{,}000)}{250{,}000(A280 - A650 \times 0.05)}$$

6. In general, a labeling efficiency of 1–2 fluorophores per antibody is ideal. In the case of under-labeling, repeat **steps 2–4** with adjusted Cy5 added (*see* **Note** 7).

3.2.2 Measuring the Antibody Penetration Across BBB

One needs to determine the ideal dilution of the Cy5 labeled IgG before using the IgG for measuring its ability to penetrate the BBB. First, prepare a 50-fold dilution of Cy5 labeled IgG in Phenol Red-free medium. Then prepare twofold serial dilutions using Phenol Red-free medium to generate eight different twofold serial dilutions. Measure the fluorescence emission using a BioTek Cytation 5 Cell Imaging Multimode Reader or other spectrofluorometer. Plot the fluorescence values on the *x* axis and the concentration of the labeled IgG (μg/μL) on the *y* axis. From the graph, determine the concentration at which 400 units or higher fluorescence emission is obtained. This allows one to determine how many microliters of the original labeled antibody solution needs to be added in the 200 μL that is placed in the upper chamber.

Place 200 μL of Phenol Red-free medium containing Cy5 labeled antibody (around 400 units of fluorescence) in the upper chamber of a trans-well insert that has been previously determined to show a Pe value between 5 and 10 P_e (cm/s). An insert without BBB with Cy5 labeled antibody is used as blank control. Collect the media that diffuses to the lower chamber at 10, 20, and 30 min separately and measure the fluorescence as before using BioTek Cytation 5 Cell Imaging Multimode Reader or other spectrofluorometer. Calculate the P_e values as described above in Subheading 3.1.5.

A typical pair of Pe values of BBB penetrating and non-penetrating IgGs obtained in our experiments are plotted in Fig. 3.

3.3 Establishing a Reverse BBB Model to Study Retention of IgG in the "CNS" Compartment

One can establish a reverse BBB model to study whether a given labled IgG species freely diffuses back into blood circulation or is retained in the CNS compartment. If one can achieve high enough specific labeling efficiency for the IgG on hand, this can be tested by simply adding the labeled IgG in the lower chamber and measuring the Pe as before. But since the volume of the lower chamber is three times that of the upper chamber (600 μL) this requires three to five times higher labeling efficiency. Alternately, one can reverse the

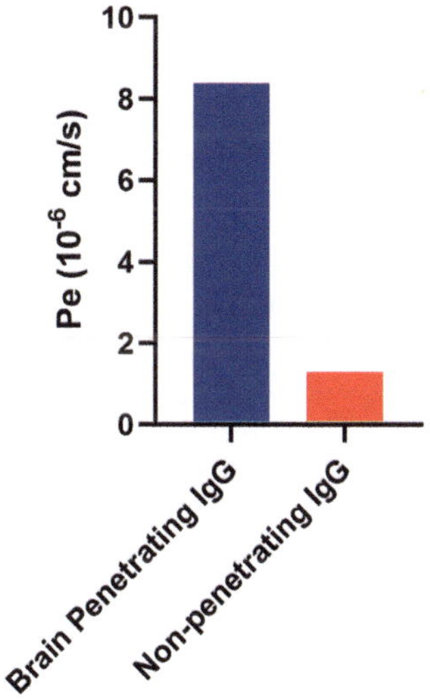

Fig. 3 Pe value comparison between different antibodies. A pair of antibodies, BBB penetrating IgG and a non-BBB-penetrating IgG, were tested on an in vitro BBB dual-cell model to measure the efficiency of their penetration across BBB. The BBB penetrating IgG shows ~8 times higher Pe value compared with non-BBB-penetrating IgG, demonstrating that brain penetrating IgG has much better ability to cross the BBB

placement of cell layers—astrocytes on the upper surface of the inserts (within the inserts) and the HBMECs on the lower surface of the inserts. HBMECs adhere poorly—we have solved this problem by using a combination of (80–100) μg/mL Type I Rat tail collagen, 3 μg/mL of Fibronectin, 3 μg/mL poly-L-Lysine. We employed 3×10^5 HBMECs and 3×10^4 astrocytes. One needs to verify the integrity of the BBB established over 2-week period before conducting their experiments.

3.4 Utilizing a Single Cell In Vitro BBB Model to Measure the Impact of HIV-1 on Alterations in Barrier Permeability

HIV-1 infection of myeloid cells in the periphery can alter the permeability of the BBB. A single cell in vitro BBB setup (using HBMECs) can be used to measure the alterations in permeability caused by HIV-1 and other infectious agents. The BBB is established on the trans-well side of the cell culture insert with HBMECs as mentioned in Subheading 3.1. Documented below are the details pertaining to infection of cells of myeloid origin with HIV-1, incubation of HIV-1 infected supernatant on the in vitro BBB, and measurement of TEER as an indicator for permeability of the BBB.

3.4.1 Infection of Cells of Myeloid Origin with HIV-1

1. Plate and differentiate 5×10^4 primary human monocytes into macrophages in a 24-well tissue culture plate by incubating them at 37 °C for 5–10 days in 1 mL DMEM supplemented with 10% human serum, PenStrep, L-glutamine, and 6.6 ng/mL MCSF (Sigma). Change the media every other day. On day

5, begin to observe cells under a microscope to ensure that the majority of the plated cells have adhered and have gained a macrophage morphology. Upon confirmation of monocyte differentiation into macrophages through adherence and morphological indicators, they are poised for infection.

2. Post-differentiation, infect the macrophages with HIV-1 ADA isolate (10 ng p24). For infection with full-length infectious molecular clones, use the virus supernatants obtained from 293T cell transfection. For clinical isolates, employ infectious PBMC supernatants with similar p24 concentrations.
3. Incubate the infected macrophages between 5 and 10 days. Perform half-media changes every third day.
4. From day 5 through day 10, collect macrophage viral supernatants and measure p24 values using ABL HIV-p24 ELISA kit.

3.4.2 Incubation of HIV-1 Infected Supernatant on the In Vitro BBB to Study the Impact of HIV-1 on BBB Permeability

1. The in vitro BBB is prepared as mentioned above, by seeding HBMEC cells onto the upper chamber of the trans-well inserts at a density of 2×10^4 cells per insert, followed by an incubation period of 14 days in a CO_2 incubator.
2. These trans-wells are then placed in a new 24-well plate with fresh media in both the apical and basal chambers.
3. Following quantitation of p24, the HIV-infected macrophage supernatants are added at a ratio of 1:2 (Supernatant: Media) to the upper chamber of the trans-well inserts. Typically, 1 ng, 10 ng, and 100 ng (p24) of infected macrophage virus supernatants are used to titer the impact of HIV on the in vitro BBB. The results presented in Fig. 4 are from 250 μL of supernatant containing 100 ng p24 of HIV-1 ADA.
4. These trans-wells with HIV-infected supernatants are then incubated in a CO_2 incubator for a period of a predetermined time, typically 24 or 48 h to assess the impact on barrier integrity.
5. Trans-wells incubated with the supernatants of uninfected macrophages serve as controls.

3.4.3 Assess the Integrity and Barrier Function of the HBMEC Cell Monolayer by Measuring Its Transendothelial Electrical Resistance (TEER) Using an EVOM Volt-Ohmmeter

1. Ensure that the EVOM volt-ohmmeter is calibrated according to the manufacturer's instructions.
2. Sterilize EVOM electrodes in Isopropanol for 15 min and place them in PBS.
3. Remove the HBMEC cell monolayer cultured on trans-well inserts treated with HIV infected and uninfected supernatants from the incubator.
4. Allow the cell monolayer to equilibrate to room temperature for approximately 15–20 min.

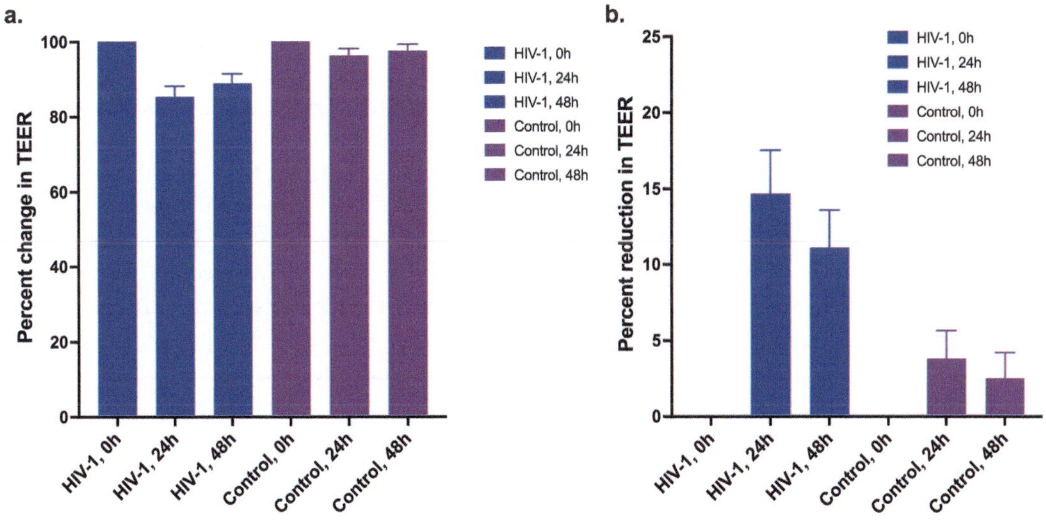

Fig. 4 Reduction in permeability of the in vitro BBB model upon exposure to HIV. The in vitro BBB model was exposed to supernatant obtained from HIV-infected macrophages. (Panel **a**) Illustrates percent change in transendothelial electrical resistance (TEER) upon exposure to supernatant from HIV-infected and uninfected macrophages (Control) at 24 h and 48 h timepoints. (Panel **b**) Illustrates the percent reduction in TEER in the same experiment

5. Remove the media from the trans-well inserts and move them to a new 24-well plate with 500 μL PBS per well. Add 100 μL of PBS to the upper chamber of the trans-well insert and allow the chambers equilibrate for 5 min.
6. Have a blank trans-well insert without any cells with PBS to measure background resistance by using the EVOM electrodes.
7. Gently place the EVOM electrodes in the treated trans-well chamber, with the longer electrode being placed in the bottom chamber, and the shorter electrode in the upper chamber of the insert.
8. Record the resistance value displayed on the EVOM volt-ohmmeter.
9. Calculations: Subtract the baseline resistance (from **step 6**) from the measured resistance (from **step 8**) to obtain the net resistance value. Multiply the net resistance value by the surface area of the trans-well insert to obtain the TEER value (0.33 cm^2 for a 24-well plate insert). The result is typically expressed in Ohm*cm^2.
10. Data Analysis: Compare the TEER values of treated HBMEC monolayers with control monolayers to assess the impact of any treatments or conditions on barrier function. Analyze the data statistically to identify any significant differences between groups.

    As documented in the protocol above, an in vitro BBB model was exposed to 250 μL of supernatant derived from HIV-infected macrophages and uninfected macrophage supernatant (control). Figure 4a depicts the percentage change in Transendothelial Electrical Resistance (TEER) at 24-h and 48-h timepoints, comparing the effects of supernatant from HIV-infected macrophages to uninfected macrophages (control). TEER was calculated by multiplying the net resistance value (actual value- baseline) obtained from EVOM measurements by the surface area of the trans-well insert. Figure 4b showcases the percentage reduction in TEER, providing insights into the impact of the HIV-infected macrophage supernatant on BBB integrity over the specified time intervals.

4 Notes

1. Collagen stock concentration will vary from batch to batch. The collagen solution, once prepared, is stable for 3 months at 4 °C.
2. Collagen gel is known to form at a wide range of concentrations. We routinely use 50 μg/mL concentration. One can test a range of concentrations around this if the two cell types are being replaced by immortalized cell lines of either astrocytes or HBMECs—which may display altered properties in adherence to the plastic surfaces.
3. Astrocytes and HBMECs can be passaged up to 50 times without the loss of morphology, rate of proliferation, or BBB integrity. Although HBMECs and astrocytes can be passaged even longer and potentially used successfully in the preparation and use of in vitro BBB, it is advisable to restart with a fresh vial after 50 passages.
4. When reviving astrocytes should never be centrifuged (even at low speed) to minimize cell death. The yields will be considerably higher when the centrifugation is avoided.
5. During the incubation, the media flows through the trans-well membrane to the other side, therefore 50 μL of media must be replenished every 5–15 min on the apical surface of each trans-well insert, and one should not allow the insert to dry.
6. In order to be doubly sure that no free dye remains, one can run the eluate in an SDS page gel and visualize on LiCOR Odyssey to ensure the absence of free dye.
7. If labeled antibody needs to be sterile, perform labeling in a tissue culture hood and filter sterilize contents through a .45 um filter after labeling.
8. Ensure that the EVOM electrodes are fully submerged in the medium when measuring TEER but not in direct contact with the insert or chamber walls.

References

1. Saloner R, Cysique LA (2017) HIV-associated neurocognitive disorders: a global perspective. J Int Neuropsychol Soc 23(9–10):860–869. https://doi.org/10.1017/S1355617717001102
2. Heaton RK, Clifford DB, Franklin DR Jr, Woods SP, Ake C, Vaida F, Ellis RJ, Letendre SL, Marcotte TD, Atkinson JH, Rivera-Mindt-M, Vigil OR, Taylor MJ, Collier AC, Marra CM, Gelman BB, McArthur JC, Morgello S, Simpson DM, McCutchan JA, Abramson I, Gamst A, Fennema-Notestine C, Jernigan TL, Wong J, Grant I, Group C (2010) HIV-associated neurocognitive disorders persist in the era of potent antiretroviral therapy: CHARTER study. Neurology 75(23): 2087–2096. https://doi.org/10.1212/WNL.0b013e318200d727
3. Davis LE, Hjelle BL, Miller VE, Palmer DL, Llewellyn AL, Merlin TL, Young SA, Mills RG, Wachsman W, Wiley CA (1992) Early viral brain invasion in iatrogenic human immunodeficiency virus infection. Neurology 42(9): 1736–1739. https://doi.org/10.1212/wnl.42.9.1736
4. Sillman B, Woldstad C, McMillan J, Gendelman HE (2018) Neuropathogenesis of human immunodeficiency virus infection. Handb Clin Neurol 152:21–40. https://doi.org/10.1016/B978-0-444-63849-6.00003-7
5. Ginhoux F, Greter M, Leboeuf M, Nandi S, See P, Gokhan S, Mehler MF, Conway SJ, Ng LG, Stanley ER, Samokhvalov IM, Merad M (2010) Fate mapping analysis reveals that adult microglia derive from primitive macrophages. Science 330(6005):841–845. https://doi.org/10.1126/science.1194637
6. Conant K, Garzino-Demo A, Nath A, McArthur JC, Halliday W, Power C, Gallo RC, Major EO (1998) Induction of monocyte chemoattractant protein-1 in HIV-1 Tat-stimulated astrocytes and elevation in AIDS dementia. Proc Natl Acad Sci USA 95(6): 3117–3121. https://doi.org/10.1073/pnas.95.6.3117
7. Weiss JM, Nath A, Major EO, Berman JW (1999) HIV-1 Tat induces monocyte chemoattractant protein-1-mediated monocyte transmigration across a model of the human blood-brain barrier and up-regulates CCR5 expression on human monocytes. J Immunol 163(5):2953–2959
8. Rao VR, Ruiz AP, Prasad VR (2014) Viral and cellular factors underlying neuropathogenesis in HIV associated neurocognitive disorders (HAND). AIDS Res Ther 11:13. https://doi.org/10.1186/1742-6405-11-13
9. Brightman MW, Reese TS (1969) Junctions between intimately apposed cell membranes in the vertebrate brain. J Cell Biol 40(3): 648–677. https://doi.org/10.1083/jcb.40.3.648
10. Bernas MJ, Cardoso FL, Daley SK, Weinand ME, Campos AR, Ferreira AJ, Hoying JB, Witte MH, Brites D, Persidsky Y, Ramirez SH, Brito MA (2010) Establishment of primary cultures of human brain microvascular endothelial cells to provide an in vitro cellular model of the blood-brain barrier. Nat Protoc 5(7):1265–1272. https://doi.org/10.1038/nprot.2010.76
11. Daneman R, Prat A (2015) The blood-brain barrier. Cold Spring Harb Perspect Biol 7(1): a020412. https://doi.org/10.1101/cshperspect.a020412
12. Deli MA, Abraham CS, Kataoka Y, Niwa M (2005) Permeability studies on in vitro blood-brain barrier models: physiology, pathology, and pharmacology. Cell Mol Neurobiol 25(1): 59–127. https://doi.org/10.1007/s10571-004-1377-8
13. Audus KL, Borchardt RT (1986) Characterization of an in vitro blood–brain barrier model system for studying drug transport and metabolism. Pharm Res 3:81–87
14. Dehouck MP, Jolliet-Riant P, Brée F, Fruchart JC, Cecchelli R, Tillement JP (1992) Drug transfer across the blood-brain barrier: correlation between in vitro and in vivo models. J Neurochem 58(5):1790–1797
15. Dehouck MP, Meresse S, Dehouck B, Fruchart JC, Cecchelli R (1992) In vitro reconstituted blood brain barrier. J Control Release 21(1–3):81–91
16. Schneider JR, Carias AM, Bastian AR, Cianci GC, Kiser PF, Veazey RS, Hope TJ (2017) Long-term direct visualization of passively transferred fluorophore-conjugated antibodies. J Immunol Methods 450:66–72. https://doi.org/10.1016/j.jim.2017.07.009

Part VII

HIV-1 Preclinical Studies

Chapter 20

Generation of Anti-HIV CAR-T Cells for Preclinical Research

Hang Su, Kim Anthony-Gonda, Rimas J. Orentas, Boro Dropulić, and Harris Goldstein

Abstract

The inability of people living with HIV (PLWH) to eradicate human immunodeficiency virus (HIV) infection is due in part to the inadequate HIV-specific cellular immune response. The antiviral function of cytotoxic CD8+ T cells, which are crucial for HIV control, is impaired during chronic viral infection because of viral escape mutations, immune exhaustion, HIV antigen downregulation, inflammation, and apoptosis. In addition, some HIV-infected cells either localize to tissue sanctuaries inaccessible to CD8+ T cells or are intrinsically resistant to CD8+ T cell killing. The novel design of synthetic chimeric antigen receptors (CARs) that enable T cells to target specific antigens has led to the development of potent and effective CAR-T cell therapies. While initial clinical trials using anti-HIV CAR-T cells performed over 20 years ago showed limited anti-HIV effects, the improved CAR-T cell design, which enabled its success in treating cancer, has reinstated CAR-T cell therapy as a strategy for HIV cure with notable progress being made in the recent decade.

Effective CAR-T cell therapy against HIV infection requires the generation of anti-HIV CAR-T cells with potent in vivo activity against HIV-infected cells. Preclinical evaluation of anti-HIV efficacy of CAR-T cells and their safety is fundamental for supporting the initiation of subsequent clinical trials in PLWH. For these preclinical studies, we developed a novel humanized mouse model supporting in vivo HIV infection, the development of viremia, and the evaluation of novel HIV therapeutics. Preclinical assessment of anti-HIV CAR-T cells using this mouse model involves a multistep process including peripheral blood mononuclear cells (PBMCs) harvested from human donors, T cell purification, ex vivo T cell activation, transduction with lentiviral vectors encoding an anti-HIV CAR, CAR-T cell expansion and infusion in mice intrasplenically injected with autologous PBMCs followed by the determination of CAR-T cell capacity for HIV suppression. Each of the steps described in the following protocol were optimized in the lab to maximize the quantity and quality of the final anti-HIV CAR-T cell products.

Key words HIV, Immunotherapy, Cytotoxic CD8+ T cell, CAR-T cell, Humanized mouse model

1 Introduction

Antiretroviral therapy (ART) suppresses human immunodeficiency virus (HIV) infection to undetectable levels and prevents viral transmission to uninfected individuals [1]. However, ART does not cure HIV. Despite years of ART, a population of

Vinayaka R. Prasad and Ganjam V. Kalpana (eds.), *HIV Protocols*, Methods in Molecular Biology, vol. 2807,
https://doi.org/10.1007/978-1-0716-3862-0_20,

HIV-infected cells (mainly CD4+ T and myeloid-derived cells) function as HIV reservoirs, which persist with a long half-life of approximately 44 months [2]. The disseminated anatomical distribution of HIV reservoirs in the peripheral blood, lymphoid, and other tissues including the less accessible central nervous system makes them extremely difficult to locate and eradicate [3]. Upon cessation of ART, HIV replication in latent infected cells occurs, enabling the emergence of HIV from the reservoir to mediate recurrent disseminated infection. The long-term remission of HIV in the absence of ART will likely require continued effective immune surveillance and elimination. While CD8+ T cells play a major role in killing HIV-infected cells, they are unable to fully suppress HIV expansion as HIV can quickly generate immune escape mutations in the HIV-derived peptide epitopes recognized by T cell receptors (TCRs), enabling infected cells to evade immune surveillance. As infection progresses, HIV-specific CD8+ T cells recognizing epitopes that are not mutated become exhausted after repetitive engagement of their TCR with viral antigens, and their exposure to inflammatory cytokines further impairs their antiviral function [4]. To overcome these limitations of HIV-specific T cell function, we and other groups have developed chimeric antigen receptor (CAR)-T cells targeting HIV-infected cells by engineering T cells with a CAR that targets highly conserved HIV antigens. The intrinsic attributes of CAR-T cells such as long half-life, broad tissue infiltration, and major histocompatibility complex (MHC) independence make them ideal therapeutic candidates for contributing to long-term HIV remission as indicated by the successful development and application of CAR-T cells for treating cancer [5]. Indeed, over the past decade, remarkable innovation has been achieved on developing more effective HIV-specific CAR-T cells, which strongly supports their role as a promising strategy for enabling HIV cure [6–10].

In current approved applications, clinical anti-HIV CAR-T cell therapeutic trials utilize molecular engineered autologous T cells collected from patients rather than allogeneic donor T cells, which would require further engineering to prevent elimination by the host immune system. After T cells are collected and isolated from the peripheral blood of a patient, they undergo ex vivo activation, transduction with the lentiviral vector (LV) encoding the anti-HIV CAR gene(s), and expansion of the CAR-T cells. Optimizing the conditions for each step of the procedure to generate ex vivo CAR-T cells is crucial to maximize the quantity and quality of anti-HIV CAR-T cells and eventually, their therapeutic efficacy [11]. After confirming successful ex vivo generation of functional anti-HIV CAR-T cells, they are infused back to the patient. This process, from T cell collection to anti-HIV CAR-T cell infusion, may take several weeks. After infusion, the therapeutic efficacy and side effects of anti-HIV CAR-T cells are closely monitored. Preclinical

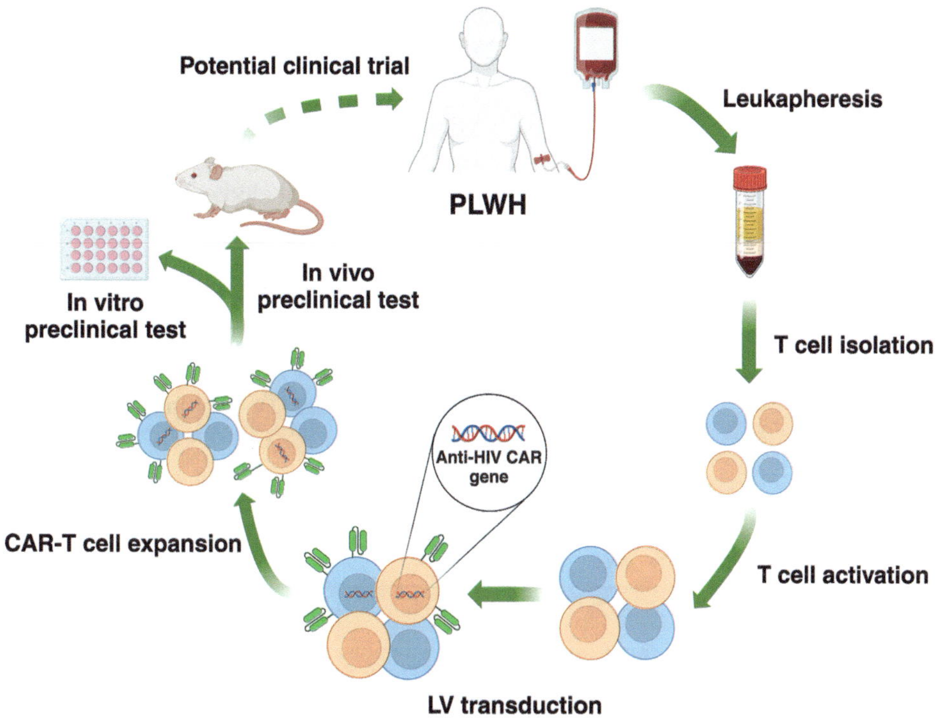

Fig. 1 Schematic of anti-HIV CAR-T cell engineering and production for preclinical research. Total PBMCs were isolated from people living with HIV (PLWH) via leukapheresis. T cells were then enriched using CD4 and CD8 microbeads and subsequently activated by anti-CD3 and anti-CD28 antibodies. After 24–72 h, activated T cells were transduced by the lentiviral vector (LV) encoding the anti-HIV CAR gene and converted to anti-HIV CAR-T cells. Anti-HIV CAR-T cells were expanded for 1–2 weeks and subjected to downstream in vitro and in vivo evaluation of their safety and antiviral function. Candidates displaying highly potent anti-HIV activities may proceed to the clinical trials

research that generates crucial efficacy and safety data is required to advance primary anti-HIV CAR-T cells into clinical trials. In this chapter, we present the techniques for the efficient generation of anti-HIV for in vitro and in vivo preclinical studies (Fig. 1). In addition, we have developed a high-throughput humanized mouse model system employing immune cells derived from human donors, including from PLWH, for evaluating the in vivo function of anti-HIV CAR-T cells [12].

2 Materials

1. Ficoll-Paque Plus density gradient media for PBMC isolation.
2. Cell culture media: Iscove's Modification of DMEM supplemented with 10% fetal bovine serum (FBS), penicillin (100 IU/mL), streptomycin (100 μg/mL), HEPES (25 mM), GlutaMAX™ (2 mM).

3. Trypan Blue solution (0.4%) for cell counting.
4. CD4 and CD8 microbeads (Miltenyi Biotec) for T cell enrichment.
5. LS Columns (Miltenyi Biotec) composed of ferromagnetic spheres to amplify magnetic fields for binding and isolation of superparamagnetic microbead-labeled cells.
6. Anti-CD3 and anti-CD28 antibodies for T cell activation.
7. Recombinant human IL-2 for cell culture.
8. Lentiviral vector encoding the anti-HIV .
9. 24-well plates and tissue culture flasks (e.g. T-25, T-75).
10. Darunavir (2 μM) to suppress HIV production in CAR-T cell culture from PLWH donors.
11. Freezing media: 90% FBS and 10% dimethyl sulfoxide (DMSO).

3 Methods

1. Isolate PBMCs from donor's peripheral blood by the Ficoll density gradient centrifugation method. Cryopreserve the isolated PBMCs in freezing media and aliquot into cryovials. Slow-freeze cryovials in a −80 °C freezer in a cryofreezing container overnight before placing them in liquid nitrogen for future use (*see* **Note 1**).
2. On the day before PBMCs usage, quickly thaw the cells in a 37 °C water bath and wash the cells with cell culture media to remove DMSO-containing cryopreservation media.
3. Perform cell count and viability determination using the Trypan Blue exclusion method.
4. Resuspend PBMCs in cell culture media supplemented with IL-2 (200 IU/mL) and incubate cells overnight in 37 °C incubator with 5% CO_2 (*see* **Note 2**).
5. The following day, perform cell count and viability determination using the Trypan Blue exclusion method and isolate CD4+ and CD8+ T cells from PBMCs using both the CD4 and CD8 microbeads together followed by passage through an LS column (*see* **Note 3**).
6. Resuspend isolated CD4+ and CD8+ T cells (1×10^6 cells/mL) in cell culture media supplemented with IL-2 (200 IU/mL). Activate CD4+ and CD8+ T cells by incubating with anti-CD3 and anti-CD28 antibodies (1 μg/mL each) and culture the cells at 37 °C for 48–72 h (*see* **Note 4**).

7. Adjust T cell concentration to 1×10^6 cells/mL without removing the anti-CD3 and anti-CD28 antibodies (*see* **Note 5**).
8. Transduce activated T cells with anti-HIV CAR lentiviral vector at a multiplicity of infection (MOI) of 40 (*see* **Notes 6** and **7**) and distribute cells into a 24-well plate (0.5×10^6 cells per well) (*see* **Note 8**).
9. Briefly and gently shake the plate. Place the plate in the 37 °C incubator for 15 min to allow the lentiviral vector to "settle on the T cells" (*see* **Note 9**).
10. Centrifuge the plate at 2500 rpm for 60 min at room temperature to amplify vector-cell contact and promote lentiviral vector transduction.
11. Add 500 μL of cell culture media supplemented with IL-2 (200 IU/mL) to each well and incubate the plate in the 37 ° C incubator for 48 h (*see* **Note 10**).
12. Harvest the transduced T cells and wash with 1× sterile PBS. Perform cell count after staining with Trypan Blue solution.
13. Resuspend the transduced cells at the density of 0.5×10^6 cells/mL in cell culture media supplemented with IL-2 (200 IU/mL). Culture the transduced T cells in a tissue culture flask appropriate for the selected T cell density and placed upright in the 37 °C incubator.
14. Examine the T cell culture daily by microscope. Every 2–3 days, or when cell confluence exceeds 70%, or when cell culture media turns yellow, change the cell culture media of the transduced T cells by performing **steps 12** and **13**.
15. Expand anti-HIV CAR-T cells for about 7–14 days to obtain sufficient numbers of CAR-T cells for in vitro or in vivo evaluation.
16. At the end of the anti-HIV CAR-T cell expansion, collect the T cells and count viable T cells after Trypan Blue solution staining. Use flow cytometry to quantify the anti-HIV CAR-T cell population by measuring the expression of CAR or a marker gene (Fig. 2 and *see* **Note 11**).
17. Preserve anti-HIV CAR-T cells using freezing media and store them in liquid nitrogen for future use as described above.
18. To investigate the in vivo antiviral function of anti-HIV CAR-T cells, a high-throughput humanized mouse model developed by our lab was employed where donor autologous PBMCs infected ex vivo with HIV or obtained from PLWH (10–100 × 10^6 cells) were intrasplenically injected into the immunodeficient NOD.Cg-*Prkdc*^{scid} *Il2rg*^{tm1Wjl}/SzJ (NSG) mice and anti-HIV CAR-T cells were administered

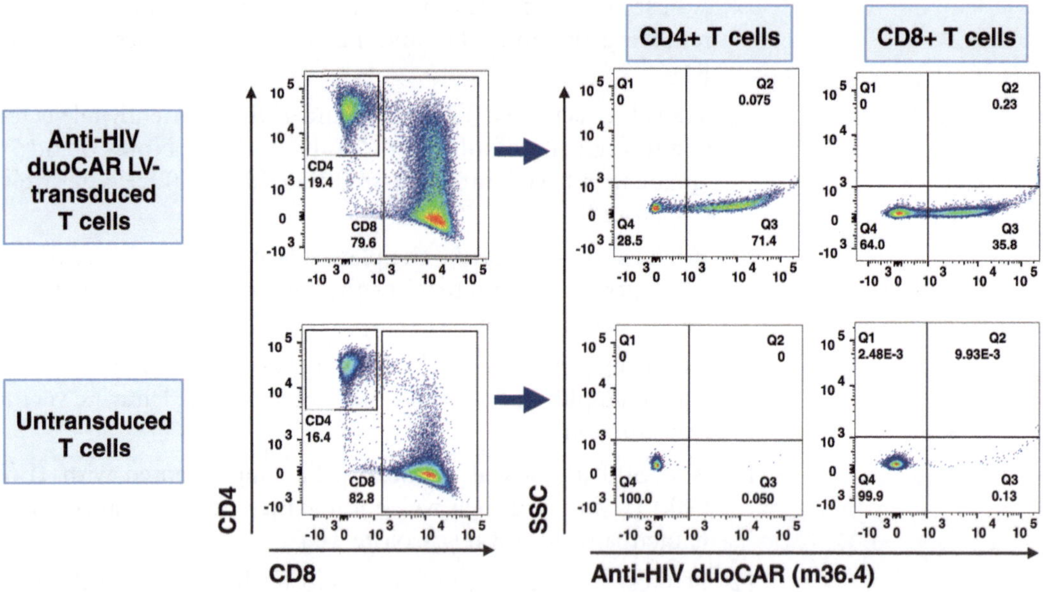

Fig. 2 Quantification of anti-HIV duoCAR expression on T cells by flow cytometry. Viable anti-HIV duoCAR lentiviral vector (LV)-transduced T cells (upper panel) were first gated on CD4 and CD8 expression (left). The increased CD4 expression on both transduced CD4+ and CD8+ T cells as compared to that on untransduced T cells (lower panel) was due to the mD1.22-CAR expression, which was derived from a modified D1 domain of CD4 receptor. CD4+ (middle) and CD8+ (right) T cells were then analyzed for the anti-HIV duoCAR (m36.4) expression. Untransduced T cells were tested as negative control

intrasplenically or intravenously ([7, 9, 12, 13], Fig. 3). These humanized mice can be maintained for weeks to months to quantify the in vivo HIV replication and to evaluate the antiviral function of anti-HIV CAR-T cells (*see* **Note 12**).

19. After animal sacrifice, harvest humanized mouse plasma and splenocytes and quantify the level of HIV infection and number of CD4+ T cells, to evaluate the anti-HIV activity of anti-HIV CAR-T cells. As shown in a representative experiment (Fig. 4), humanized mice were injected with HIV-infected autologous PBMCs and left untreated (red), or treated with untransduced T cells (UTD, blue), or treated with anti-HIV duoCAR-T cells (CAR-T, green), and maintained for 3 weeks before animal sacrifice. As mentioned in **Note 11**, duoCAR was designed to contain two CAR molecules binding to highly conserved regions of the HIV-1 envelope glycoprotein (mD1.22 and m36.4) [7, 9]. Levels of HIV viremia (Fig. 4a) and spleen HIV DNA (Fig. 4b) were significantly lower in anti-HIV duoCAR-T cell-treated humanized mice as compared to HIV control and UTD-treated humanized mice. The population of spleen CD4+ T cells was preserved in anti-HIV duoCAR-T cell-treated humanized mice as compared to the other two groups (Fig. 4c). Together, in the preclinical studies, we

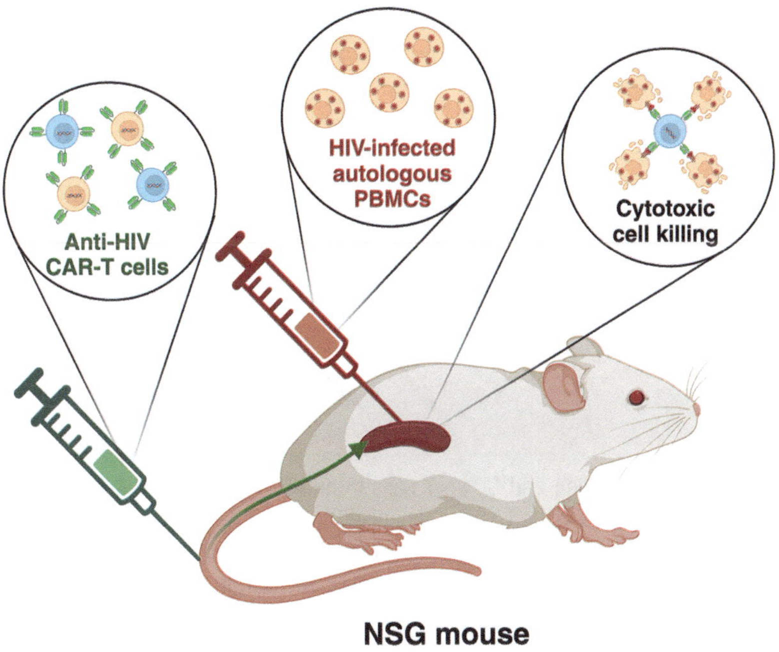

Fig. 3 In vivo evaluation of HIV inhibition by anti-HIV CAR-T cells using a high-throughput humanized mouse model. Adult NSG mice were first anesthetized. HIV-infected autologous PBMCs were then injected into the spleen while anti-HIV CAR-T cells were injected either into the spleen or via the tail vein according to the experimental design. After surgery, humanized mice were closely monitored until sacrifice. Mouse plasma and splenocytes were harvested to quantify HIV infection and determine the in vivo cytotoxicity of anti-HIV CAR-T cells

demonstrate that the anti-HIV duoCAR-T cells engineered and produced following the above protocol display potent in vivo suppression of HIV infection (*see* **Note 13**).

4 Notes

1. The entire protocol should be conducted in a Biosafety level 2+ (BSL-2+) lab, and biosafety precautions should be performed as required.
2. It has been reported that overnight resting of T cells from freeze-thaw process facilitates the recovery of T cell polyfunctionality [14].
3. The presence of CD4+ CAR-T cells enhances in vivo survival and function of CD8+ CAR-T cells [15]. In addition, cytotoxic CD4+ CAR-T cells play an important role in the long-term control of diseases [16, 17].
4. If the PBMC donor is a PLWH, antivirals should be added to the media to prevent CD4+ T cell loss from HIV infection.

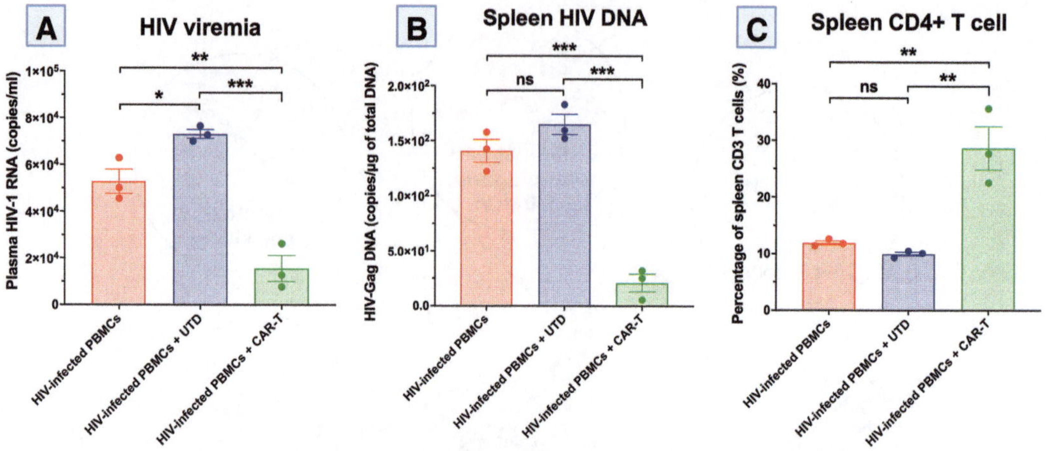

Fig. 4 Quantification of HIV infection and CD4+ T cells in anti-HIV duoCAR-T cell-treated humanized mice. Plasma and splenocytes were harvested from humanized mice that were administered with HIV-infected autologous PBMCs and left untreated (red), or treated with untransduced T cells (UTD, blue), or treated with anti-HIV duoCAR-T cells (CAR-T, green). Three animals were included in each group. (**a**) HIV viremia (plasma HIV RNA) and (**b**) spleen HIV DNA were quantified by RT-PCR and digital droplet PCR (ddPCR), respectively. (**c**) Spleen CD4+ T cells were quantified by flow cytometry and gated on CD3+ T cells. Data were presented as mean ± SEM. One-way ANOVA followed by Tukey's multiple comparisons post-test were used for comparisons between two groups for statistical analyses. $^{*}P < 0.05$, $^{**}P < 0.01$, $^{***}P < 0.001$. ns, not significant

5. Visualize activated T cells by microscope examination. Activated T cells should form clumps, have an irregular morphology, and increase in diameter.

6. Avoid freezing and thawing the lentiviral vector repeatedly, which reduces the lentiviral vector titer. Lentiviral vector should be stored in aliquots. During transduction, lentiviral vector stocks should be quickly thawed and added to the T cell culture at the appropriate MOI.

7. Using a higher MOI can increase the transduction efficiency of lentiviral vectors. However, high MOI can have cytotoxic effects on T cells. In addition, the lentiviral vector transduction efficiency may vary between donors. Therefore, for each individual donor, a pilot experiment for MOI optimization should be performed by testing a range of MOIs and the corresponding anti-HIV CAR expression of T cells. The minimum MOI at which a comparable level of anti-HIV CAR is expressed on T cells should be selected.

8. As an alternative to bulk transduction, activated T cells can be distributed into the 24-well plate first and then transduced individually. Our experience indicated that there was no significant difference between the two methods on anti-HIV CAR transduction efficiency.

9. To improve lentiviral vector transduction efficiency, transduction enhancers such as ViralEntry™ Transduction Enhancer (abm) or LentiBlast Premium Transduction Enhancer (OZBiosciences) can be used. Transduction enhancers improve lentiviral vector transduction efficiency by reducing electrostatic repulsions between the cellular membrane and lentiviral vector particles and enhancing lentiviral vector fusion with the cellular membrane.
10. While IL-2 supplemented in the cell culture media plays a major role in driving CAR-T cell proliferation, the CAR-T cell products grown in IL-2 may display a more differentiated and effector phenotype and display reduced in vivo survival [18]. The combination of IL-7 and IL-15 facilitates the production of CAR-T cells that are more activated, proliferative, and less exhausted as compared to CAR-T cells generated with IL-2 [19]. In addition, IL-7/IL-15 treatment enhances the preservation of less differentiated naïve/T stem memory CAR-T cells that are correlated with more extended in vivo CAR-T cell survival and enhanced cytotoxic responses [20, 21]. The effects of other cytokines added during CAR-T cell manufacture have been recently reviewed [11, 22].
11. One flow cytometry example of anti-HIV CAR detection is where we developed an anti-HIV CAR construct (named anti-HIV duoCAR) that targets two highly conserved binding sites on the HIV-1 envelope glycoprotein (mD1.22 and m36.4) [7, 9]. We employed a fluorochrome-labeled antibody that specifically detects the m36.4-CAR idiotype, which was used to quantify anti-HIV duoCAR expression on the LV-transduced T cells. In our studies, among different donors, the anti-HIV duoCAR expression on LV-transduced CD4+ and CD8+ T cells ranged from ~50- to 80% and from ~30- to 60%, respectively. Additional studies can be performed to characterize the cellular phenotypes of anti-HIV CAR-T cells, such as activation, exhaustion, and differentiation. In addition, ex vivo and in vivo cytotoxicity assays can be conducted to quantify the capacity of anti-HIV CAR-T cells to kill autologous HIV-infected cells by determining the lowest effector to target cell ratio that provides maximal killing or inhibition of HIV infection.
12. HIV is a human-specific virus and does not infect mice. To establish the in vivo animal model supporting HIV infection, humanized mice have been developed by reconstituting the human immune system in mice engineered by gene deletion to severely compromise their immune function and prevent rejection of transferred human cells to enable the study of HIV pathogenesis and evaluation of novel antiviral therapeutics [6, 23–26]. Using the humanized mouse model as

described in the current protocol, a wide range of donor HIV-infected PBMCs (from 10 to 100 million) can be injected into NSG mouse spleen to enable the establishment of sustained in vivo HIV infection for months. HIV viremia can be quantified at different time points to depict the temporal dynamics of in vivo HIV replication. Tissue HIV amplification, specifically in the spleen, can also be continuously monitored if using the *Renilla reniformis* luciferase HIV (HIV-LucR) for infection and quantified by the in vivo imaging system (IVIS) spectrum imager as described before [12]. Effector anti-HIV molecules/cells of interest can be administered once or multiple times according to the treatment design, via different routes (e.g. intrasplenic, intravenous, intraperitoneal, and subcutaneous) to evaluate their impact on in vivo HIV infection.

13. Valid animal models and rigorous preclinical tests enabled us to identify the CAR candidate with the most potent anti-HIV activity and then evaluate the impact of additional genetic modifications of the CAR to increase its anti-HIV function. To this end, the anti-HIV duoCAR-T cell therapy as introduced above that was developed and preclinically tested following the above protocol has proceeded to an ongoing phase I/IIa clinical trial (NCT04648046) to evaluate its safety and anti-HIV efficacy in ART-treated PLWH.

References

1. The Lancet H (2017) U=U taking off in 2017. Lancet HIV 4(11):e475. https://doi.org/10.1016/S2352-3018(17)30183-2
2. Finzi D, Blankson J, Siliciano JD, Margolick JB, Chadwick K, Pierson T, Smith K, Lisziewicz J, Lori F, Flexner C, Quinn TC, Chaisson RE, Rosenberg E, Walker B, Gange S, Gallant J, Siliciano RF (1999) Latent infection of CD4+ T cells provides a mechanism for lifelong persistence of HIV-1, even in patients on effective combination therapy. Nat Med 5(5):512–517. https://doi.org/10.1038/8394
3. Chaillon A, Gianella S, Dellicour S, Rawlings SA, Schlub TE, De Oliveira MF, Ignacio C, Porrachia M, Vrancken B, Smith DM (2020) HIV persists throughout deep tissues with repopulation from multiple anatomical sources. J Clin Invest 130(4):1699–1712. https://doi.org/10.1172/JCI134815
4. Fenwick C, Joo V, Jacquier P, Noto A, Banga R, Perreau M, Pantaleo G (2019) T-cell exhaustion in HIV infection. Immunol Rev 292(1):149–163. https://doi.org/10.1111/imr.12823
5. Hong M, Clubb JD, Chen YY (2020) Engineering CAR-T cells for next-generation cancer therapy. Cancer Cell 38(4):473–488. https://doi.org/10.1016/j.ccell.2020.07.005
6. Maldini CR, Claiborne DT, Okawa K, Chen T, Dopkin DL, Shan X, Power KA, Trifonova RT, Krupp K, Phelps M, Vrbanac VD, Tanno S, Bateson T, Leslie GJ, Hoxie JA, Boutwell CL, Riley JL, Allen TM (2020) Dual CD4-based CAR T cells with distinct costimulatory domains mitigate HIV pathogenesis in vivo. Nat Med 26(11):1776–1787. https://doi.org/10.1038/s41591-020-1039-5
7. Anthony-Gonda K, Bardhi A, Ray A, Flerin N, Li M, Chen W, Ochsenbauer C, Kappes JC, Krueger W, Worden A, Schneider D, Zhu Z, Orentas R, Dimitrov DS, Goldstein H, Dropulic B (2019) Multispecific anti-HIV duoCAR-T cells display broad in vitro antiviral activity and potent in vivo elimination of HIV-infected cells in a humanized mouse model. Sci Transl Med 11(504). https://doi.org/10.1126/scitranslmed.aav5685
8. Liu B, Zhang W, Xia B, Jing S, Du Y, Zou F, Li R, Lu L, Chen S, Li Y, Hu Q, Lin Y, Zhang Y, He Z, Zhang X, Chen X, Peng T,

Tang X, Cai W, Pan T, Li L, Zhang H (2021) Broadly neutralizing antibody-derived CAR T cells reduce viral reservoir in individuals infected with HIV-1. J Clin Invest 131(19). https://doi.org/10.1172/JCI150211

9. Anthony-Gonda K, Ray A, Su H, Wang Y, Xiong Y, Lee D, Block A, Chilunda V, Weiselberg J, Zemelko L, Wang YY, Kleinsorge-Block S, Reese JS, de Lima M, Ochsenbauer C, Kappes JC, Dimitrov DS, Orentas R, Deeks SG, Rutishauser RL, Berman JW, Goldstein H, Dropulic B (2022) In vivo killing of primary HIV-infected cells by peripheral-injected early memory-enriched anti-HIV duoCAR T cells. JCI Insight 7(21). https://doi.org/10.1172/jci.insight.161698
10. York J, Gowrishankar K, Micklethwaite K, Palmer S, Cunningham AL, Nasr N (2022) Evolving strategies to eliminate the CD4 T cells HIV viral reservoir via CAR T cell immunotherapy. Front Immunol 13:873701. https://doi.org/10.3389/fimmu.2022.873701
11. Stock S, Schmitt M, Sellner L (2019) Optimizing manufacturing protocols of chimeric antigen receptor T cells for improved anticancer immunotherapy. Int J Mol Sci 20(24). https://doi.org/10.3390/ijms20246223
12. Thomas T, Seay K, Zheng JH, Zhang C, Ochsenbauer C, Kappes JC, Goldstein H (2016) High-throughput humanized mouse models for evaluation of HIV-1 therapeutics and pathogenesis. Methods Mol Biol 1354: 221–235. https://doi.org/10.1007/978-1-4939-3046-3_15
13. Li M, Garforth SJ, O'Connor KE, Su H, Lee DM, Celikgil A, Chaparro RJ, Seidel RD, Jones RB, Arav-Boger R, Almo SC, Goldstein H (2021) T cell receptor-targeted immunotherapeutics drive selective in vivo HIV- and CMV-specific T cell expansion in humanized mice. J Clin Invest 131(23). https://doi.org/10.1172/JCI141051
14. Kutscher S, Dembek CJ, Deckert S, Russo C, Korber N, Bogner JR, Geisler F, Umgelter A, Neuenhahn M, Albrecht J, Cosma A, Protzer U, Bauer T (2013) Overnight resting of PBMC changes functional signatures of antigen specific T- cell responses: impact for immune monitoring within clinical trials. PLoS One 8(10):e76215. https://doi.org/10.1371/journal.pone.0076215
15. Walker RE, Bechtel CM, Natarajan V, Baseler M, Hege KM, Metcalf JA, Stevens R, Hazen A, Blaese RM, Chen CC, Leitman SF, Palensky J, Wittes J, Davey RT Jr, Falloon J, Polis MA, Kovacs JA, Broad DF, Levine BL, Roberts MR, Masur H, Lane HC (2000) Long-term in vivo survival of receptor-modified syngeneic T cells in patients with human immunodeficiency virus infection. Blood 96(2):467–474
16. Melenhorst JJ, Chen GM, Wang M, Porter DL, Chen C, Collins MA, Gao P, Bandyopadhyay S, Sun H, Zhao Z, Lundh S, Pruteanu-Malinici I, Nobles CL, Maji S, Frey NV, Gill SI, Loren AW, Tian L, Kulikovskaya I, Gupta M, Ambrose DE, Davis MM, Fraietta JA, Brogdon JL, Young RM, Chew A, Levine BL, Siegel DL, Alanio C, Wherry EJ, Bushman FD, Lacey SF, Tan K, June CH (2022) Decade-long leukaemia remissions with persistence of CD4(+) CAR T cells. Nature 602(7897):503–509. https://doi.org/10.1038/s41586-021-04390-6
17. Anderson ND, Birch J, Accogli T, Criado I, Khabirova E, Parks C, Wood Y, Young MD, Porter T, Richardson R, Albon SJ, Popova B, Lopes A, Wynn R, Hough R, Gohil SH, Pule M, Amrolia PJ, Behjati S, Ghorashian S (2023) Transcriptional signatures associated with persisting CD19 CAR-T cells in children with leukemia. Nat Med. https://doi.org/10.1038/s41591-023-02415-3
18. Pipkin ME, Sacks JA, Cruz-Guilloty F, Lichtenheld MG, Bevan MJ, Rao A (2010) Interleukin-2 and inflammation induce distinct transcriptional programs that promote the differentiation of effector cytolytic T cells. Immunity 32(1):79–90. https://doi.org/10.1016/j.immuni.2009.11.012
19. Gargett T, Brown MP (2015) Different cytokine and stimulation conditions influence the expansion and immune phenotype of third-generation chimeric antigen receptor T cells specific for tumor antigen GD2. Cytotherapy 17(4):487–495. https://doi.org/10.1016/j.jcyt.2014.12.002
20. Xu Y, Zhang M, Ramos CA, Durett A, Liu E, Dakhova O, Liu H, Creighton CJ, Gee AP, Heslop HE, Rooney CM, Savoldo B, Dotti G (2014) Closely related T-memory stem cells correlate with in vivo expansion of CAR.CD19-T cells and are preserved by IL-7 and IL-15. Blood 123(24):3750–3759. https://doi.org/10.1182/blood-2014-01-552174
21. Arcangeli S, Bove C, Mezzanotte C, Camisa B, Falcone L, Manfredi F, Bezzecchi E, El Khoury R, Norata R, Sanvito F, Ponzoni M, Greco B, Moresco MA, Carrabba MG, Ciceri F, Bonini C, Bondanza A, Casucci M (2022) CAR T cell manufacturing from naive/stem memory T lymphocytes enhances antitumor responses while curtailing cytokine release syndrome. J Clin Invest 132(12). https://doi.org/10.1172/JCI150807

22. Silveira CRF, Corveloni AC, Caruso SR, Macedo NA, Brussolo NM, Haddad F, Fernandes TR, de Andrade PV, Orellana MD, Guerino-Cunha RL (2022) Cytokines as an important player in the context of CAR-T cell therapy for cancer: their role in tumor immunomodulation, manufacture, and clinical implications. Front Immunol 13:947648. https://doi.org/10.3389/fimmu.2022.947648
23. Dash PK, Kaminski R, Bella R, Su H, Mathews S, Ahooyi TM, Chen C, Mancuso P, Sariyer R, Ferrante P, Donadoni M, Robinson JA, Sillman B, Lin Z, Hilaire JR, Banoub M, Elango M, Gautam N, Mosley RL, Poluektova LY, McMillan J, Bade AN, Gorantla S, Sariyer IK, Burdo TH, Young WB, Amini S, Gordon J, Jacobson JM, Edagwa B, Khalili K, Gendelman HE (2019) Sequential LASER ART and CRISPR treatments eliminate HIV-1 in a subset of infected humanized mice. Nat Commun 10(1):2753. https://doi.org/10.1038/s41467-019-10366-y
24. McBrien JB, Mavigner M, Franchitti L, Smith SA, White E, Tharp GK, Walum H, Busman-Sahay K, Aguilera-Sandoval CR, Thayer WO, Spagnuolo RA, Kovarova M, Wahl A, Cervasi B, Margolis DM, Vanderford TH, Carnathan DG, Paiardini M, Lifson JD, Lee JH, Safrit JT, Bosinger SE, Estes JD, Derdeyn CA, Garcia JV, Kulpa DA, Chahroudi A, Silvestri G (2020) Robust and persistent reactivation of SIV and HIV by N-803 and depletion of CD8(+) cells. Nature 578(7793):154–159. https://doi.org/10.1038/s41586-020-1946-0
25. Dash PK, Chen C, Kaminski R, Su H, Mancuso P, Sillman B, Zhang C, Liao S, Sravanam S, Liu H, Waight E, Guo L, Mathews S, Sariyer R, Mosley RL, Poluektova LY, Caocci M, Amini S, Gorantla S, Burdo TH, Edagwa B, Gendelman HE, Khalili K (2023) CRISPR editing of CCR5 and HIV-1 facilitates viral elimination in antiretroviral drug-suppressed virus-infected humanized mice. Proc Natl Acad Sci USA 120(19): e2217887120. https://doi.org/10.1073/pnas.2217887120
26. Su H, Cheng Y, Sravanam S, Mathews S, Gorantla S, Poluektova LY, Dash PK, Gendelman HE (2019) Immune activations and viral tissue compartmentalization during progressive HIV-1 infection of humanized mice. Front Immunol 10:340. https://doi.org/10.3389/fimmu.2019.00340

Chapter 21

Highly Sensitive Analysis of Cervical Mucosal HIV-1 Infection Using Reporter Viruses Expressing Secreted Nanoluciferase

Dana F. Indihar, Jennifer J. Jones, Christina Ochsenbauer, and John C. Kappes

Abstract

Ex vivo cervical tissue explant models offer a physiologically relevant approach for studying virus-host interactions that underlie mucosal HIV-1 transmission to women. However, the utility of cervical explant tissue (CET) models has been limited for both practical and technical reasons. These include assay variation, inadequate sensitivity for assessing HIV-1 infection and replication in tissue, and constraints imposed by the requirement for using multiple replica samples of CET to test each experimental variable and assay parameter. Here, we describe an experimental approach that employs secreted nanoluciferase (sNLuc) and current HIV-1 reporter virus technologies to overcome certain limitations of earlier ex vivo CET models. This method augments application of the CET model for investigating important questions involving mucosal HIV-1 transmission.

Key words HIV-1, Transmission, Cervix, Cervical mucosa, Ex vivo model, Reporter virus, Nanoluciferase

1 Introduction

HIV-1 infection of the genital mucosa represents the principal route of all transmission events worldwide [1, 2]. Tissue explant models may offer major advantages for studying disease and virus–host interactions in humans since they recapitulate the cytoarchitecture and temporal spatial dynamics and functional interactions that exist in vivo. Unique insights into key aspects of mucosal HIV-1 transmission have come from multi-parametric imaging studies of mucosal tissue following simian immunodeficiency virus (SIV) vaginal challenge of rhesus macaques, including the identification of early infected target cells in vivo, early virus-host dynamics, and patterns of viral acquisition and dissemination [3–6]. Virus

Vinayaka R. Prasad and Ganjam V. Kalpana (eds.), *HIV Protocols*, Methods in Molecular Biology, vol. 2807,
https://doi.org/10.1007/978-1-0716-3862-0_21,

challenge studies utilizing human mucosal tissue models have helped to understand early pathogen-host interactions [7, 8].

Cervical explant tissue (CET) has been utilized to create ex vivo model systems of mucosal HIV-1 transmission. A standardized protocol using CET to study human pathogens that cause infection through mucosal routes was described in 2009 by Margolis and Grivel [9]. This ex vivo cervical tissue model has notable features pertinent to the assessment of early virus-host interactions in a physiologically relevant context, including that the tissue cytoarchitecture and spatial distribution of HIV-target cell populations is preserved during short-term culture, and exogenous activation is not required for infection of tissue-resident CD4 T lymphocytes [9–14]. Notably, the CET model recapitulates infection by CCR5-tropic HIV-1, while it is refractory to infection by CXCR4-tropic virus [10, 15, 16]. In other studies utilizing the ex vivo CET model, transmitted/founder (T/F) viruses were shown to productively infect cervical tissue and replicate in CD4 T cells [17]. As our understanding of HIV transmission biology improved, these models evolved to address more complex questions. In 2018, Introini et al. described ex vivo CET culture methods using gelatin sponges for non-polarized CET infection that facilitate access of the cultured tissue to medium nutrients and improve viability [16]. These and other ex vivo cervical tissue culture systems have been successfully used to investigate early disease pathogenesis, cell targets of acute mucosal HIV infection [18, 19], infection and replication efficiencies of mucosally transmitted HIV [15, 17, 19], and efficacy of anti-HIV therapeutics [20–23].

Intact virus-infected CET can be analyzed by various methods to assess HIV-1 infection and virus-host interactions. The temporal-spatial dynamics of virus replication and spread among target cells within the tissue can be evaluated using techniques such as multi-parametric imaging analysis of virus antigen (e.g., p24 antigen) and cell-specific differentiation markers. Using cells isolated from the tissue, multicolor flow cytometric analyses have helped inform the identity and dynamics of target cell infection. While these ex vivo CET models offer certain advantages over monotypic bi-dimensional cell cultures, the models are technologically demanding. Obtaining meaningful research results is further complicated by intra- and inter-donor variation and the progressive loss of tissue integrity and functional architecture during ex vivo culture. Established CET models analyze multiple replica tissue samples for assessment of each experimental condition to augment sensitivity and minimize assay variation [9, 10, 16, 24]. Practical constraints imposed by requiring multiple replica samples of dissected tissue for the assessment of each experimental variable and assay parameter have further limited usefulness of the CET model.

Here, we describe a method that employs HIV-1 reporter virus technologies using physiologically relevant HIV-1 strains engineered to encode secreted nanoluciferase (sNLuc) that overcome certain limitations of the ex vivo CET transmission model of HIV-1 infection. The method offers unprecedented sensitivity for detecting and quantifying *de novo* HIV-1 infection and replication in situ within the first few days after virus challenge. Notably, using the highly sensitive sNLuc readout, quantitative assessment of infection and replication kinetics of HIV-1 reporter viruses in individual 2-mm^2 mucosal explant tissue blocks is highly reproducible and minimizes the intra- and inter-donor variation that has been reported with previous CET models [9, 10, 16, 24]. Thus, this method, which expresses sNLuc from replication-competent HIV-1 reporter viruses engineered in a manner intended to retain wild-type-like expression and function of all viral proteins with physiologically relevant HIV-1 virus strains, augments the utility of the CET model. In this chapter, we describe experimental methods that utilize sNLuc reporter viruses to analyze HIV-1 infection and replication in human CET. We describe applicable HIV-1 reporter virus technologies, the generation and titration of viral stocks, the acquisition and dissection of CET for ex vivo culture, an efficient non-polarized approach for virus inoculation, and the sampling and analysis of CET culture supernatants for sNLuc activity as a surrogate of productive HIV-1 infection. Finally, we show representative results to demonstrate performance in terms of assay sensitivity and reproducibility. The methodology can be adapted to augment research on correlates of mucosal HIV-1 transmission and to help understand early virus-host interactions that underlie that acquisition of mucosal HIV-1 infection.

2 Materials

2.1 Human Tissue Acquisition

The use of human tissues must adhere to all Institutional Review Board (IRB) protocols and guidelines, including those for informed patient consent. Remnant cervical tissue may be obtained from individuals undergoing medically necessary hysterectomies in hospitals associated with the research institution or sourced from a tissue bank. Standard clinical protocol requires assessment of the resected tissue by the surgical pathology laboratory for abnormalities and disease. Clear and precise communication with individuals directly involved with tissue procurement from the surgical pathology lab is vital for obtaining useful remnant tissue (*see* **Note 1**).

2.2 HIV-1 Reporter Infectious Molecular Clones

The methods described herein utilize replication-competent HIV-1 reporter viruses engineered to express secreted nanoluciferase (sNLuc) in a manner intended to retain wild-type-like expression and function of all viral proteins, including Nef. HIV-1 infectious

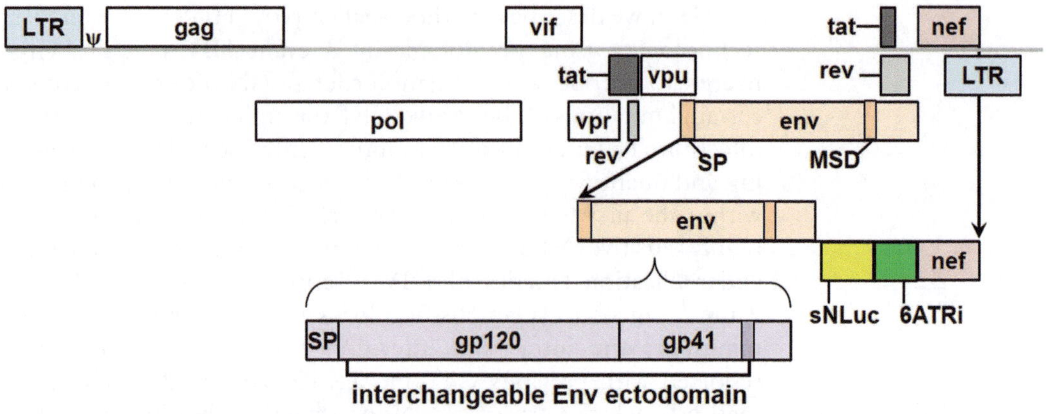

Fig. 1 Schematic of HIV-1 reporter virus construct. Schematic representation of the insertion of the coding sequence for a reporter gene (shown in yellow), followed by a modified, truncated ECMV internal ribosome entry site (IRES) sequence (6ATRi, shown in green), into an HIV-1 proviral genome between the *env* and *nef* orfs. First described in Alberti et al. [31] for the *Renilla* (LucR) reporter 6ATRi cassette, this approach has been adapted to express sNLuc [44], used herein, as well as other reporter genes (*see* **Note 2**). The modified 6ATRi IRES has been shown to maintain native-like levels of Nef expression [31, 32, 37, 39]. The reporter cassette can be inserted into T/F-IMC or HIV-1 references strains like NL4-3. The schematic also depicts how the coding sequence for the NL4-3 Env ectodomain, encompassing essentially all of gp120 and the external and membrane-spanning domain (MSD) of gp41, can be exchanged for heterologous *env* sequences. As illustrated, this avoids any overlap with *vpu*, *tat,* and *rev* orfs and mimics preferential in vivo recombination sites in the C1 and membrane spanning domain coding regions of *env* [56, 57]. SP: signal peptide; MSD: membrane spanning domain

molecular clones (IMCs) modified to express sNLuc can comprise either full-length proviral genomes of primary viruses (e.g. T/F HIV-1 IMC) or an isogenic NL4-3 proviral backbone modified to encode the Env ectodomain of heterologous *env* genes of interest (Env-IMC). Notably, the latter strategy facilitates studies involving the virologic and biological properties of different Env strains of HIV-1 while retaining the entire NL4-3 *vpu*, *tat,* and *rev* open reading frames, thus avoiding inter-strain chimerisms [25]. Figure 1 depicts the molecular genetic structure of the two complementary strategies: sNLuc-IMCs and sNLuc-Env-IMCs. We have extensively reported on the rationale, molecular design, and virologic properties of different reporter virus approaches (*see* **Note 2**).

1. sNLuc-Env-IMC and sNLuc-IMC proviral plasmids. These plasmid DNAs are prepared using appropriate midi- or maxi-prep methods to ensure high-quality, endotoxin-free, and sufficiently concentrated (ideally >0.5 μg/μl) DNA master stocks (*see* **Note 3**). For consideration of virus strains to utilize, *see* **Note 4**.

2.3 Generation of Virus Stocks

1. HIV proviral DNA plasmid clones to generate infectious virus by transfection.
2. 293T cells. These human embryonic kidney cells contain the simian virus 40 (SV40) large tumor antigen (T-antigen), which enhances expression from transfected plasmid DNA containing the SV40 origin of replication sequence [26, 27].
3. Dulbecco's Modified Eagle's Medium (DMEM) (*see* **Note 5**).
4. D10++ Medium: DMEM comprising 10% heat-inactivated fetal bovine serum (FBS), 25 mM HEPES, 2 mM L-glutamine, 100 μg/ml streptomycin, 100 U/ml of penicillin.
5. FuGENE™ HD Transfection Reagent, or equivalent.
6. 2 ml screw-cap microtubes.
7. Standard 6-well mammalian cell culture plates: clear, flat-bottom, sterile, polystyrene.

2.4 Virus Stock Titration

1. TZM-bl reporter cells [28], available from the NIH HIV Reagent Program (managed by the ATCC) and contributed by Drs. John C. Kappes and Xiaoyun Wu. In response to HIV-1 infection and Tat expression, TZM-bl cells express both firefly luciferase and β-galactosidase (β-gal). In this chapter, we use the β-gal reporter to determine virus titers: TZM-bl infectious units per ml (IU/ml).
2. D10++ Medium: As described above.
3. D1++ Medium: DMEM comprising 1% heat-inactivated fetal bovine serum (FBS), 25 mM HEPES, 2 mM L-glutamine, 100 μg/ml streptomycin, 100 U/ml of penicillin.
4. DEAE-Dextran stock solution: Sterile deionized water containing 10 mg/ml diethylaminoethyl-dextran hydrochloride. Aliquot 500–1000 μl into microtubes and store at −20 °C.
5. Standard 24-well mammalian cell culture plates: Clear, flat-bottom, sterile, polystyrene, treated for cell culture.
6. Cell fixation buffer: Combine 500 ml of 1× PBS with 4 ml glutaraldehyde solution (Grade I; 25% in H_2O) and 11 ml formaldehyde solution (37 wt. % in H_2O; contains 10–15% methanol as stabilizer). Store in the dark at 4 °C.
7. X-gal (5-Bromo-4-Chloro-3-Indolyl-AY-D-galactopyranoside) stock solution: 100 ml Dimethyl Sulfoxide (DMSO) containing 40 mg/ml X-gal.
8. TZM-bl cell β-gal staining solution: For 100 ml of staining solution, combine 94.8 ml of 1× PBS, 2 ml 0.2 M Potassium Ferricyanide, 2 ml 0.2 M Potassium Ferrocyanide, 200 μl 1M $MgCl_2$, and 1 ml of 40 mg/ml X-gal stock solution. Store the first two components in the dark at 4 °C, and store X-gal stock

solution at -20°C ; combine individual components to make the X-gal staining solution fresh when needed.

9. Brightfield or phase-contrast microscope.
10. Manual tally counter for enumerating β-gal+ cells.

2.5 Cervical Explant Medium (CEM)

1. Cervical explant medium (CEM): RPMI 1640 medium comprising 1× Modified Eagle's medium (MEM) non-essential amino acids solution, 1 mM sodium-pyruvate, 50 μg/ml gentamicin, 15% fetal bovine serum (FBS), 2 mM L-glutamine, 100 U/ml penicillin, 100 ug/ml streptomycin. Store at 4 °C. Just prior to use for cervical explant culture, one may supplement CEM with 30 U/ml IL-2.

2.6 Tissue Dissection

1. The materials and reagents described in this section should be readily available to ensure that tissue can be processed without delay after it is received.
2. Class II biological safety cabinet.
3. CEM: Described in Subheading 2.5.
4. The surgical instruments listed below are recommended for tissue dissection. These instruments (a–e) should be autoclaved before dissection in order to reduce risk of microbial contamination.
   - (a) #4 stainless-steel scalpel blade handle.
   - (b) Standard stainless-steel dissecting scissors.
   - (c) General-purpose extra-long stainless-steel tweezers/forceps.
   - (d) Precision tweezers/forceps with curved tips and fine serrated points.
   - (e) Precision straight and tapered tweezers/forceps.
   - (f) Disposable, sterile surgical blades fitting the #4 stainless-steel scalpel blade handle.
   - (g) Mopec BladeFLASK Blade Remover (or equivalent) to safely remove scalpel blades from the handle.
5. Sterile WypAll™ L40 Wipers, or equivalent. These are thick wipes/pads that help delineate a sterile workspace during the dissection process. Prepare and store in a reusable vessel that will withstand steam sterilization.
6. Laminated 2-mm × 2-mm dissection guide. This is a reusable template made to facilitate the tissue dissection into uniform 2 mm^3 blocks. Go to https://gridzzly.com and in the white box with the text "Make your own grid paper with Gridzzly," click on the "□," and then type "2" into the first space. This will generate a 2-mm × 2-mm grid pattern. Print on 8.5 × 11″ paper, cut into four quadrants, and laminate.

7. Standard tissue culture dish: sterile, 10 cm diameter.
8. Vinyl ruler: ~15 cm in length with 1 mm demarcations. Choose one that is flat, with low profile that can be positioned under a cell culture dish for measuring dimensions of remnant tissue samples.
9. Surgical face masks.
10. Powder-free nitrile gloves.
11. Optional tools that may facilitate tissue dissection: A focused-beam light source, a dissection microscope, surgical/dental loupes.

2.7 CET Infection and Culture

1. 2 ml screw-cap microtubes.
2. Tube revolver/rotator with 2 ml microtube paddle attachments or equivalent device.
3. Humidified CO_2 incubator with CO_2 levels at 5% and temperature at 37 °C.
4. Standard 48-well tissue culture plates, clear, flat-bottom, sterile, polystyrene, treated for cell culture.
5. Standard tissue culture dishes: sterile, 10 cm diameter.
6. SURGIFOAM Absorbable Gelatin Sponge purchased in sterile pieces measuring 2 cm × 6 cm × 0.7 cm, referred to as "gelatin sponge" throughout the text.
7. 12-channel reagent reservoirs with lids: sterile, contains up to 5 ml per channel.

2.8 Secreted Nanoluciferase Analysis

1. Promega Nano-Glo™ Luciferase Assay System. This assay kit includes Nano-Glo™ Luciferase assay Substrate, and Nano-Glo™ Luciferase Assay Buffer (*see* **Note 6**). This Nano-Glo™ System generates a continuous glow-type luminescence with a half-life of approximately 120 min.
2. Luminometer (Promega GloMax 96 Microplate Luminometer or equivalent).
3. 96-well white plates: Non-sterile, flat-bottom.
4. Computer-based data analysis and graphing software.

3 Methods

3.1 Generation of Infectious HIV-1 Stocks

The generation of HIV-1 stocks involves replication competent HIV-1 molecular clones and recombinant DNA. Therefore, appropriate biosafety practices must be followed, including the necessary biocontainment infrastructure. This includes consulting applicable Occupational Safety and Health Administration (OSHA) regulations and implementing proper biosafety level (BSL) containment procedures and practices, and review of the NIH Guidelines for research involving recombinant or synthetic nucleic acid molecules [29], and the CDC report on Biosafety in Microbiological and Biomedical Laboratories (BMBL) [30]. Herein, we describe how stocks (master banks) of HIV-1 reporter viruses are generated, aliquoted into multiple vials, stored at −80 °C, and titered for TZM-bl infectious units. Working from master banks of characterized virus augments the reproducibility of results between experiments and ensures that the viruses are ready for immediate use when tissue becomes available.

1. Seed 4×10^5 293T cells in 3 ml of D10++ medium per well in 6-well plates. The next day, the cells should be adherent and evenly distributed on the bottom of each well. Suboptimal virus titers will be obtained if the wells contain cell aggregates or regions of high cell density around the perimeter or the center of the wells.
2. Between 18 and 24 h after seeding, transfect the 293T cells with proviral DNA to generate infectious virus (*see* **Note 7**).
3. Collect the culture supernatant 60–72 h after transfection, and centrifuge it at 200 g for 10 min to pellet cells. Transfer the supernatant into a new tube. A second centrifugation step at higher speed can be performed to remove subcellular debris.
4. Aliquot the clarified supernatant into pre-labeled 2 ml screw-cap microtubes and store at −80 °C. We generally recommend aliquots of 250 μl. Each well of transfected cells will yield ~3 ml of supernatant, which represents 12 vials containing 250 μl each. Depending on the scope of the research, it may be beneficial to transfect additional cells/wells to procure greater amounts of virus, both in terms of volume per vial and number of vials.

3.2 Titration of HIV-1 Stocks

Viruses should be titered using TZM-bl cells [28] by β-gal staining to quantify infectious units (IU) per milliliter. The efficiency of proviral DNA transfection, condition of the TZM-bl cell culture, and the viral strain are all factors that will influence the virus titer.

1. Seed 7×10^4 TZM-bl cells in each well of a 24-well plate with 500 μl of D10++ medium per well, and incubate at 37 °C with 5% CO_2. The titration of each virus will require 10 wells.
2. The next day, prepare serial dilutions of the virus for titration. Using a 96-well plate, mix 20 μl of the virus stock with 180 μl of D1++ medium supplemented with 40 μg/ml DEAE-dextran. Then prepare five 5-fold serial dilutions by transferring 40 μl into 160 μl of D1++ medium containing 40 μg/ml DEAE-dextran.

3. Gently remove the cell culture supernatant from the 24-well plate and replace it with 200 μl per well of D1++ medium containing 40 μg/ml of DEAE-dextran. Add 50 μl of each of the 5-fold virus dilutions to two replica wells (50 μl each) of the 24-well plate, and mix by gentle horizontal rotation and side-to-side movement. Incubate at 37 °C for 4 h, and then add 500 μl per well of pre-warmed D10++ medium per well. Continue to incubate at 37 °C for 48 h.
4. Remove the supernatant from each well and add 250 μl of TZM-bl cell fixation buffer being careful to not dislodge the cells in these and subsequent steps, and incubate for 5–8 min at room temperature. Gently remove the cell fixation buffer, and wash each well three times with 500 μl of PBS. After the final wash, add 250 μl of TZM-bl β-gal staining solution per well and incubate at 37 °C for 2 h.
5. Remove the β-gal staining solution and add 250 μl PBS per well. The β-gal+ blue-stained cells can be enumerated immediately or the 24-well plate can be stored in a sealed plastic bag at 4 °C for counting at a later time.
6. To calculate the TZM-bl IU-titer from the β-gal+ blue-stained cells, use a microscope to visually inspect each well at each dilution. Identify two replica-dilution wells with 50–200 blue-stained cells each. With a grid marked on the underside of the wells to aid counting, use a manual tally counter and enumerate the number of blue-stained cells in each of the replica wells. Calculate the average and divide it by the dilution factor to determine IU per ml. (*see* equation below).
7. Equation to calculate IU/ml: $\left(\frac{(\#\text{ of well 1 blue cells} + \#\text{ of well 2 blue cells}) \div 2}{\text{virus stock input per well } [\mu\text{l}]}\right) \times 1000 = \text{IU/ml}.$
8. Example calculation: $\left(\frac{(58\text{ cells} + 62\text{ cells}) \div 2}{0.0016\ \mu\text{l}}\right) \times 1000 =$ 3.75E + 07 IU/ml

3.3 Preparation of CETs for Dissection

The laboratory location for CET dissection and HIV infection should be considered prior to receiving tissue. We recommend both procedures be performed in a Class II biosafety cabinet. Once tissue is received, it should be immediately dissected and used for HIV infection (*see* **Note 8**).

1. Take 1–2 autoclaved WypAll™ wipes, unfold them, and lay them flat inside of the biosafety cabinet. This area is the "sterile working space," and all preparation and dissection steps should occur within this area.
2. Place one sterile tissue culture dish (10 cm) and its lid atop the WypAll™ wipes. Add 4–5 ml of fresh CEM into the lid and set it aside. Add 10–15 ml of CEM to the base of the culture dish.

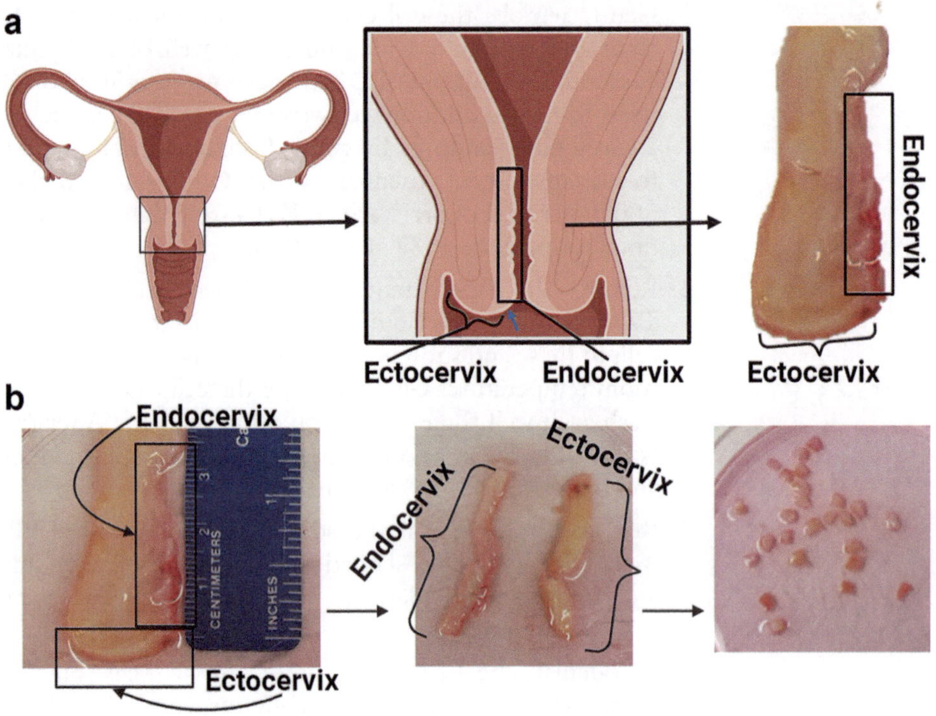

Fig. 2 Anatomical origin, features, and dissection of CET specimens. (**a**) A diagram of the human female reproductive track delineating the position of the cervix (left) with the ecto- and endocervical tissue sites identified (center). The blue arrow indicates the transformation zone at the transition of ectocervix and endocervix. On the right, an excised donor tissue specimen is shown with the ecto- and endocervical tissue sites indicated. (**b**) After assessment and release of a section of the cervix by the surgical pathology laboratory, the remnant tissue should be carefully evaluated, giving particular attention to the overall size and relative amounts and integrity of the ectocervical and endocervical mucosal surfaces (*see* also **Note 1** and Fig. 6). The images show a tissue specimen of optimal size (>2 cm in length) that was received with discernable regions of endo- and ectocervical mucosae (left), the mucosal surfaces of the endo- and ectocervix after dissection form the submucosal tissue (center), and ~2-mm^3 blocks of tissue prepared by further dissection (right). (Part of the illustration was created with biorender.com)

3. Use the curved tip tweezers to place the explant tissue into the CEM-containing culture dish base. Gently swirl the media over the tissue. Carefully use the tweezers to manipulate and evaluate the general condition of the tissue (*see* **Note 9**).
4. Visually identify the anatomical features of the tissue, in particular the mucosal surfaces comprising the ectocervix and endocervix (Fig. 2).
5. Slide a vinyl ruler under the tissue culture dish to measure the length of the ectocervix and endocervix. These measurements allow for an estimation of the total number of tissue blocks that can be obtained by dissection. Document the appearance of the tissue specimen by photographing it.

3.4 Dissection of the CET

1. With the CET in the base of the tissue culture dish, use sterile curved tip tweezers to grip the tissue, then use a sterile surgical blade attached to a #4 scalpel blade handle to excise the endocervix and ectocervix surfaces from the remaining part of the tissue. Remove as much of the underlying submucosal tissue as possible and discard (*see* **Note 10**).
2. Place the excised ectocervix and endocervix at opposite ends of the culture dish base. Remove and dispose of the remaining tissue.
3. Position the laminated 2-mm × 2-mm dissection grid under the culture dish.
4. Gripping the excised piece of endocervix or ectocervix with the curve-tip tweezers, cut the tissue in half, then cut those halves in half. Cut away any additional submucosal tissue to leave only the mucosal surface. Using the tissue dissection grid as a reference, continue until the tissue has been dissected into uniform ~2-mm^3 blocks. Transfer the blocks from the culture dish base into the tissue culture dish lid.
5. Repeat the dissection steps for the other excised tissue pieces and transfer those blocks into the culture plate lid, keeping the endocervix and ectocervix blocks separated (Fig. 2).
6. Count and record the number of tissue blocks for each tissue piece: ectocervix and endocervix. A standard remnant tissue specimen of 2-cm length can yield between 20 and 40 blocks each of ecto- and endocervix, for a total of 40–80 CET blocks.
7. If necessary, remove excess mucus from the surface of the dissected tissue blocks (*see* **Note 11**).

3.5 Non-polarized CET Infection

1. Add 1 ml of CEM to 2 ml screw-cap microtubes. The number of tubes required will depend on the total number of tissue blocks that were obtained by dissection and research-specific experimental parameters, including controls and the number of blocks per infection condition (see below). Use straight-tip tweezers to transfer up to eight dissected tissue blocks into each microtube. Here too, the number of tissue blocks per tube will depend on the overall experimental design.
2. Thaw the required amount of sNLuc virus stock from −80 °C cryostorage. Add a desired amount of virus to the screw-cap microtubes containing tissue blocks (*see* **Note 12**).
3. Because sNLuc is present in the virus inoculum, it may cause relatively high background RLU levels and an appropriate negative control must be included. We suggest including one 2 ml screw-cap microtube containing 1 ml of medium and 2–3 tissue blocks without virus as a negative control, and one inoculated with non-infectious virus stock derived from transfection of 293T cells with *env*-minus sNLuc-IMC virus.

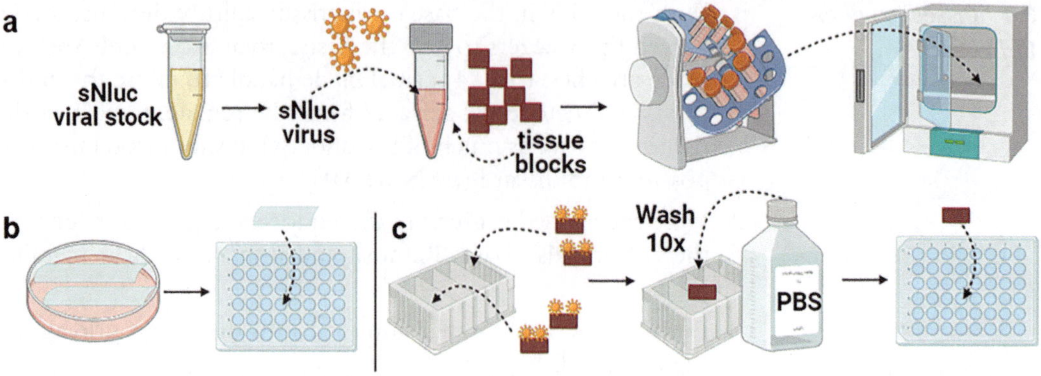

Fig. 3 Workflow for virus inoculation and preparation for culture. (**a**) sNLuc-IMC or sNLuc-Env-IMC reporter virus stocks maintained at −80 °C are thawed at room temperature (left). The dissected CET blocks are placed into 2 ml screw-cap tubes and inoculated with a desired amount of virus (total volume of 1 ml) for 18–24 h in a 37 °C incubator while rotating end-over-end at 10 rpm. (**b**) Gelatin sponge is submerged in CEM until fully saturated, cut into uniform 5-mm × 5-mm squares, and then individually placed into wells of a 48-well plate (left). (**c**) Following the virus inoculation step (a), the tissue blocks are repeatedly washed to minimize sNLuc present in the virus inoculum. Then each CET block is placed into a well of a 48-well plate and gently nestled atop of the gelatin sponge. (Illustration created with biorender.com)

4. Additional controls may include tissue blocks treated with antiretroviral drugs at time points before or after virus inoculation, depending on drug mechanism and experimental design.
5. Securely tighten the cap of each microtube, place the tubes into the paddle attachment of a tube revolver/rotator, and rotate at approximately 10 rpm for 18–24 h in a 37 °C incubator. An overview of the steps described in this section is illustrated in Fig. 3.

3.6 CET HIV Culture

1. Add 8–10 ml of CEM to a sterile 10 cm tissue culture dish.
2. Cut the gelatin sponge into 1 cm wide strips using sterile dissecting scissors.
3. Use tweezers to gently place the gelatin sponge strips into the CEM, and submerge using the blunt side of the tweezers. The gelatin sponge has sufficiently absorbed CEM when it turns from brittle and white to soft and semi-translucent. This may take up to 1 h.
4. While the gelatin sponge is hydrating, add 2–4 ml of cold PBS to each channel of a 12-channel reagent reservoir.
5. Remove the virus-inoculated tissue blocks from the incubator, and use tweezers to transfer all tissue blocks from one microtube into one channel of the 12-channel reservoir. Label each channel to distinguish the contents from different microtubes.

6. Repeat **step 5** for all microtubes, disinfecting the tweezers between microtubes with 70% ethanol. When complete, set the reservoir with tissue blocks aside while performing the next two steps.
7. Using sterile tweezers and dissecting scissors, cut the CEM-saturated gelatin sponge strips into ~0.5-cm × 0.5-cm squares, and for each tissue block, place one gelatin sponge square at the bottom of each well of a 48-well plate.
8. Add 400 μl of fresh CEM to each gelatin sponge-containing well.
9. Continue with washing of the tissue blocks by removing the PBS in each reservoir channel and replacing it with ~4 ml of fresh PBS (*see* **Note 13**).
10. After washing, transfer each tissue block atop of the gelatin sponge in a well of the 48-well plate with gentle manipulation so it is positioned securely atop the gelatin sponge (*see* **Note 14**).
11. Place the 48-well plate into a humidified incubator calibrated at 37 °C and 5% CO_2.
12. After 24 h (Day 1), move the plate to the biosafety cabinet, gently remove as much of the CEM as possible from each well, and replenish with 450 μl of pre-warmed, fresh CEM.
13. Perform an 80% medium change anytime the medium color changes from pink/red to yellow or at least every 5 days (*see* **Note 15**). Ensure that the tissue remains submersed and atop the gelatin sponge.

3.7 Sample Collection and sNLuc Analysis

1. Move the 48-well culture plate to a Class II biosafety cabinet, collect 20 μl of supernatant immediately adjacent to each tissue block, and place into a white 96-well luminometer plate for immediate analysis (*see* **Note 16**).
2. Return the 48-well culture plate to incubator.
3. Prepare the Nano-Glo™ luciferase assay buffer and substrate solution according to the manufacturer's recommendations.
4. Add 20 μl of the luciferase assay solution to each well containing culture supernatant and mix well.
5. Analyze sNLuc relative light units (RLU) using a 96-well microplate luminometer programmed with an integration time of 1 s/well. The sNLuc activity should be analyzed within 10 min.
6. Repeat **steps 1–6** daily, or as desired, for up to 14 days.
7. Import the raw RLU values into appropriate analytical software for data analysis.

3.8 Representative Performance Characteristics and Data Analysis Considerations

Selected data generated using these methods are presented here to demonstrate expected outcomes and important performance parameters of the sNLuc reporter virus mucosal model, in particular; assay sensitivity, reproducibility, intra- and inter-donor variation, and potential limitations. To augment translation and application of this mucosal model, example experiments are described below with data that were analyzed and presented to emphasize salient assay performance characteristics.

3.8.1 Detection of Productive HIV-1 infection of CET

1. Tissue blocks prepared from CET were inoculated with a sNLuc reporter virus in either the absence or presence of nevirapine to inhibit HIV-1 infection. Over a period of 2 weeks, supernatant samples from each individual CET block culture were collected and analyzed for sNLuc RLU.
2. Raw RLU values plotted against time are shown in Fig. 4, demonstrating the following: (i) Residual sNLuc RLU from the virus inoculum persists despite extensive washing after virus inoculation (*see* **Note 13**). This is apparent at day 2 by

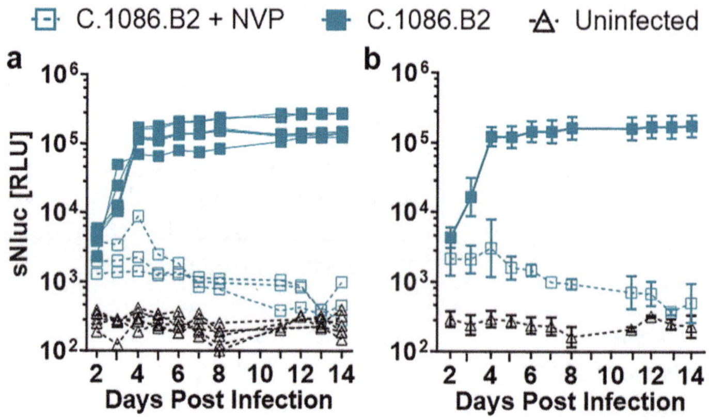

Fig. 4 Productive HIV-1 infection of CET and intra-donor reproducibility. (**a**) Nine CET blocks prepared by dissection from a single tissue donor were inoculated with sNLuc-Env-IMC reporter virus comprising the C.1086.B2 *env*. Three of the nine CET blocks were both inoculated and cultured in the presence of 20 ng/ml of nevirapine (NVP) (open squares). Six of the nine CET blocks were not treated with drug (solid squares). Note that because of similarities between the RLU values from each of the six CET blocks, not all six curves shown on the plot can be readily distinguished. Five additional CET blocks were mock-infected (open triangles). The CET blocks were washed and then cultured individually in wells of a 48-well plates as described above. Supernatant samples collected from day 2 through day 14 were analyzed to quantify sNLuc RLU. Samples collected from wells with uninfected CET blocks served as a control to help assess residual sNLuc RLU in the samples collected from virus inoculated CET blocks. The raw RLU values (*y*-axis) were plotted against time (*x*-axis). (**b**) The geometric mean RLU values were calculated for each group of replica blocks and plotted with error bars representing geomean standard deviation

comparing RLU values of samples collected from uninfected versus drug treated controls. (ii) Strong evidence of virus infection is shown by a marked increase in RLU values for the non-drug-treated CET blocks versus those treated with RT inhibitor. (iii) Each of six replica virus-inoculated CET blocks exhibits a similar sNLuc RLU profile of HIV-1 infection and replication.

3.8.2 Profiles of HIV-1 infection and Replication in CET

1. Tissue blocks from two different CET donors were prepared and inoculated with sNLuc reporter virus representing two different strains of subtype B virus in either the absence or presence of drug (Fig. 5). Over a period of 2 weeks, supernatant was collected from the CET block cultures and analyzed for sNLuc RLU.
2. The raw RLU values, generated from each CET donor (a and b respectively), are plotted against time (Fig. 5), demonstrating the following: (i) Similar to Fig. 4a, there is a marked difference in the kinetics of sNLuc expression in CET infected without drug versus with drug. (ii) The RLU kinetic profile of the two different HIV subtype B strains is similar. (iii) RLU values from CET treated with protease inhibitor are indistinguishable from CET treated with RT inhibitor. This result has two notable implications for understanding the model system: First, it

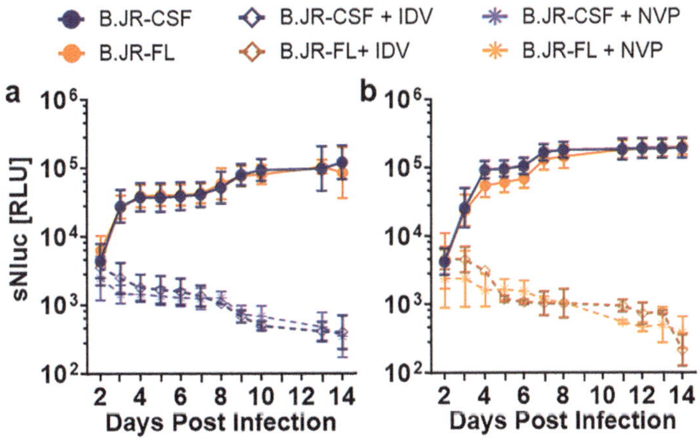

Fig. 5 Inter-donor reproducibility of HIV-1 infection and replication profiles in CET. CET blocks prepared from remnant tissue specimens obtained from two donors (depicted in panels **a** and **b**, respectively) were divided into four groups each and inoculated with sNLuc-Env-IMC reporter viruses comprising the B.JR-FL and the B.JR-CSF *envs*, respectively. Each group included at least five CET blocks that were virus inoculated and subsequently cultured in CEM containing either no inhibitor, 20 μM of NVP, or 2 μM of Indinavir (IDV). Samples collected from day 2 through 14 were analyzed for sNLuc RLU. The sNLuc RLU values in each group are presented as geometric means calculated from replica samples. The data were plotted with error bars representing geomean standard deviation

shows that sNLuc expressed from cells infected in the first round by virus in the inoculum (i.e., prior to secondary infection by *de novo* expressed virus) is not quantifiable over the background RLU; and second, it indicates that the RLU detected subsequent to day 2 represents ongoing HIV replication and spreads to secondary target cells in the tissue.

3.9 Concluding Remarks

The sNLuc-Env-IMC and sNLuc-IMC reporter viruses provide exceptional sensitivity for detecting HIV-1 infection in CET. Moreover, the sNLuc enables simple and repeated sampling from culture supernatants and quantitative analysis of virus infection and replication over several days following virus inoculation. Importantly, the viral dynamics exhibited among individual 2-mm^3 CET blocks from a given tissue specimen are highly reproducible. This underpinning enables meaningful comparative analyses of virus infection and replication in the CET model. The methods described herein may be applied to study determinants of viral fitness, to further examine the fitness properties of different HIV-1 subtypes, to test the effectiveness of neutralizing antibodies, or to examine effects of cytokine/chemokine expression. The methods may also be useful for correlative analyses of early virus-host interactions or factors affecting risk of HIV acquisition. Finally, the methods may be adapted for other mucosal tissue studies, including those involving rectal, tonsillar, foreskin, vaginal, or endometrial tissues.

4 Notes

1. Access to useful remnant CET is essential, and therefore, considerations of the logistics for effectively accomplishing this routinely are particularly relevant. We highly recommend gaining an understanding of the administrative and practical procedures used by both the tissue procurement service and the surgical pathology lab for postoperative assessment of the tissue. It is imperative to educate the tissue procurement staff of your specific tissue requirement needs. Furthermore, it is critically important to precisely identify and communicate what anatomical features must be retained for your research on the tissue sections released by the surgical pathology lab (i.e., remnant tissue that retains a portion of the endo- and ectocervix mucosae).
2. The experimental needs of any given application determine the specific molecular design of an HIV-1 reporter virus that will be most useful. To give some broader context for the sNLuc reporter viruses used herein, we provide references that will allow the interested reader to gain more in-depth information on the features of the reporter IMCs. Usually, when HIV-1 is

modified with reporter genes they are positioned within the 3′ end of the genome, including insertion into or adjacent to the *nef* or *env* open reading frames (orfs). The HIV-1 reporter virus approaches developed in our laboratory have found application in various immune-monitoring assays and in other transmission and pathogenesis-related research, including those in mucosal explant tissue and humanized mice [25, 31–51]. Our initial design [25] was refined to maintaining Nef protein expression at physiological levels of the parental non-reporter provirus [31] by inserting reporter genes in frame with the *nef* start codon followed by the modified "6ATRi" IRES illustrated in Fig. 1. Nef function, including CD4 and MHC I down-regulation, is preserved. This has demonstrated advantages for utilization of reporter IMC encoding *Renilla* luciferase (LucR), sNLuc, NLuc, mouse CD24, or Green Fluorescent Protein (GFP) in several research applications [31, 32, 37, 39, 44, 52–54]. We have employed the reporter 6ATRi cassette approach both in the context of Env-IMC [31, 32, 37, 44, 52–54] and IMC, including TF-IMC [37, 39, 53, 54]. The rationale for introducing the *env* ectodomain coding region (spanning from the C1 region of gp120, through the membrane-spanning domain (MSD) of gp41) of heterologous HIV-1 strains of interest, rather than the entire *env* orf, into an isogenic Env-IMC backbone [25, 55] is backed by findings that inter-strain and inter-clade recombination sites at these locations are also selected for in vitro and in vivo [56, 57]. This approach avoids creating chimeric *tat, rev,* and *vpu* genes that may alter the replication properties of Env-IMC, and thus confound *env* comparative analyses. Furthermore, such "Env. ecto"-recombinant Envs maintain the neutralization phenotype of the *env* from which the ectodomain coding region was derived [58]; we consistently found that Env-IMC with full-length and "Env-ecto" heterologous sequences have the same neutralization profile in several different target cell types (Ochsenbauer et al., unpublished). The stable retention of reporter activity is also to be considered since it is critical for experimental applications that require multiple rounds of replication, including tissue explant models. We demonstrated retention of luciferase activity in PBMC using our first- and second-generation HIV-1 reporter virus design [25, 31]. Of interest, the work of Roy et al. showed that reporter gene retention in and replication of IMC in humanized mice are improved further by removal of CG dinucleotides in the reporter coding regions [59].

3. Care should be taken when growing bacterial cultures transformed with HIV-1 proviral plasmids. We recommend using STBL2 or STBL3 competent cells and growth temperatures

between 30 and 35 °C. This helps with preventing deletions from occurring during culture for DNA miniprep and maxiprep. In our experience, selection pressure for such deletions seems to predominantly come from HIV-1 virus strain specific sequences in the 3′ portion of the genome (including *env*). In NL4-3-based Env-IMC without or with reporter gene cassettes, problems with proviral plasmid deletions occur only occasionally and are *env* stain dependent. For proviral plasmids encoding primary T/F IMC genomes, we observe with higher frequency that parts of the proviral clone become deleted. Therefore, we routinely grow bacteria transformed with such plasmids at 30 °C, accepting potentially lower yields of maxiprep DNA in favor of high DNA quality. We occasionally clone T/F IMC genomes into low-copy-number plasmids for increased genetic stability [60].

4. Consideration should be given for the selection of HIV-1 strains. Strains that utilize the CXCR4 coreceptor for cell entry either do not replicate or replicate poorly in the CET model [10, 15]. The R5-tropic BaL strain of HIV-1 replicates well [10, 12, 16, 18, 20] and is a useful positive control. Biologically relevant HIV-1 strains to consider include those representing T/F viruses. We previously described the generation of sNLuc-Env-IMC, pNL-sNLuc.6ATRi-B.BaL.ecto, which expresses the Env ectodomain of HIV-1_{BaL} within the NL4-3-derived proviral backbone [44]. In addition, we found that proviral clones of sNLuc-Env-IMC comprising *env* sequences derived from several Clade A, B, C, and D T/F HIV strains, and full-length T/F sNLuc-IMCs are infectious and replication competent and thus, well suited for application in the CET model.

5. Certain research studies utilizing the CET model may want to avoid medium that contains phenol-red due to its estrogen-like effect on the mucosa. Phenol-red-free DMEM and RPMI are available from various manufacturers. The use of charcoal stripped FBS may be considered for similar reasons as a substitute for normal FBS.

6. If using the sNLuc gene provided by Promega Inc., users are required to use Promega Nano-Glo reagents, per Promega's Limited Use Label License, available on the company website.

7. While several different methods can be used for DNA transfection, we suggest using the Fugene HD transfection reagent, according to the manufacture's recommendations with 2 μg of proviral plasmid-cloned DNA for each well of a 6-well plate.

8. It is strongly recommended to complete the dissection and infection of the CET as soon as it is received. Since the resected donor tissue will require assessment by surgical pathology, it is

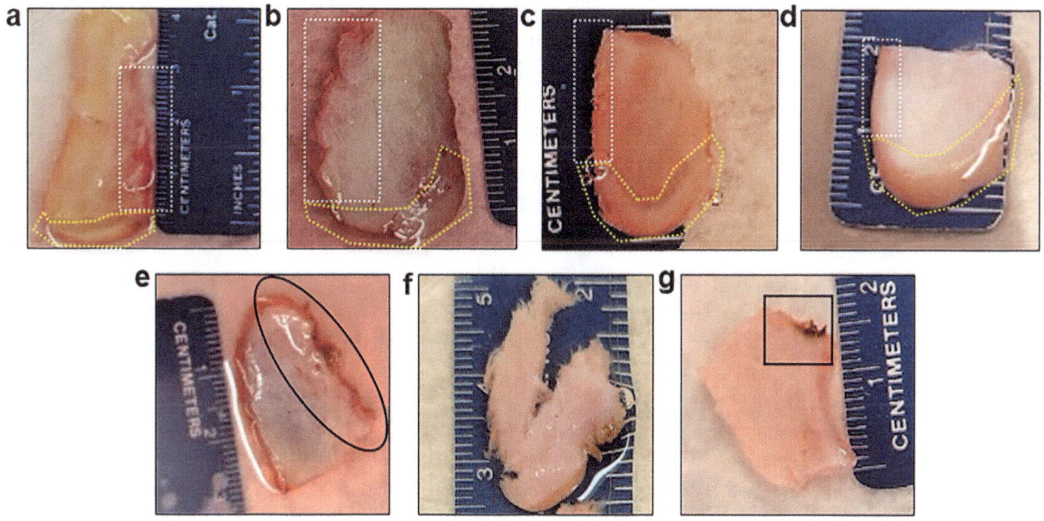

Fig. 6 Examples of remnant cervical tissue specimens. (**a–d**) Four examples are shown of donor specimens considered at or above standard for preparation of CET blocks. All four specimens are ~2 cm in length or longer, which is considered optimal; they contain undamaged mucosal surfaces, both ecto- and endocervical mucosae are present, and specimens have been cleanly excised from the donor tissue. There are no "burnt" or "charred" areas resulting from cauterization of the donor tissue. (**e–g**) Three examples of donor specimens considered substandard: The tissue on the left (**e**) is excessively bloody, and the mucosal surface is "shredded." The center tissue (**f**) is abnormally shaped and lacks the endocervical mucosa. The tissue on the right (**g**) is missing the ectocervical mucosa, is less than 2 cm in length, and is partially "charred" from cauterization during surgical resection. White and yellow dotted lines delineate endocervical and ectocervical mucosae, respectively. Black outlines draw attention to substandard characteristics in **e** and **g**

often not available until relatively late in the day. If the tissue cannot be used the day it is received, place it into ~30 ml of fresh CEM at 4 °C. Maintaining the CETs at 4 °C longer than 24 h will compromise viability.

9. First, visually evaluate the tissue to assess its general condition and anatomical features. Ideal CET specimens are greater than 2 cm in length and exhibit a healthy tissue appearance and structure, particularly along the mucosal surface of the ecto- and endocervix (Figs. 2 and 6). Substandard tissue specimens are small, have non-discernable ecto- and/or endocervix mucosal surfaces, and/or have visible injury. Injury may include "shredding," where the mucosal surface becomes detached or torn from the underlying submucosa, or "burning/charring," where the mucosal surface is blackened from cauterization during surgery (Fig. 6e–g). Tissue of poor quality should not be used.

10. It is critical to understand the anatomical features of human cervix, both to identify useful tissue specimens and to guide the dissection. Ectocervix and endocervix can be distinguished by their physical appearance and relative proximity. Ectocervix is covered by a pink stratified squamous epithelium, consisting of

multiple layers of cells, and it has a smooth, pearly, opaque appearance. Endocervix comprises columnar epithelium consisting of a single layer of cells that appears reddish in color from the underlying vasculature in the stroma that is visible (Figs. 2 and 6). Remove these mucosal surfaces from the underlying tissue by manipulating the scalpel in a guillotine-like motion, with long, unidirectional cuts. Dissection should not use a "sawing" motion as this creates jagged, uneven edges.

11. Endocervix will often continue to secrete mucus after dissection. Excess mucus may by temporally removed using tweezers to gently blot the CET blocks against the surface of the WypAll™ wipe.
12. A virus inoculum of 1×10^6 IU is generally recommended; however, this may vary depending on the question and experimental design. An inoculum of 1×10^6 IU should reproducibly result in productive HIV-1 infection of ~80% of the virus inoculated tissue blocks.
13. The virus stocks contain an abundance of sNLuc originating from the 293T cells transfected with reporter virus proviral DNA. Therefore, to maximize the detection of sNLuc produced from *de novo* infection of the CET, the tissue blocks must be washed extensively. The tissue blocks should be washed 8–10 times with 2–4 ml of PBS per wash. Virus stocks prepared by ultracentrifugation will also help reduce RLU background; however, at least two washes are still recommended. Insufficient washing will result in high RLU values, which may obscure the detection of virus infection, particularly at the early time points after inoculation.
14. Surgifoam and similar products are sterile, water-insoluble, malleable, porcine gelatin sponges intended for hemostatic use. The beneficial effects for CET culture include physical support for the tissue that helps retain tissue cytoarchitecture, slowing the rate of degradation, minimizing cell loss, and facilitating access to culture medium nutrients and oxygen through the sponge capillaries [16].
15. If phenol-red-free medium is used to satisfy specific experimental criteria, we recommend to culture one uninfected tissue block on gelatin sponge in phenol-red-containing CEM to help identify when medium changes will be necessary.
16. As an alternative to contemporaneous analysis of sNLuc, samples can be collected and stored at −80 °C, in which case it is recommended to collect samples of a larger volume into appropriate tubes or 96-well plates for storage, and to transfer 20 μl into white luminometer plates immediately before analysis. Be aware that volume loss by sampling medium from the wells should be replaced, albeit this is not necessary each time unless an appreciable volume has been removed.

References

1. Burgener A, McGowan I, Klatt NR (2015) HIV and mucosal barrier interactions: consequences for transmission and pathogenesis. Curr Opin Immunol 36:22–30. https://doi.org/10.1016/j.coi.2015.06.004
2. Global AIDS Update 2016 (2016) UNAIDS, UNAIDS
3. Haase AT (2011) Early events in sexual transmission of HIV and SIV and opportunities for interventions. Annu Rev Med 62:127–139. https://doi.org/10.1146/annurev-med-080709-124959
4. Keele BF, Estes JD (2011) Barriers to mucosal transmission of immunodeficiency viruses. Blood 118(4):839–846. https://doi.org/10.1182/blood-2010-12-325860
5. Smedley J, Turkbey B, Bernardo ML, Del Prete GQ, Estes JD, Griffiths GL, Kobayashi H, Choyke PL, Lifson JD, Keele BF (2014) Tracking the luminal exposure and lymphatic drainage pathways of intravaginal and intrarectal inocula used in nonhuman primate models of HIV transmission. PLoS One 9(3):e92830. https://doi.org/10.1371/journal.pone.0092830
6. Stieh DJ, Maric D, Kelley ZL, Anderson MR, Hattaway HZ, Beilfuss BA, Rothwangl KB, Veazey RS, Hope TJ (2014) Vaginal challenge with an SIV-based dual reporter system reveals that infection can occur throughout the upper and lower female reproductive tract. PLoS Pathog 10(10):e1004440. https://doi.org/10.1371/journal.ppat.1004440
7. Dezzutti CS (2015) Animal and human mucosal tissue models to study HIV biomedical interventions: can we predict success? J Int AIDS Soc 18(1):20301. https://doi.org/10.7448/ias.18.1.20301
8. Rollenhagen C, Lathrop MJ, Macura SL, Doncel GF, Asin SN (2014) Herpes simplex virus type-2 stimulates HIV-1 replication in cervical tissues: implications for HIV-1 transmission and efficacy of anti-HIV-1 microbicides. Mucosal Immunol 7(5):1165–1174. https://doi.org/10.1038/mi.2014.3
9. Grivel J-C, Margolis L (2009) Use of human tissue explants to study human infectious agents. Nat Protoc 4(2):256–269. https://doi.org/10.1038/nprot.2008.245
10. Saba E, Grivel JC, Vanpouille C, Brichacek B, Fitzgerald W, Margolis L, Lisco A (2010) HIV-1 sexual transmission: early events of HIV-1 infection of human cervico-vaginal tissue in an optimized ex vivo model. Mucosal Immunol 3(3):280–290. https://doi.org/10.1038/mi.2010.2
11. Estes JD, Legrand R, Petrovas C (2018) Visualizing the immune system: providing key insights into HIV/SIV infections. Front Immunol 9. https://doi.org/10.3389/fimmu.2018.00423
12. Introini A, Boström S, Bradley F, Gibbs A, Glaessgen A, Tjernlund A, Broliden K (2017) Seminal plasma induces inflammation and enhances HIV-1 replication in human cervical tissue explants. PLoS Pathog 13(5):e1006402. https://doi.org/10.1371/journal.ppat.1006402
13. Merbah M, Introini A, Fitzgerald W, Grivel J-C, Lisco A, Vanpouille C, Margolis L (2011) Cervico-vaginal tissue ex vivo as a model to study early events in HIV-1 infection. Am J Reprod Immunol 65(3):268–278. https://doi.org/10.1111/j.1600-0897.2010.00967.x
14. Introini A, Vanpouille C, Lisco A, Grivel JC, Margolis L (2013) Interleukin-7 facilitates HIV-1 transmission to cervico-vaginal tissue ex vivo. PLoS Pathog 9(2):e1003148. https://doi.org/10.1371/journal.ppat.1003148
15. Abraha A, Nankya IL, Gibson R, Demers K, Tebit DM, Johnston E, Katzenstein D, Siddiqui A, Herrera C, Fischetti L, Shattock RJ, Arts EJ (2009) CCR5- and CXCR4-tropic subtype C human immunodeficiency virus type 1 isolates have a lower level of pathogenic fitness than other dominant group M subtypes: implications for the epidemic. J Virol 83(11): 5592–5605. https://doi.org/10.1128/jvi.02051-08
16. Introini A, Vanpouille C, Fitzgerald W, Broliden K, Margolis L (2018) Ex vivo infection of human lymphoid tissue and female genital mucosa with human immunodeficiency virus 1 and histoculture. J Vis Exp (140). https://doi.org/10.3791/57013
17. Merbah M, Arakelyan A, Edmonds T, Ochsenbauer C, Kappes JC, Shattock RJ, Grivel JC, Margolis LB (2012) HIV-1 expressing the envelopes of transmitted/founder or control/reference viruses have similar infection patterns of CD4 T-cells in human cervical tissue ex vivo. PLoS One 7(12):e50839. https://doi.org/10.1371/journal.pone.0050839
18. Peters PJ, Gonzalez-Perez MP, Musich T, Moore Simas TA, Lin R, Morse AN, Shattock RJ, Derdeyn CA, Clapham PR (2015)

Infection of ectocervical tissue and universal targeting of T-cells mediated by primary non-macrophage-tropic and highly macrophage-tropic HIV-1 R5 envelopes. Retrovirology 12: 48. https://doi.org/10.1186/s12977-015-0176-2

19. Klein K, Nankya I, Nickel G, Ratcliff AN, Meadows AAJ, Hathaway N, Bailey JA, Stieh DJ, Cheeseman HM, Carias AM, Lobritz MA, Mann JFS, Gao Y, Hope TJ, Shattock RJ, Arts EJ (2021) Deep gene sequence cluster analyses of multi-virus-infected mucosal tissue reveal enhanced transmission of acute HIV-1. J Virol 95(3). https://doi.org/10.1128/jvi.01737-20
20. Rohan LC, Moncla BJ, Kunjara Na Ayudhya RP, Cost M, Huang Y, Gai F, Billitto N, Lynam JD, Pryke K, Graebing P, Hopkins N, Rooney JF, Friend D, Dezzutti CS (2010) In vitro and ex vivo testing of tenofovir shows it is effective as an HIV-1 microbicide. PLoS One 5(2): e9310. https://doi.org/10.1371/journal.pone.0009310
21. Cummins JE Jr, Guarner J, Flowers L, Guenthner PC, Bartlett J, Morken T, Grohskopf LA, Paxton L, Dezzutti CS (2007) Preclinical testing of candidate topical microbicides for anti-human immunodeficiency virus type 1 activity and tissue toxicity in a human cervical explant culture. Antimicrob Agents Chemother 51(5): 1770–1779. https://doi.org/10.1128/aac.01129-06
22. Buffa V, Stieh D, Mamhood N, Hu Q, Fletcher P, Shattock RJ (2009) Cyanovirin-N potently inhibits human immunodeficiency virus type 1 infection in cellular and cervical explant models. J Gen Virol 90(Pt 1): 234–243. https://doi.org/10.1099/vir.0.004358-0
23. Fletcher P, Kiselyeva Y, Wallace G, Romano J, Griffin G, Margolis L, Shattock R (2005) The nonnucleoside reverse transcriptase inhibitor UC-781 inhibits human immunodeficiency virus type 1 infection of human cervical tissue and dissemination by migratory cells. J Virol 79(17):11179–11186. https://doi.org/10.1128/jvi.79.17.11179-11186.2005
24. Introini A, Vanpouille C, Grivel JC, Margolis L (2014) An ex vivo model of HIV-1 infection in human lymphoid tissue and cervico-vaginal tissue. Bio Protoc 4(4). https://doi.org/10.21769/bioprotoc.1047
25. Edmonds TG, Ding H, Yuan X, Wei Q, Smith KS, Conway JA, Wieczorek L, Brown B, Polonis V, West JT, Montefiori DC, Kappes JC, Ochsenbauer C (2010) Replication competent molecular clones of HIV-1 expressing Renilla luciferase facilitate the analysis of antibody inhibition in PBMC. Virology 408(1):1–13. https://doi.org/10.1016/j.virol.2010.08.028
26. 293T CRL-3216™. ATCC. https://www.atcc.org/products/crl-3216. 2023
27. Pear WS, Nolan GP, Scott ML, Baltimore D (1993) Production of high-titer helper-free retroviruses by transient transfection. Proc Natl Acad Sci USA 90(18):8392–8396. https://doi.org/10.1073/pnas.90.18.8392
28. Wei X, Decker JM, Liu H, Zhang Z, Arani RB, Kilby JM, Saag MS, Wu X, Shaw GM, Kappes JC (2002) Emergence of resistant human immunodeficiency virus type 1 in patients receiving fusion inhibitor (T-20) monotherapy. Antimicrob Agents Chemother 46(6): 1896–1905. https://doi.org/10.1128/aac.46.6.1896-1905.2002
29. Policy NOoS (2019) NIH guidelines for research involving recombinant or synthetic nucleic acid molecules (NIH guidelines). National Institutes of Health, https://osp.od.nih.gov/
30. Chosewood LC, Wilson DE (2009) Biosafety in microbiological and biomedical laboratories (BMBL), 5th edn. US Department of Health and Human Services
31. Alberti MO, Jones JJ, Miglietta R, Ding H, Bakshi RK, Edmonds TG, Kappes JC, Ochsenbauer C (2015) Optimized replicating Renilla luciferase reporter HIV-1 utilizing novel internal ribosome entry site elements for native Nef expression and function. AIDS Res Hum Retrovir 31(12):1278–1296. https://doi.org/10.1089/aid.2015.0074
32. Cavrois M, Banerjee T, Mukherjee G, Raman N, Hussien R, Rodriguez BA, Vasquez J, Spitzer MH, Lazarus NH, Jones JJ, Ochsenbauer C, McCune JM, Butcher EC, Arvin AM, Sen N, Greene WC, Roan NR (2017) Mass cytometric analysis of HIV entry, replication, and remodeling in tissue CD4+ T cells. Cell Rep 20(4):984–998. https://doi.org/10.1016/j.celrep.2017.06.087
33. deCamp A, Hraber P, Bailer RT, Seaman MS, Ochsenbauer C, Kappes J, Gottardo R, Edlefsen P, Self S, Tang H, Greene K, Gao H, Daniell X, Sarzotti-Kelsoe M, Gorny MK, Zolla-Pazner S, LaBranche CC, Mascola JR, Korber BT, Montefiori DC (2014) Global panel of HIV-1 Env reference strains for standardized assessments of vaccine-elicited neutralizing antibodies. J Virol 88(5): 2489–2507. https://doi.org/10.1128/JVI.02853-13
34. Freel SA, Lamoreaux L, Chattopadhyay PK, Saunders K, Zarkowsky D, Overman RG, Ochsenbauer C, Edmonds TG, Kappes JC,

Cunningham CK, Denny TN, Weinhold KJ, Ferrari G, Haynes BF, Koup RA, Graham BS, Roederer M, Tomaras GD (2010) Phenotypic and functional profile of HIV-inhibitory CD8 T cells elicited by natural infection and heterologous prime/boost vaccination. J Virol 84(10):4998–5006. https://doi.org/10.1128/JVI.00138-10

35. Naarding MA, Fernandez N, Kappes JC, Hayes P, Ahmed T, Icyuz M, Edmonds TG, Bergin P, Anzala O, Hanke T, Clark L, Cox JH, Cormier E, Ochsenbauer C, Gilmour J (2014) Development of a luciferase based viral inhibition assay to evaluate vaccine induced CD8 T-cell responses. J Immunol Methods 409: 161–173. https://doi.org/10.1016/j.jim.2013.11.021

36. Pollara J, McGuire E, Fouda GG, Rountree W, Eudailey J, Overman RG, Seaton KE, Deal A, Edwards RW, Tegha G, Kamwendo D, Kumwenda J, Nelson JA, Liao HX, Brinkley C, Denny TN, Ochsenbauer C, Ellington S, King CC, Jamieson DJ, van der Horst C, Kourtis AP, Tomaras GD, Ferrari G, Permar SR (2015) Association of HIV-1 envelope-specific breast milk IgA responses with reduced risk of postnatal mother-to-child transmission of HIV-1. J Virol 89(19): 9952–9961. https://doi.org/10.1128/JVI.01560-15

37. Prévost J, Richard J, Medjahed H, Alexander A, Jones J, Kappes JC, Ochsenbauer C, Finzi A (2018) Incomplete downregulation of CD4 expression affects HIV-1 Env conformation and antibody-dependent cellular cytotoxicity responses. J Virol 92(13). https://doi.org/10.1128/jvi.00484-18

38. Sarzotti-Kelsoe M, Daniell X, Todd CA, Bilska M, Martelli A, Labranche C, Perez LG, Ochsenbauer C, Kappes JC, Rountree W, Denny TN, Montefiori DC (2014) Optimization and validation of a neutralizing antibody assay for HIV-1 in A3R5 cells. J Immunol Methods 409:147–160. https://doi.org/10.1016/j.jim.2014.02.013

39. Ventura JD, Beloor J, Allen E, Zhang T, Haugh KA, Uchil PD, Ochsenbauer C, Kieffer C, Kumar P, Hope TJ, Mothes W (2019) Longitudinal bioluminescent imaging of HIV-1 infection during antiretroviral therapy and treatment interruption in humanized mice. PLoS Pathog 15(12):e1008161. https://doi.org/10.1371/journal.ppat.1008161

40. Anthony-Gonda K, Bardhi A, Ray A, Flerin N, Li M, Chen W, Ochsenbauer C, Kappes JC, Krueger W, Worden A, Schneider D, Zhu Z, Orentas R, Dimitrov DS, Goldstein H, Dropulic B (2019) Multispecific anti-HIV duoCAR-T cells display broad in vitro antiviral activity and potent in vivo elimination of HIV-infected cells in a humanized mouse model. Sci Transl Med 11(504). https://doi.org/10.1126/scitranslmed.aav5685

41. Astronomo RD, Lemos MP, Narpala SR, Czartoski J, Fleming LB, Seaton KE, Prabhakaran M, Huang Y, Lu Y, Westerberg K, Zhang L, Gross MK, Hural J, Tieu HV, Baden LR, Hammer S, Frank I, Ochsenbauer C, Grunenberg N, Ledgerwood JE, Mayer K, Tomaras G, McDermott AB, McElrath MJ (2021) Rectal tissue and vaginal tissue from intravenous VRC01 recipients show protection against ex vivo HIV-1 challenge. J Clin Invest 131(16). https://doi.org/10.1172/jci146975

42. Astronomo RD, Santra S, Ballweber-Fleming-L, Westerberg KG, Mach L, Hensley-McBain-T, Sutherland L, Mildenberg B, Morton G, Yates NL, Mize GJ, Pollara J, Hladik F, Ochsenbauer C, Denny TN, Warrier R, Rerks-Ngarm S, Pitisuttithum P, Nitayapan S, Kaewkungwal J, Ferrari G, Shaw GM, Xia SM, Liao HX, Montefiori DC, Tomaras GD, Haynes BF, McElrath JM (2016) Neutralization takes precedence over IgG or IgA isotype-related functions in mucosal HIV-1 antibody-mediated protection. EBioMedicine 14:97–111. https://doi.org/10.1016/j.ebiom.2016.11.024

43. Bardhi A, Wu Y, Chen W, Li W, Zhu Z, Zheng JH, Wong H, Jeng E, Jones J, Ochsenbauer C, Kappes JC, Dimitrov DS, Ying T, Goldstein H (2017) Potent in vivo NK cell-mediated elimination of HIV-1-infected cells mobilized by a gp120-bispecific and hexavalent broadly neutralizing fusion protein. J Virol 91(20). https://doi.org/10.1128/JVI.00937-17

44. Gornalusse GG, Vojtech LN, Levy CN, Hughes SM, Kim Y, Valdez R, Pandey U, Ochsenbauer C, Astronomo R, McElrath J, Hladik F (2021) Buprenorphine increases HIV-1 infection in vitro but does not reactivate HIV-1 from latency. Viruses 13(8). https://doi.org/10.3390/v13081472

45. Mielke D, Stanfield-Oakley S, Borate B, Fisher LH, Faircloth K, Tuyishime M, Greene K, Gao H, Williamson C, Morris L, Ochsenbauer C, Tomaras G, Haynes BF, Montefiori D, Pollara J, deCamp AC, Ferrari G (2022) Selection of HIV envelope strains for standardized assessments of vaccine-elicited antibody-dependent cellular cytotoxicity-mediating antibodies. J Virol 96(2):

e0164321. https://doi.org/10.1128/JVI.01643-21

46. Seay K, Church C, Zheng JH, Deneroff K, Ochsenbauer C, Kappes JC, Liu B, Jeng EK, Wong HC, Goldstein H (2015) In vivo activation of human NK cells by treatment with an interleukin-15 superagonist potently inhibits acute in vivo HIV-1 infection in humanized mice. J Virol 89(12):6264–6274. https://doi.org/10.1128/JVI.00563-15
47. Seay K, Khajoueinejad N, Zheng JH, Kiser P, Ochsenbauer C, Kappes JC, Herold B, Goldstein H (2015) The vaginal acquisition and dissemination of HIV-1 infection in a novel transgenic mouse model is facilitated by coinfection with herpes simplex virus 2 and is inhibited by microbicide treatment. J Virol 89(18): 9559–9570. https://doi.org/10.1128/JVI.01326-15
48. Seay K, Qi X, Zheng JH, Zhang C, Chen K, Dutta M, Deneroff K, Ochsenbauer C, Kappes JC, Littman DR, Goldstein H (2013) Mice transgenic for CD4-specific human CD4, CCR5 and cyclin T1 expression: a new model for investigating HIV-1 transmission and treatment efficacy. PLoS One 8(5):e63537. https://doi.org/10.1371/journal.pone.0063537
49. Sung JA, Pickeral J, Liu L, Stanfield-Oakley SA, Lam CY, Garrido C, Pollara J, LaBranche C, Bonsignori M, Moody MA, Yang Y, Parks R, Archin N, Allard B, Kirchherr J, Kuruc JD, Gay CL, Cohen MS, Ochsenbauer C, Soderberg K, Liao HX, Montefiori D, Moore P, Johnson S, Koenig S, Haynes BF, Nordstrom JL, Margolis DM, Ferrari G (2015) Dual-Affinity Re-Targeting proteins direct T cell-mediated cytolysis of latently HIV-infected cells. J Clin Invest 125(11):4077–4090. https://doi.org/10.1172/JCI82314
50. Pollara J, Hart L, Brewer F, Pickeral J, Packard BZ, Hoxie JA, Komoriya A, Ochsenbauer C, Kappes JC, Roederer M, Huang Y, Weinhold KJ, Tomaras GD, Haynes BF, Montefiori DC, Ferrari G (2011) High-throughput quantitative analysis of HIV-1 and SIV-specific ADCC-mediating antibody responses. Cytometry A 79(8):603–612. https://doi.org/10.1002/cyto.a.21084
51. Mielke D, Bandawe G, Zheng J, Jones J, Abrahams MR, Bekker V, Ochsenbauer C, Garrett N, Abdool Karim S, Moore PL, Morris L, Montefiori D, Anthony C, Ferrari G, Williamson C (2021) ADCC-mediating non-neutralizing antibodies can exert immune pressure in early HIV-1 infection. PLoS Pathog 17(11):e1010046. https://doi.org/10.1371/journal.ppat.1010046
52. Neidleman JA, Chen JC, Kohgadai N, Müller JA, Laustsen A, Thavachelvam K, Jang KS, Stürzel CM, Jones JJ, Ochsenbauer C, Chitre A, Somsouk M, Garcia MM, Smith JF, Greenblatt RM, Münch J, Jakobsen MR, Giudice LC, Greene WC, Roan NR (2017) Mucosal stromal fibroblasts markedly enhance HIV infection of CD4+ T cells. PLoS Pathog 13(2): e1006163. https://doi.org/10.1371/journal.ppat.1006163
53. Fernandez N, Hayes P, Makinde J, Hare J, King D, Xu R, Rehawi O, Mezzell AT, Kato L, Mugaba S, Serwanga J, Chemweno J, Nduati E, Price MA, Osier F, Ochsenbauer C, Yue L, Hunter E, Gilmour J, investigators IpC (2022) Assessment of a diverse panel of transmitted/founder HIV-1 infectious molecular clones in a luciferase based CD8 T-cell mediated viral inhibition assay. Front Immunol 13:1029029. https://doi.org/10.3389/fimmu.2022.1029029
54. Hayes P, Fernandez N, Ochsenbauer C, Dalel J, Hare J, King D, Black L, Streatfield C, Kakarla V, Macharia G, Makinde J, Price M, Hunter E, Gilmour J, investigators IpC (2021) Breadth of CD8 T-cell mediated inhibition of replication of diverse HIV-1 transmitted-founder isolates correlates with the breadth of recognition within a comprehensive HIV-1 Gag, Nef, Env and Pol potential T-cell epitope (PTE) peptide set. PLoS One 16(11):e0260118. https://doi.org/10.1371/journal.pone.0260118
55. Ochsenbauer C, Kappes JC (2009) New virologic reagents for neutralizing antibody assays. Curr Opin HIV AIDS, 4(5):418–425. https://doi.org/10.1097/COH.0b013e32832f011e
56. Simon-Loriere E, Galetto R, Hamoudi M, Archer J, Lefeuvre P, Martin DP, Robertson DL, Negroni M (2009) Molecular mechanisms of recombination restriction in the envelope gene of the human immunodeficiency virus. PLoS Pathog 5(5):e1000418. https://doi.org/10.1371/journal.ppat.1000418
57. Balinda SN, Kapaata A, Xu R, Salazar MG, Mezzell AT, Qin Q, Herard K, Dilernia D, Kamali A, Ruzagira E, Kibengo FM, Song H, Ochsenbauer C, Salazar-Gonzalez JF, Gilmour J, Hunter E, Yue L, Kaleebu P (2022) Characterization of near full-length transmitted/founder HIV-1 subtype D and A/D recombinant genomes in a heterosexual Ugandan population (2006–2011). Viruses 14(2). https://doi.org/10.3390/v14020334
58. Binley JM, Lybarger EA, Crooks ET, Seaman MS, Gray E, Davis KL, Decker JM, Wycuff D, Harris L, Hawkins N, Wood B, Nathe C, Richman D, Tomaras GD, Bibollet-Ruche F,

Robinson JE, Morris L, Shaw GM, Montefiori DC, Mascola JR (2008) Profiling the specificity of neutralizing antibodies in a large panel of plasmas from patients chronically infected with human immunodeficiency virus type 1 subtypes B and C. J Virol 82(23): 11651–11668. https://doi.org/10.1128/JVI.01762-08

59. Roy CN, Benitez Moreno MA, Kline C, Ambrose Z (2021) CG dinucleotide removal in bioluminescent and fluorescent reporters improves HIV-1 replication and reporter gene expression for dual imaging in humanized mice. J Virol 95(19):e0044921. https://doi.org/10.1128/JVI.00449-21

60. Ochsenbauer C, Edmonds TG, Ding H, Keele BF, Decker J, Salazar MG, Salazar-Gonzalez JF, Shattock R, Haynes BF, Shaw GM, Hahn BH, Kappes JC (2012) Generation of transmitted/founder HIV-1 infectious molecular clones and characterization of their replication capacity in CD4 T lymphocytes and monocyte-derived macrophages. J Virol 86(5): 2715–2728. https://doi.org/10.1128/jvi.06157-11

Chapter 22

Multiparameter Flow Cytometry Monitoring of T Cell Responses

Holden T. Maecker

Abstract

Multiparameter flow cytometry is a common tool for assessing responses of T, B, and other cells to pathogens or vaccines. Such responses are likely to be important for predicting the efficacy of an HIV vaccine, despite the elusive findings in HIV vaccine trials to date. Fortunately, flow cytometry has evolved to be capable of readily measuring 30–40 parameters, providing the ability to dissect detailed phenotypes and functions that may be correlated with disease protection. Nevertheless, technical hurdles remain, and standardization of assays is still largely lacking. Here an optimized protocol for antigen-specific T cell monitoring is presented, with specific variations for particular markers. It covers the analysis of multiple cytokines, cell surface proteins, and other functional markers such as CD107, CD154, CD137, etc. References are given to published panels of 8–28 colors.

Key words Flow cytometry, Cytokines, T cells, Antigen-specific, Intracellular, Multicolor, HIV/ AIDS, Vaccine, Monitoring

1 Introduction

HIV vaccine researchers were among the first to adopt polychromatic flow cytometry as a tool to dissect T cell responses to HIV infection and to HIV vaccine candidates. While interest in the HIV vaccine field has fluctuated between generation of polyfunctional T cells and broadly neutralizing antibodies, it remains likely that both will be important at some level [1]. As such, detailed phenotyping of cellular responses, as afforded by newer flow cytometry instrumentation, is likely to be critical to defining vaccine correlates of protection.

Early studies in the SIV model, as well as in HIV+ long-term non-progressors, implicated specific types of T cell responses in protection from disease. These included CD8+ central memory T cells [2] with maintained proliferative capacity and polyfunctionality, including IL-2 production [3, 4]. Similarly, CD4+ T cells with

Vinayaka R. Prasad and Ganjam V. Kalpana (eds.), *HIV Protocols*, Methods in Molecular Biology, vol. 2807,
https://doi.org/10.1007/978-1-0716-3862-0_22,

good proliferative capacity were implicated in protection [5, 6]. The concept of polyfunctionality of T cell responses took hold, with studies not only in HIV but also in multiple other diseases showing the importance of maintaining polyfunctional T cells [7–12].

Unfortunately, the first HIV vaccine to demonstrate at least partial protection in a phase 3 clinical trial, RV144, did not show a correlation of polyfunctional T cells (or any T cell responses) with protection. Instead, two antibody-based correlates were identified [13], and the field turned to broadly neutralizing antibodies as the holy grail to be sought by a successful HIV vaccine. Still, antibody responses depend upon CD4 T cells, in particular, T follicular helper cells (TFH), as demonstrated in influenza vaccination [14, 15]. Furthermore, antibody levels are tied to circulating levels of plasmablasts, as again shown in the influenza system [16]. Thus, measurement of T cell responses to HIV vaccines will likely be important even if antibodies provide the major route of disease protection.

Conversely, progressive HIV disease can be correlated with a number of cellular markers. These include broad CD38 expression on CD8 T cells [17–19] and PD-1 expression on HIV-specific T cells [20, 21]. Altered differentiation of HIV-specific T cells has also been demonstrated by staining for markers such as CCR7, CD62L, CD27, CD28, CD45RA, and CD127 [22–28].

Finally, innate immune cell responses to vaccines have become a topic of recent interest. It is likely that trained immunity within cells such as NK cells and innate lymphoid cells (ILC) in general will be an important consequence of successful vaccines, as shown in tuberculosis and other systems [29, 30]. For identification of all of these cell types and their responses, especially in combination, multiparameter flow cytometry is the tool of choice.

1.1 Instrumentation Considerations

The number of parameters that can be simultaneously measured in flow cytometry has grown in the last 30 years due to multiple factors. One class of factors are specific developments in instrumentation, beginning with digital processing in the 1990s. Digital processing is important in that it allows for more precise calculation of optical spillover between detectors, and thus, more precise compensation than was possible with analog systems [31, 32]. This was matched with software routines that automate the compensation process when presented with a set of single-color fluorescent controls.

Recently, spectral cytometry has emerged as a more precise method for dealing with optical spillover between dyes. This in turn enables the use of more colors, including dyes that have very similar emission peaks but different overall emission spectra [33, 34]. Spectral cytometry measures the full spectrum of light emitted by the analyzed cells, then mathematically deconvolutes the signal contributed by individual fluorochromes, by reference to single-color controls. This is in contrast to conventional fluorescence flow cytometry, which attempts to isolate the signal from individual fluorochromes using an optimized bandpass filter for each one. Spectral cytometry has provided well-resolved data with up to 40 fluorescent parameters [35].

Another development in instrumentation is mass cytometry, or CyTOF, which uses heavy metal ion labels to tag antibodies, with a time-of-flight mass spectrometry readout (reviewed in [36]). This technology allows for even more simultaneously analyzed parameters (50+), with little to no spillover between detector channels (depending mainly on the purity of the metal isotopes used). The CyTOF has a distinct advantage for highly multiparameter protocols involving intracellular staining (e.g., for cytokines), in that the metal ion labels are not degraded by fixation/permeabilization protocols as some fluorescent tandem dyes are (*see* [32] and discussion under Subheading 1.2 below). However, this advantage is counter-balanced by the slow acquisition rate of the CyTOF (at least 10× slower than fluorescence flow cytometry). Together with the loss of ~40% of cells on injection, this makes CyTOF potentially more difficult to implement when analyzing rare events such as antigen-specific T cells. In any case, the workflow and considerations for CyTOF are sufficiently different that they will not be considered in this chapter. A protocol for CyTOF with intracellular cytokine staining of antigen-specific T cells is published elsewhere [37].

Large, multicenter clinical studies will require standardization not only of the instruments used but also of their setup. Fortunately, highly multicolor instrumentation has moved from largely customized hardware to more standard products. Automated setup routines using standardized beads are now commonplace, but experiment-specific settings may still need to be determined based on running fully stained cell samples with the panel of interest [38]. Finally, monitoring of certain cytometer parameters over time is an important quality control tool to ensure consistent data and to anticipate and/or identify potential problems that might require service [39].

1.2 Reagent Considerations

Together with instrument developments, there has been a rise in the number of available fluorochromes for flow cytometry. These include synthetic polymer dyes with spectrally different excitation wavelengths and relatively narrow emission peaks, such as the "brilliant" dyes [40]. There has also been a proliferation of "tandem" dyes, which use the emission from one fluorochrome to excite another, covalently coupled fluorochrome. Some of these can pose unique problems of degradation in the presence of light, heat, and fixation, which affect their spillover properties [32].

The basic elements that require consideration in designing a multicolor flow cytometry panel include brightness and spillover. In general, one should select a set of fluorochromes that offer the greatest brightness, within the constraints of the user's instrument, while minimizing spectral overlap. Antibodies with the dimmest staining are then assigned to the brightest fluorochromes, but with an eye toward minimizing spillover issues. For example, antibodies for two markers that are co-expressed on the same cell population should not be used in channels with a large amount of spillover between them. Rather, such channels could more effectively be used for antibodies that bind to mutually exclusive cell populations, thus rendering the spillover moot.

The choice of antibody specificities to include in a multicolor panel is somewhat daunting, given that there are literally hundreds of possible markers to include. While the number of markers that can be simultaneously used in multicolor flow cytometry is now larger than ever, choices still need to be made. Besides the basic lineage and memory subset markers, there are many possible activation markers and cytokines to consider and even more potential homing receptors and markers of differentiation and exhaustion. Table 1 organizes some of these into classes for consideration.

Once an antibody panel is designed, titration of certain reagents is often required to achieve an optimal ratio of signal: noise. While many reagents are sold pre-titered, this does not always mean that the specified titer is optimal in a given application. Since polychromatic experiments already compromise sensitivity due to spillover between detectors, the need for optimal titration is critical. It should also be noted that the optimal titer for cell surface staining with an antibody is often higher than that for intracellular staining (post-fixation and permeabilization). For some antibodies, resolution is compromised so drastically by fixation and permeabilization that these antibodies need to be used prior to application of a fixative.

Given the many considerations in optimizing a reagent panel for polychromatic flow cytometry, it makes sense to take advantage, wherever possible, of panels already created and proven by others. The OMIP publications in Cytometry Part A are useful resources for well-described and tested panels. For example, a recently published 21-color panel from Nogimori et al. [41] probes eight cytokines, CD107 and CD154, and multiple memory, homing, and differentiation markers, along with CD3, CD4, and CD8. Similarly, Swanson and Seder published a 28-color panel that includes eight cytokines, CD69 and CD154, and a large number of differentiation and homing receptors, plus antibodies to multiple TCR γ and δ families, CD3, CD4, and CD8. A more modest eight-color panel was used by our group to demonstrate HIV-specific T cell differentiation states [28], and the gating scheme for this panel is shown in Fig. 1. Panels such as these can at least be taken as a

Table 1
Types of markers to consider

Class	Example Marker(s)	Comments	Reference (s)
T Lineage	CD3, CD4, CD8, γδTCR	All are needed for best discrimination of T cell lineages	
Dump	CD14, CD33, CD19, CD20, CD16, CD62P	Some or all of these can be combined in a single channel, to make T cell gating cleaner, unless parallel analysis of these cell lineages is desired	[52]
T Memory Subsets	CCR7, CD45RA	While these are canonical for defining naïve, central memory, effector memory, and TEMRA, there can be modifications (e.g., CD27 or CD62L for CCR7)	[55]
T differentiation	CD28	Costimulatory protein lost in immune aging	[5, 56, 57]
	CD57	Marker of effector differentiation	[58]
	ICOS	Costimulatory and TFH marker	[14, 15]
	PD-1	Potential exhaustion marker	[20, 21]
Activation induced	CD69	Rapidly induced activation marker	[59]
	CD154	Marker of activation for CD4 T cells	[60]
	CD25, OX-40, CD137	Markers of activation for CD4 and CD8 T cells	[60, 61]
Degranulation	CD107	Marker of CD8 T cell degranulation	[62]
Homing	CCR4, CXCR3, CCR6, CXCR5, CLA, b7 integrin	Chemokines and other tissue homing markers	
Cytokines	IL-2, IL-4, IL-5, IL-10, IL-13, IL-17, IL-21, IFNγ, MIP-1β, TNF	T cell cytokines used in intracellular staining	
	IL-6, IL-8, IL-12	Non-T cell cytokines used in intracellular staining	

starting point, streamlining the optimization process compared to starting from "scratch."

Multiparameter intracellular cytokine staining has been reviewed elsewhere [42], and standardization and tips for optimizing protocols have also been published [43, 44]. The main focus of the protocol presented here is to show how intracellular cytokine staining of antigen-specific T cells can be adapted to a polychromatic format, which may include readout not only of phenotypic markers but also of functional markers such as CD107 or CD154, in addition to cytokines. Variables in the stimulation and processing steps that apply to these markers and others are summarized in Table 2.

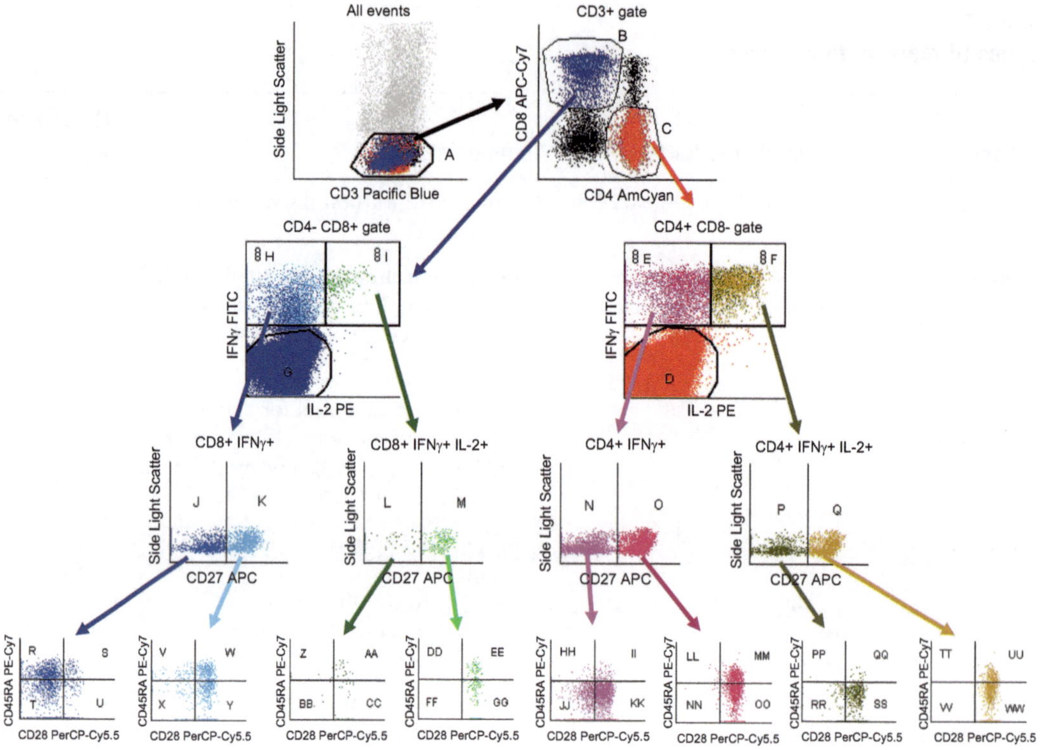

Fig. 1 Gating scheme for eight-color ICS study (reproduced from ref. 28). An acquisition threshold was used to exclude CD3-negative cells. Dynamic gates were used for defining CD3+, CD4+, and CD8+ lymphocytes, tethered regions were used to define the cytokine-positive cells, and conventional (static) regions or quadrants were used to define all other (rare) subsets. Note that all possible combinatorial subsets of CD27, CD28, and CD45RA were reported for each subset of cytokine-positive cells, resulting in a set of 32 populations whose frequencies describe a response "fingerprint" for that sample

2 Materials

2.1 Reagents

1. Freshly isolated or cryopreserveds PBMCs, isolated by Ficoll gradient centrifugation or via cell Preparation Tubes (CPT) or equivalent.
2. Complete RPMI: RPMI-1640 medium with 20 mM HEPES, 10% fetal bovine serum, and antibiotic/antimycotic solution (cRPMI-10).
3. Stimulation antigens, e.g., commercial peptide mixes (resuspended at 500 μg/mL per peptide and stored in aliquots at −80 C) [45], and PMA/ionomycin (Sigma) or Staphylococcal enterotoxin B (SEB, Sigma) as a positive control. Optional: Preconfigured plates containing lyophilized stimulation reagents and secretion inhibitor(s) can be purchased [e.g., BD Lyoplate, BD Biosciences, San Jose, CA [46]].

Table 2
Procedural Variables for Different Functional Markers

Variable	Covered by this protocol: IL-2, IL-4, IL-5, IL-10[b], IL-13, IL-17, IL-21, IFNγ, MIP-1β, TNF	Covered by this protocol: CD107, CD154	Not covered[a]: TGFβ
Stimulation conditions	6–12 h	5–6 h in the presence of staining antibodies for these markers	16–24 h in serum-free medium[c]
Secretion inhibitor	Brefeldin A	Monensin[d]	Monensin
Fixation/ permeabilization system	FACS Lysing Solution, FACS Permeabilizing Solution 2[e]	FACS Lysing Solution, FACS Permeabilizing Solution 2[e]	Cytofix, Cytoperm[f]

[a]This marker would be difficult to combine with those in the first two columns, mainly due to the longer optimal stimulation time
[b]IL-10 positive cells are very rare, and detection may be increased with up to 24 h incubation
[c]Serum-free medium (e.g., AIM V, Invitrogen, Grand Island, NY) produces much stronger TGFβ responses, presumably because serum contains free TGFβ that blocks staining for this marker
[d]CD107 and CD154 are taken up by endocytic vesicles and degraded. Monensin blocks this degradation by preventing acidification of these vesicles. When doing combined assays with cytokines, a monensin+brefeldin A combination is recommended (*see* **Note 5**)
[e]BD Biosciences, San Jose, CA. Note that other fixation/permeabilization systems such as Cytofix/Cytoperm (BD Biosciences, San Diego, CA) are also used successfully for these markers by many investigators
[f]BD Biosciences, San Diego, CA

(a) For use of SEB as a positive control: Prepare a 50× SEB stock of 50 μg/mL in sterile PBS. Store this stock at 4 °C (aliquoting is not necessary).

(b) For use of PMA + ionomycin as a positive control: Prepare a 1 mg/ml stock of PMA in DMSO, and store in small aliquots at −80 C. Dilute an aliquot 1:1000 in PBS just before use to yield a 100× (1 μg/ml) stock. Also prepare a 1 mg/ml stock of ionomycin in DMSO, and store in small aliquots at −80 C. Dilute a small aliquot 1:10 in PBS just before use to yield a 100× (100 μg/ml) stock. Use 2 μl of 100× PMA + 2 μl of 100× ionomycin for each positive control well.

4. Optional: costimulatory antibodies to CD28 and CD49d, 0.1 mg/mL each in sterile PBS (FastImmune, BD Biosciences).
5. 5 mg/mL brefeldin A, in DMSO, stored in aliquots at −80 C. Assays involving CD154 or CD107a staining also require: 5 mg/ml monensin in methanol, stored at −80 C (aliquoting not necessary).
6. EDTA, 20 mM in PBS (pH 7.4).

7. Cell fixation reagent, e.g., BD FACS Lysing Solution (BD Biosciences) or equivalent.
8. Cell permeabilizing reagent, e.g., BD FACS Permeabilizing Solution 2 (BD Biosciences) or equivalent.
9. Fluorescent-labeled antibodies. Optional: Preconfigured plates containing lyophilized antibody cocktails, available from vendors such as Biolegend and BD Biosciences.
10. Wash buffer (0.5% bovine serum albumin +0.1% NaN_3 in PBS).
11. Antibody capture beads (e.g., BD CompBeads (BD Biosciences)), for creating single-color compensation controls.
12. Optional, to reduce biohazard potential, or if samples will be stored >24 h prior to acquisition: 1% paraformaldehyde in PBS [dilute 10% paraformaldehyde (EM Science, Gibbstown, NJ) 1:10 in PBS]; or BD Stabilizing Fixative (BD Biosciences) when tandem dyes are used.

2.2 Equipment

1. 96-well conical bottom polypropylene plates with lids (*see* **Note 1**).
2. 12-channel aspiration manifold with 7 mm prongs (V&P Scientific, San Diego, CA).
3. Plate holders for table-top centrifuge.
4. Polychromatic (3–5 laser), preferably spectral, flow cytometer.
5. Optional: 96-well plate loader for flow cytometer.

3 Methods

3.1 Sample Collection

1. (a) For fresh PBMC (*see* **Note 2**): Resuspend at 5×10^6 to 1×10^7 viable lymphocytes/mL in warm (37 °C) cRPMI-10 (*see* **Note 3**).

   (b) For cryopreserved PBMC (*see* **Note 4**): Thaw briefly in a 37 °C water bath, then slowly dilute up to 10 mL with warm (37 °C) cRPMI-10, and centrifuge for approximately 7 min at 250 × g. Resuspend in a small volume of warm cRPMI-10, perform a viable cell count, and dilute to a final concentration of 5×10^6 to 1×10^7 viable lymphocytes/mL (*see* **Note 3**).
2. Add 200 μL of cell suspension per well to a 96-well plate. For cryopreserved PBMC, incubate at 37 °C for 6–18 h prior to stimulation (*see* **Note 4**).

3.2 Cell Activation

1. (a) For assays not involving CD107 or CD154: Thaw an aliquot of 5 mg/mL brefeldin A stock (*see* **Note 5**). Dilute 1:10 in sterile PBS to make a 50× working stock.

   (b) For assays measuring CD107 and/or CD154: Thaw an aliquot of 5 mg/mL brefeldin A and add an equal volume of 5 mg/mL monensin stock (*see* **Note 5**). Dilute 1:10 in sterile PBS to make a 50× working stock.

   (c) For assays using preconfigured lyophilized stimulation reagents in plates: Add 200 μL of cell suspension directly to the appropriate wells, let sit for a few minutes, then pipet up and down thoroughly to mix. Skip to Subheading 3.2, **step** 7.
2. Thaw and dilute peptide stocks in sterile PBS, if necessary, to achieve a 50× working stock that is between 50 and 100 μg/mL/peptide (when diluted 1:50, this will yield a final concentration of 1–2 μg/mL/peptide) (*see* **Note 6**).
3. For each stimulation condition, prepare a "master mix" of the 50× working stocks and costimulatory antibodies as follows:

   4 μL/well peptides, positive control(s), or PBS (negative control)

   4 μL/well brefeldin A or brefeldin A + monensin

   4 μL/well CD28 + CD49d Ab stock (*see* **Note** 7).
4. Pipet 12 μL of the appropriate master mix into each well containing cells. Mix by gently pipetting. Always separate positive control wells from antigen stimulation wells to avoid possible cross-contamination during stimulation and acquisition.
5. For assays involving CD107 and/or CD154, also add the recommended titer of the antibody conjugate(s) to each well. Minimize exposure to light, particularly for tandem dye conjugates (*see* **Note 8**).
6. Incubate covered plate for 6–24 h at 37 °C (*see* **Notes 9** and **10**).

3.3 Sample Processing

1. To halt activation and detach adherent cells, add 20 μL per well of 20 mM EDTA in PBS and mix by pipetting.
2. Incubate 15 min at room temperature, then mix again by vigorous pipetting to fully resuspend adhered cells.
3. Centrifuge plate at 250 × g for 5 min. Aspirate supernatant with 7 mm vacuum manifold (*see* **Note 11**).
4. For assays using amine-reactive dye for staining non-viable cells: Resuspend the amine dye at optimum concentration in PBS (usually around 2.5 μg/mL, but this should be determined for individual lots of dye). Resuspend each well with 100 μL of this solution, incubate 20 min at room temperature, then add 100 μL wash buffer, and wash as in Subheading 3.3, **step 3** above.

5. (a) For assays using liquid reagents and cell-surface markers other than CD3, CD4, and CD8: Resuspend each well in 100 μL wash buffer and add optimal titers of all Abs to cell-surface markers (*see* **Note 12**), incubate 30–60 min at room temperature, then add 100 μL wash buffer, and wash as in Subheading 3.3, **step 3** above.

   (b) For assays using preconfigured lyophilized cell-surface staining Abs: Resuspend the appropriate wells of the surface Ab plate with 50 μL of wash buffer. Let it sit for a few minutes, then pipet up and down thoroughly to mix. Transfer the solution to appropriate wells of the cell plate, incubate 30–60 min at room temperature in the dark, then add 100 μL wash buffer, and wash as in Subheading 3.3, **step 3** above.
6. Resuspend cell pellets with 100 μL of 1× BD FACS Lysing Solution per well. Incubate at room temperature for 10 min (*see* **Notes 13** and **14**).
7. Add 100 μL wash buffer to each well, then centrifuge plate at 500 × g for 5 min (*see* **Note 15**). Aspirate supernatant with 7 mm vacuum manifold.
8. Resuspend cell pellets with 200 μL of 1× BD FACS Permeabilizing Solution 2 per well. Incubate at room temperature for 10 min (*see* **Note 14**).
9. Centrifuge plate at 500 × g for 5 min (*see* **Note 15**). Aspirate supernatant with 7 mm vacuum manifold.
10. Add 200 μL wash buffer to each well and wash as in Subheading 3.3, **step 9** above.
11. Again add 200 μL wash buffer to each well, and wash as in Subheading 3.3, **step 9** above.
12. (a) For assays using liquid reagents: Resuspend pellet in an optimally titrated staining cocktail of all Abs to intracellular markers. Incubate in the dark at room temperature for 60 min, mixing by pipetting or gentle agitation every 15–20 min.

    (b) For assays using preconfigured lyophilized intracellular staining reagents: Resuspend the appropriate wells of the intracellular Ab plate with 50 μL of wash buffer. Let it sit for a few minutes, then pipet up and down thoroughly to mix. Transfer the solution to the appropriate wells of the cell plate, mix to resuspend cell pellets, and incubate at room temperature in the dark for 60 min, mixing by pipetting or gentle agitation every 15–20 min.
13. Add 200 μL of wash buffer to each well, and wash as in Subheading 3.3, **step 9** above.
14. Again add 200 μL wash buffer to each well, and wash as in Subheading 3.3, **step 9** above.

15. Resuspend pellets with 150 μL wash buffer. Store at 4 °C in the dark until ready for data acquisition, which should be performed within 24 h. Optional: resuspend pellets with 150 μL of 1% paraformaldehyde in PBS or BD Stabilizing Fixative (*see* **Note 16**).

3.4 Data Acquisition and Analysis

1. First determine optimal instrument settings for the reagent panel in question.
   (a) Establish baseline PMT settings using vendor's setup beads and software.
   (b) Run a sample stained with the full reagent cocktail in question, and adjust baseline PMT settings as needed so that events are mostly above zero but do not register in the highest fluorescence channel. Save these experiment-specific settings if allowed by the software.
2. Create a set of compensation controls consisting of single-stained cells or beads (*see* **Note 17**). Acquire these controls and use the software's automated algorithm to calculate compensation (*see* **Note 18**).
3. Create a template for acquisition that displays the relevant parameters in the test samples in the form of dot plots. This template need not be the same as that used for analysis, i.e., it does not need to specify all gates or regions of interest. However, the template should show any gates used to define the saved population of cells or the stopping criteria (e.g., CD3+ cells).
4. Set an appropriate threshold, usually on forward scatter, to eliminate small debris, and set the stopping and storage criteria. It is usually safest to store all events (rather than a gated subset) to allow exploration of all data on analysis. However, sometimes a threshold or gate on CD3+ cells may be employed in order to reduce file sizes (*see* **Note 19**). When using a plate loader, be sure to set a stopping criterion based on time, so that samples with insufficient cells will not run dry.
5. Acquire data from a fully stained sample to verify that the settings chosen are appropriate. If any changes to PMT voltages need to be made, be sure to rerun compensation samples and recalculate compensation based on the new voltages.
6. Record data from samples. Be sure to collect positive controls as the last sample for each donor and follow by thorough washing, to avoid cell carryover that could result in a false-positive result.
7. Analyze data using the acquisition software or compatible third-party software. Be sure to define all regions of interest and report the desired statistics on these (*see* **Note 20** and

Fig. 1). Where possible, use a batch analysis function to analyze all samples from a given experiment or study and export the statistical data to a spreadsheet (*see* **Note 21**).

8. For large studies, it is helpful to create a database to accept the statistical output files from batch analysis. This database can then be queried to create data tables from subsets of the data, allowing rapid graphing, statistical analysis, background subtraction, conversion to absolute counts, etc.
9. Consider performing unbiased clustering and/or dimensionality reduction in addition to manual gating. Analysis packages such as UMAP [47] or FlowSOM [48] are available in premium versions of the online Cytobank software (Beckman Coulter).

4 Notes

1. Plates versus tubes: Cells can also be stimulated in 15 mL conical polypropylene tubes, with staining in 12 × 75 mm polystyrene tubes (BD Falcon). However, plates are preferred for ease of handling multiple samples, and results for human PBMCs are equivalent to those in tubes [49].
2. Fresh PBMCs: If PBMCs are not to be cryopreserved, they should ideally be prepared on the day of blood draw, then either stimulated the same day, or rested at 37 °C in cRPMI-10 overnight and stimulated the following day. Overnight resting at 37 °C increases the staining intensity of cytokines, but the effect is more pronounced with cryopreserved samples. Overnight shipping of whole blood or PBMCs at ambient temperatures can cause a variable decrease in cell function and should be avoided if possible.
3. Higher cell concentrations (1×10^7/mL, 2×10^6/well) should be used when possible, especially when response levels are low and/or there are many cell subsets to enumerate.
4. Cryopreserved PBMCs: If cells cannot be stimulated within 24 h of blood draw, they should be cryopreserved by a validated protocol [50]. Upon thawing, recoveries of >60% and viabilities of >80% should usually be obtained; values much lower than these will be accompanied by functional losses as well, and acceptance criteria may be set to avoid acquisition of non-functional samples. The method of thawing is equally as important as that of cryopreservation [50]. Thawed cells should be rested in cRPMI-10 for 6–18 h at 37 °C to maximize cytokine staining intensity [49]. Some cell loss may occur during this period, so plating a slight excess of cells is desirable. Alternately, the cells can be rested in bulk (e.g., in a slanted

15 mL conical polypropylene tube), then recounted and resuspended at the desired concentration after resting.

5. Brefeldin A versus monensin: Secretion of most cytokines of interest (IFNγ, IL-2, etc.) is best inhibited by brefeldin A at 10 μg/mL cell suspension. However, CD107 and CD154 are transiently expressed on the cell surface. Therefore, staining Abs to CD107 and/or CD154 are added to the stimulation culture to bind the antigen(s) as soon as they are expressed. Monensin increases the intensity of staining under these conditions by preventing the acidification and degradation of lysosomal vesicles that contain the recycled CD107 and CD154. Thus, for combined cytokine and CD107 or CD154 detection, 5 μg/mL each of brefeldin A and monensin is recommended.
6. Peptide mixes: Peptide mixes can be prepared and lyophilized as premixed pools of up to several hundred peptides [45]. These can then be resuspended in DMSO at high concentration per peptide, to avoid DMSO toxicity to cells. The total concentration of DMSO in the assay should be kept at $<0.5\%$.
7. Costimulatory antibodies: Antibodies to CD28 and CD49d can increase the cytokine response to protein antigens, peptides, and SEB by amplifying the signal for low-affinity T cells [51]. In occasional donors, they increase cytokine production in the absence of antigen (TNF is usually most affected).
8. Adding staining Abs during stimulation: As described in **Note 5**, staining Abs to CD107 and CD154 are best added during stimulation, to capture the transiently expressed antigen. Fluorochrome conjugated Abs are sensitive to light exposure, so they should be handled in low light and, once added, the samples should be incubated in the dark. Certain tandem dyes such as APC-Cy7 and PE-Cy7 are particularly sensitive to light and temperature [32] and are not optimal choices for use in stimulation cultures.
9. Stimulation time: A minimum of 5–6 h allows adequate detection of most proinflammatory cytokines like IFNγ, TNF, and IL-2 [52]. Increasing the time of incubation (in the presence of brefeldin A) increases cytokine staining intensity, but is not recommended for CD107 or CD154. For whole proteins requiring intracellular processing, a pre-incubation of 2 h prior to adding brefeldin A and/or monensin is recommended [52]. CD8 responses to whole protein antigens can sometimes be detected and are increased with longer incubation in antigen alone, but not in all donors [53].
10. Automating incubation times: A programmable heat block, incubator, or water bath can be used to time activation, cooling the samples to 4–18 °C at the end of a specified period at 37 °C, and holding them for later processing.

11. A fixed-length vacuum manifold helps achieve consistent washing without undue cell loss in microtiter plates. Because of the small wash volume, a sufficient number of washes and efficient removal of supernatant are essential.
12. CD3, CD4, and CD8 can be stained either before or after fixation and permeabilization. Down-modulation of these antigens occurs to a variable degree depending upon the stimulus. Cells that have down-modulated these antigens can be better detected by intracellular staining (post-fixation and permeabilization) [45], although the overall staining intensity is usually decreased. Most other cell-surface antigens are optimally stained before fixation.
13. Freezing of activated samples: Samples can be frozen at –80 °C directly in FACS Lysing Solution [52] or after treatment with FACS Lysing Solution or similar fixative. This allows for samples to be sent to another laboratory for processing, or for longitudinal samples to be accumulated for batch processing; but it is not recommended if a cell-surface staining step has been performed. FACS Lysing Solution simultaneously lyses erythrocytes and fixes leukocytes. While erythrocyte lysis is not required for PBMC samples, fixation is still helpful to prevent cell loss prior to permeabilization.
14. Fixation and permeabilization steps: Solutions for these steps should be stored and used at 22–25 °C.
15. Centrifugation speed: All centrifugation post-fixation should be done at higher g force (500 × g) due to increased cell buoyancy.
16. Use of paraformaldehyde is only helpful when samples are stored for more than 24 h prior to acquisition, or to ensure neutralization of potentially biohazardous samples. In addition to effects on cell scatter and autofluorescence, storage in paraformaldehyde can cause degradation of tandem dyes such as APC-Cy7 and PE-Cy7. An alternative fixative is available that protects these tandems from degradation (BD Stabilizing Fixative, BD Biosciences).
17. Compensation controls: Where possible, anti-immunoglobulin coated capture beads are preferred as compensation controls, because they provide a bright and homogeneous population of events stained with the antibody conjugate of interest. Ideally, the same lot of antibody should be used for compensation as is used in the experiment. In practice, however, this is only important for certain tandem conjugates, such as APC-Cy7 and PE-Cy7. The compensation controls should ideally be treated identically to the experimental samples in terms of fixation, etc., although this too is only important for the above tandem dyes.

18. When to apply compensation: While compensation can be calculated and changed at any time by software packages such as FlowJo (BD Biosciences) or Cytobank (Beckman Coulter), it is helpful to perform compensation before sample acquisition, so that any setup problems can be more readily detected.
19. Number of events to collect: Because multiparameter ICS assays tend to divide responding populations of cells into ever-smaller subsets, it is important to process and collect enough cells per sample to allow statistically significant differences between samples to be detected. The number of events required will depend upon the anticipated levels of responses and background, as well as the number of subsets of responding cells being identified.
20. Gating of down-modulated cells: Be sure that gates set on CD3, CD4, and CD8 parameters include dim-positive cells, since down-modulation of these markers occurs with activation. When using dynamic gating (*see* **Note 21**), set the region size to the maximum value possible without causing inclusion of neighboring populations. Some donors have a significant population of CD4 + CD8dim T cells. This population contains a disproportionate number of cells specific for chronic antigens such as CMV and HIV [54] and should be included in the CD4+ T cell gate to avoid under-reporting of responses.
21. Batch analysis: Dynamic gating tools can be used to accommodate staining differences between samples for populations such as CD3+, CD4+, and CD8+ cells (*see* Fig. 1). This in turn allows use of a single analysis template and batch analysis across multiple samples in an experiment or study. However, dynamic gates are not generally useful for rare populations, and their specifications (size and movement) need to be adjusted for the dataset being analyzed. Batch analysis and dynamic gating thus do not replace the need for visual inspection of all data.

Acknowledgments

Details of this protocol were optimized by Laurel Nomura and Maria Suni (BD Biosciences).

References

1. Tomaras GD, Plotkin SA (2017) Complex immune correlates of protection in HIV-1 vaccine efficacy trials. Immunol Rev 275:245–261
2. Sun Y, Schmitz JE, Buzby AP et al (2006) Virus-specific cellular immune correlates of survival in vaccinated monkeys after simian immunodeficiency virus challenge. J Virol 80: 10950–10956
3. Migueles SA, Laborico AC, Shupert WL et al (2002) HIV-specific CD8+ T cell proliferation is coupled to perforin expression and is maintained in nonprogressors. Nat Immunol 3: 1061–1068

4. Betts MR, Nason MC, West SM et al (2006) HIV nonprogressors preferentially maintain highly functional HIV-specific CD8+ T cells. Blood 107:4781–4789
5. Boaz MJ, Waters A, Murad S et al (2002) Presence of HIV-1 Gag-specific IFN-gamma +IL-2+ and CD28+IL-2+ CD4 T cell responses is associated with nonprogression in HIV-1 infection. J Immunol (Baltimore, Md.: 1950) 169:6376–6385
6. Younes S-A, Yassine-Diab B, Dumont AR et al (2003) HIV-1 viremia prevents the establishment of interleukin 2-producing HIV-specific memory CD4+ T cells endowed with proliferative capacity. J Exp Med 198:1909–1922
7. Boyd A, Almeida JR, Darrah PA et al (2015) Pathogen-specific T cell polyfunctionality is a correlate of T cell efficacy and immune protection. PLoS One 10:e0128714
8. Pera A, Vasudev A, Tan C et al (2017) CMV induces expansion of highly polyfunctional CD4+ T cell subset coexpressing CD57 and CD154. J Leukoc Biol 101:555–566
9. Moncunill G, De Rosa SC, Ayestaran A et al (2017) RTS,S/AS01E malaria vaccine induces memory and polyfunctional T cell responses in a pediatric african phase III trial. Front Immunol 8:1008
10. Akinsiku OT, Bansal A, Sabbaj S et al (2011) Interleukin-2 production by polyfunctional HIV-1-specific CD8 T cells is associated with enhanced viral suppression. J Acquir Immune Defic Syndr 58:132–140
11. Snyder LD, Chan C, Kwon D et al (2016) Polyfunctional T-cell signatures to predict protection from cytomegalovirus after lung transplantation. Am J Respir Crit Care Med 193: 78–85
12. Ferre AL, Hunt PW, McConnell DH et al (2010) HIV controllers with HLA-DRB1*13 and HLA-DQB1*06 alleles have strong, polyfunctional mucosal CD4+ T-cell responses. J Virol 84:11020–11029
13. Haynes BF, Gilbert PB, McElrath MJ et al (2012) Immune-correlates analysis of an HIV-1 vaccine efficacy trial. N Engl J Med 366:1275–1286
14. Spensieri F, Borgogni E, Zedda L et al (2013) Human circulating influenza-CD4+ ICOS1+-IL-21+ T cells expand after vaccination, exert helper function, and predict antibody responses. Proc Natl Acad Sci USA 110: 14330–14335
15. Bentebibel S-E, Lopez S, Obermoser G et al (2013) Induction of ICOS+CXCR3+CXCR5+ TH cells correlates with antibody responses to influenza vaccination. Sci Transl Med 5: 176ra32
16. Nakaya HI, Wrammert J, Lee EK et al (2011) Systems biology of vaccination for seasonal influenza in humans. Nat Immunol 12:786–795
17. Kestens L, Vanham G, Gigase P et al (1992) Expression of activation antigens, HLA-DR and CD38, on CD8 lymphocytes during HIV-1 infection. AIDS 6:793–797
18. Ho HN, Hultin LE, Mitsuyasu RT et al (1993) Circulating HIV-specific CD8+ cytotoxic T cells express CD38 and HLA-DR antigens. J Immunol 150:3070–3079
19. Giorgi JV, Liu Z, Hultin LE et al (1993) Elevated levels of CD38+ CD8+ T cells in HIV infection add to the prognostic value of low CD4+ T cell levels: results of 6 years of follow-up. The Los Angeles Center, Multicenter AIDS Cohort Study. J Acquir Immune Defic Syndr 6:904–912
20. Day CL, Kaufmann DE, Kiepiela P et al (2006) PD-1 expression on HIV-specific T cells is associated with T-cell exhaustion and disease progression. Nature 443:350–354
21. Trautmann L, Janbazian L, Chomont N et al (2006) Upregulation of PD-1 expression on HIV-specific CD8+ T cells leads to reversible immune dysfunction. Nat Med 12:1198–1202
22. Chen G, Shankar P, Lange C et al (2001) CD8 T cells specific for human immunodeficiency virus, Epstein-Barr virus, and cytomegalovirus lack molecules for homing to lymphoid sites of infection. Blood 98:156–164
23. Champagne P, Ogg GS, King AS et al (2001) Skewed maturation of memory HIV-specific CD8 T lymphocytes. Nature 410:106–111
24. Harari A, Rizzardi GP, Ellefsen K et al (2002) Analysis of HIV-1- and CMV-specific memory CD4 T-cell responses during primary and chronic infection. Blood 100:1381–1387
25. Ellefsen K, Harari A, Champagne P et al (2002) Distribution and functional analysis of memory antiviral CD8 T cell responses in HIV-1 and cytomegalovirus infections. Eur J Immunol 32:3756–3764
26. Appay V, Dunbar PR, Callan M et al (2002) Memory CD8+ T cells vary in differentiation phenotype in different persistent virus infections. Nat Med 8:379–385
27. Paiardini M, Cervasi B, Albrecht H et al (2005) Loss of CD127 expression defines an expansion of effector CD8+ T cells in HIV-infected individuals. J Immunol 174:2900–2909
28. Nomura LE, Emu B, Hoh R et al (2006) IL-2 production correlates with effector cell

differentiation in HIV-specific CD8+ T cells. AIDS Res Ther 3:18

29. O'Neill LAJ, Netea MG (2020) BCG-induced trained immunity: can it offer protection against COVID-19? Nat Rev Immunol 20: 335–337
30. Netea MG, Domínguez-Andrés J, Barreiro LB et al (2020) Defining trained immunity and its role in health and disease. Nat Rev Immunol 20:375–388
31. Roederer M (2001) Spectral compensation for flow cytometry: visualization artifacts, limitations, and caveats. Cytometry A 45:194–205
32. Maecker HT, Frey T, Nomura LE et al (2004) Selecting fluorochrome conjugates for maximum sensitivity. Cytometry A 62:169–173
33. Futamura K, Sekino M, Hata A et al (2015) Novel full-spectral flow cytometry with multiple spectrally-adjacent fluorescent proteins and fluorochromes and visualization of in vivo cellular movement. Cytometry A 87:830–842
34. Schmutz S, Valente M, Cumano A et al (2016) Spectral cytometry has unique properties allowing multicolor analysis of cell suspensions isolated from solid tissues. PLoS One 11: e0159961
35. Park LM, Lannigan J, Jaimes MC (2020) OMIP-069: forty-color full spectrum flow cytometry panel for deep immunophenotyping of major cell subsets in human peripheral blood. Cytometry A 97:1044–1051
36. Spitzer MH, Nolan GP (2016) Mass cytometry: single cells, many features. Cell 165:780–791
37. Lin D, Maecker HT (2018) Mass cytometry assays for antigen-specific T cells using CyTOF. Methods Mol Biol 1678:37–47
38. Maecker HT, Trotter J (2006) Flow cytometry controls, instrument setup, and the determination of positivity. Cytometry A 69:1037–1042
39. Perfetto SP, Ambrozak D, Nguyen R et al (2006) Quality assurance for polychromatic flow cytometry. Nat Protoc 1:1522–1530
40. Chattopadhyay PK, Gaylord B, Palmer A et al (2012) Brilliant violet fluorophores: a new class of ultrabright fluorescent compounds for immunofluorescence experiments. Cytometry A 81:456–466
41. Nogimori T, Moriishi E, Ikeda M et al (2021) OMIP 075: a 22-color panel for the measurement of antigen-specific T-cell responses in human and nonhuman primates. Cytometry A 99:884–887
42. Lovelace P, Maecker HT (2018) Multiparameter intracellular cytokine staining. Methods Mol Biol 1678:151–166
43. Lamoreaux L, Roederer M, Koup R (2006) Intracellular cytokine optimization and standard operating procedure. Nat Protoc 1: 1507–1516
44. Nomura L, Maino VC, Maecker HT (2008) Standardization and optimization of multiparameter intracellular cytokine staining. Cytometry A 73:984–991
45. Maecker HT, Dunn HS, Suni MA et al (2001) Use of overlapping peptide mixtures as antigens for cytokine flow cytometry. J Immunol Methods 255:27–40
46. Maecker HT, Rinfret A, D'Souza P et al (2005) Standardization of cytokine flow cytometry assays. BMC Immunol 6:13
47. Becht E, McInnes L, Healy J et al (2018) Dimensionality reduction for visualizing single-cell data using UMAP. Nat Biotechnol 37:38–44
48. Van Gassen S, Callebaut B, Van Helden MJ et al (2015) FlowSOM: using self-organizing maps for visualization and interpretation of cytometry data. Cytometry A 87:636–645
49. Suni MA, Dunn HS, Orr PL et al (2003) Performance of plate-based cytokine flow cytometry with automated data analysis. BMC Immunol 4:9
50. Disis ML, dela Rosa C, Goodell V et al (2006) Maximizing the retention of antigen specific lymphocyte function after cryopreservation. J Immunol Methods 308:13–18
51. Waldrop SL, Davis KA, Maino VC et al (1998) Normal human CD4+ memory T cells display broad heterogeneity in their activation threshold for cytokine synthesis. J Immunol (Baltimore, Md.: 1950) 161:5284–5295
52. Nomura LE, Walker JM, Maecker HT (2000) Optimization of whole blood antigen-specific cytokine assays for CD4(+) T cells. Cytometry 40:60–68
53. Maecker HT, Ghanekar SA, Suni MA et al (2001) Factors affecting the efficiency of CD8+ T cell cross-priming with exogenous antigens. J Immunol (Baltimore, Md.: 1950) 166:7268–7275
54. Suni MA, Ghanekar SA, Houck DW et al (2001) CD4(+)CD8(dim) T lymphocytes exhibit enhanced cytokine expression, proliferation and cytotoxic activity in response to HCMV and HIV-1 antigens. Eur J Immunol 31:2512–2520
55. Sallusto F, Lenig D, Förster R et al (1999) Two subsets of memory T lymphocytes with distinct homing potentials and effector functions. Nature 401:708–712
56. Eylar EH, Lefranc CE, Yamamura Y et al (2001) HIV infection and aging: enhanced

Interferon- and Tumor Necrosis Factor-alpha production by the CD8+ CD28- T subset. BMC Immunol 2:10

57. Weng N-P, Akbar AN, Goronzy J (2009) CD28(−) T cells: their role in the age-associated decline of immune function. Trends Immunol 30:306–312
58. Lee SA, Sinclair E, Hatano H et al (2014) Impact of HIV on CD8+ T cell CD57 expression is distinct from that of CMV and aging. PLoS One 9:e89444
59. Waldrop SL, Pitcher CJ, Peterson DM et al (1997) Determination of antigen-specific memory/effector CD4+ T cell frequencies by flow cytometry: evidence for a novel, antigen-specific homeostatic mechanism in HIV-associated immunodeficiency. J Clin Invest 99:1739–1750
60. Dan JM, Lindestam Arlehamn CS, Weiskopf D et al (2016) A cytokine-independent approach to identify antigen-specific human germinal center T follicular helper cells and rare antigen-specific CD4+ T cells in blood. J Immunol 197:983–993
61. Schöllhorn A, Maia A, Kimmerle F et al (2022) Staining of activated ß2-integrins in combination with CD137 and CD154 for sensitive identification of functional antigen-specific CD4+ and CD8+ T cells. Front Immunol 13: 1107366
62. Betts MR, Brenchley JM, Price DA et al (2003) Sensitive and viable identification of antigen-specific CD8+ T cells by a flow cytometric assay for degranulation. J Immunol Methods 281: 65–78

Chapter 23

An Ultrasensitive p24 Assay to Measure HIV-1 in Diverse Biological Matrixes

Callie Levinger, J. Natalie Howard, and Alberto Bosque

Abstract

Assays to study HIV persistence are crucial to evaluate therapeutic strategies aimed toward an HIV cure. Several assays have been developed to date that rely on the measurement of nucleic acids. In recent years, the advancement of ultrasensitive technologies for the detection of proteins has improved our understanding of the role of translation-competent reservoirs in HIV persistence. In this chapter, we describe the development of an ultrasensitive p24 ELISA that uses planar array technology. This assay allows for the detection of HIV-1 p24 in the low fg/ml range in different biological matrixes, including cell lysates. This assay can be used to investigate the efficacy of latency reversing agents to reactivate HIV or to evaluate the persistence of translation-competent reservoirs in people living with HIV (PWH) in cells or diverse biological fluids.

Key words HIV persistence, p24, Translation-competent HIV, Reservoirs, Latency reversal

1 Introduction

Human immunodeficiency virus-1 (HIV-1) is able to persist even in the presence of antiretroviral therapy (ART). The development of assays to evaluate HIV persistence has been mostly limited to the evaluation of viral nucleic acids, either DNA or RNA, due to their higher sensitivity compared to protein-based assays (for a review, *see* ref. 1). These assays allow for the analysis of sequence diversity, integrated virus, intactness of the HIV genome or assessment of RNA species, including splicing forms or poly-A mRNA [2–6]. However, these assays cannot evaluate the role of translation-competent viral reservoirs in HIV persistence. In the past years, several assays have been developed to evaluate expression of HIV proteins and in particular HIV Gag. These assays use flow cytometry [7, 8], enzyme-linked immunosorbent assay (ELISA) using a digital ELISA (dELISA) platform [9–15], or an enzyme-linked ImmunoSpot (ELISpot) assay called viral protein spot (VIP-Spot) [16]. We have recently developed an ultrasensitive p24

Vinayaka R. Prasad and Ganjam V. Kalpana (eds.), *HIV Protocols*, Methods in Molecular Biology, vol. 2807,
https://doi.org/10.1007/978-1-0716-3862-0_23,

immunoassay using the Quanterix Simoa planar array technology and the SP-X imaging and analysis system [17–20]. This immunoassay uses an anchor antibody microprinted in microwell plates specific for a peptide tag. The peptide tag is then conjugated to a p24 capture antibody via maleimide chemistry, enabling high-affinity binding of capture antibody to the anchor antibody. Samples are incubated directly, and bound p24 is sandwiched between the anti-p24 peptide-tagged capture antibody and biotinylated detector antibody. Next, streptavidin horseradish peroxidase (HRP) is added. Then, a mixture of chemiluminescent substrate Luminol and peroxide is added to the wells. The plate is scanned using Quanterix SP-X imager to measure the enhanced chemiluminescence signal generated from the enzymatic reaction of HRP with hydrogen peroxide and Luminol. This assay can be performed using volumes between 25 and 50 μl in a 96-well format, and it allows for the direct quantification in different matrixes without further manipulation in less than 5 h. The limit of detection of the assay varies as a function of the matrix used (Fig. 1). This planar SP-X immunoassay p24 ELISA could be an additional assay to evaluate HIV-1 cure approaches when sample volume is limited and to study HIV-1 persistence of translation-competent reservoirs in diverse cells and/or anatomical compartments.

2 Materials

All solutions must be prepared to be sterile and remain sterile throughout the procedure.

2.1 Antibody Preparation

1. Capture antibody: monoclonal anti-HIV-1/2 purified antibody cat#HIV-018-4830 (Capricorn Products LLC). Aliquot and store at −20 °C.
2. Detection antibody: anti-HIV type 1 p24 clone 39/5.4A. Aliquot and store at −20 °C.
3. Conjugation buffer: one powdered PBS pack dissolved in 500 mL ddH_20 (*see* **Note 1**). Store at 4 °C.
4. 2 mg Sulfo-SMCC (sulfosuccinimidyl 4-(N-maleimidomethyl) cyclohexane-1-carboxylate). Store at −20 °C.
5. 2 mg EZ-Link™ NHS-PEG4-Biotin. Store at −20 °C.
6. Simoa Planar Array Homebrew Tag 1. Store at −20 °C.
7. Simoa Planar Array Homebrew Diluent A. Store at 4 °C.
8. UltraPure™ 1 M Tris-HCl buffer, pH 7.5. Store at room temperature.
9. Amicon Ultra-0.5 centrifuge filter unit: Ultracel-50 regenerated cellulose membrane, 50 kDa MWCO.

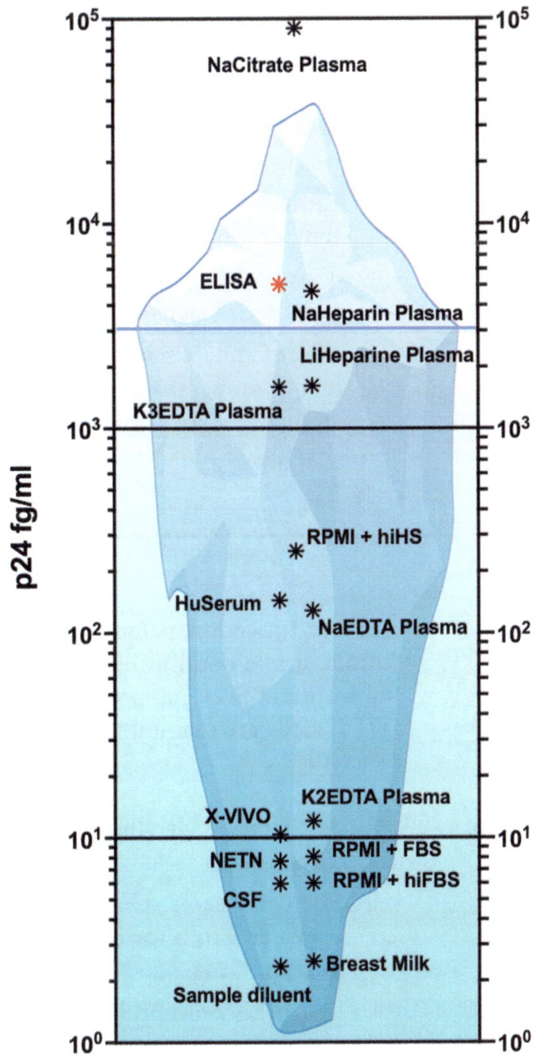

Fig. 1 Limit of detection in diverse matrixes using the ultrasensitive p24 planar array ELISA. Limit of detection of standard ELISAs is indicated in red. (Created with Biorender)

2.2 Cell Lysis Buffer

1. NETN: 100 mM NaCl, 20 mM Tris-Cl (pH 8.0) 0.5 mM EDT, 0.5% (v/v) Noidet P-40.
2. Protease inhibitors.
3. Phosphatase inhibitors.

2.3 Ultrasensitive Planar Array p24 Gag ELISA

1. Sample matrix (*see* **Note 2**).
2. Planar Array Homebrew Plate (1-spot) from Quanterix. Store at 4 °C.
3. Streptavidin HRP. Store at 4 °C.

4. SuperSignal Luminol Reagent. Store at 4 °C.
5. SuperSignal Peroxide Reagent. Store at 4 °C.
6. Purified HIV-1 p24 antigen protein to be used as standard. Store at 4 °C.
7. Wash buffer: 12.5× Concentrate of Simoa Planar Array Wash Buffer prepared using 2.85 L DI water. Store at room temperature.
8. Triton X-100. Store at room temperature. Add at 1% in each sample to inactive virus.
9. MicroClime environmental lid to reduce edge effects. Store at room temperature.
10. Blocking buffer: 5% nonfat dry milk powder in 10 mL Simoa Planar Array Diluent A.

3 Methods

All the procedures that contain live HIV have to be performed under sterile conditions in a biological safety cabinet wearing the appropriate personal protection equipment required to work with HIV. Cells are maintained in a cell culture incubator at 37 °C and 5% CO_2.

3.1 Preparation of Samples

1. All samples in biological fluids are inactivated using 1% Triton-X.
2. For cell lysates. Collect cells in 15 mL Falcon tube and spin down at 526 g for 5 min at 4 °C.
3. Aspirate media without disturbing cell pellet, resuspend pellet in 1 mL sterile PBS, and transfer to 1.5 mL Eppendorf tube.
4. Spin down cells at 935 g for 5 min at 4 °C.
5. Aspirate PBS without touching pellet.
6. Resuspend cell pellet in NETN cell lysis buffer supplemented with protease/phosphatase inhibitors at volume of 100 μL/ 1×10^6 cells.
7. Incubate cells on ice for 30 min.
8. Spin down Eppendorf tubes at 17,968 g for 20 min at 4 °C to remove insoluble fraction.
9. Collect supernatants into a new clean 1.5 mL Eppendorf tube, discard old tube with cellular debris.
10. Measure total protein concentration using a Bicinchoninic acid (BCA) assay.

3.2 Capture and Detection Antibody Buffer Exchange

1. Remove 2 mg vial of Sulfo-SMCC, 0.3 mg vial of Simoa Planar Array Homebrew Tag-1, and 2 mg NHS-PEG4-Biotin vial from freezer and let reach room temperature.
2. Place Amicon filter into Amicon microcentrifuge tube and pre-wet by adding 500 μL conjugation buffer.
3. Spin at 4 °C for 5 min at 14,000 g. Discard flow-through (*see* **Note 3**).
4. Calculate volume of antibody needed for 100 μg and add to separate microcentrifuge tube. Top off to 500 μL with conjugation buffer and vortex (*see* **Note 4**).
5. Add diluted antibody to Amicon filter unit. Spin at 4 °C for 5 min at 14,000 g. Discard flow-through.
6. Add 450 μL conjugation buffer to Amicon filter unit. Spin at 4 °C for 5 min at 14,000 g. Discard flow-through.
7. Add 450 μL conjugation buffer to Amicon filter unit. Spin at 4 °C for 5 min at 14,000 g. Discard flow-through.
8. Invert Amicon filter onto a new Amicon microcentrifuge tube (*see* **Note 5**).
9. Spin at 4 °C for 2 min at 1000 g. Do not discard flow-through (*see* **Note 6**).
10. Wash Amicon filter 5× per side of filter using 30 μL conjugation buffer (*see* **Note 7**).
11. Invert Amicon filter back into microcentrifuge tube. Spin at 4 °C for 2 min at 1000 g. Do not discard flow-through.
12. Measure the concentration of antibody and adjust the concentration to 1 mg/mL with conjugation buffer. Vortex to combine.

3.3 Capture Antibody Tag Conjugation

1. Transfer 70 μL of 1 mg/mL buffer exchanged capture antibody to fresh microcentrifuge tube.
2. Add 1 mL of diH_2O to the 2 mg Sulfo-SMCC vial (*see* **Note 8**). Let it sit at room temperature for 5 min. Vortex.
3. Add 3.1 μL of Sulfo-SMCC to the capture antibody (*see* **Note 9**). Vortex and incubate at room temperature for 30 min.
4. Add 3.65 μL of 1 M Tris-HCl pH 7.4 to stop reaction. Vortex.
5. Reconstitute Simoa Planar Array Homebrew Tag-1 with 250 μL of conjugation buffer. Vortex.
6. Add 15.4 μL of Tag-1 to the capture antibody-Sulfo-SMCC preparation. Vortex and incubate at room temperature for 30 min (can start detection antibody biotinylation during this step—*see* **Note 10**).

3.4 Detection of Antibody Biotinylation

1. Transfer 70 μL of 1 mg/mL buffer exchanged detection antibody to fresh microcentrifuge tube.
2. Reconstitute NHS-PEG4-Biotin with 170 μL of diH_2O. Vortex and let it sit at room temperature for 5 min. Vortex again.
3. Add 1.2 μL of NHS-PEG4-Biotin to buffer exchanged detection antibody (*see* **Note 11**). Vortex and incubate at room temperature for 30 min.

3.5 Purification of Capture and Detection Antibodies

1. Place Amicon filter into microcentrifuge and pre-wet by adding 500 μL conjugation buffer.
2. Spin at 4 °C for 5 min at 14,000 g. Discard flow-through (*see* **Note 3**).
3. Add 400 μL conjugation buffer to tagged capture antibody microcentrifuge tube and biotinylated detection antibody microcentrifuge tube.
4. Add diluted antibody to Amicon filter unit. Spin at 4 °C for 5 min at 14,000 g. Discard flow-through.
5. Add 450 μL conjugation buffer to Amicon filter unit. Spin at 4 °C for 5 min at 14,000 g. Discard flow-through.
6. Add 450 μL conjugation buffer to Amicon filter unit. Spin at 4 °C for 5 min at 14,000 g. Discard flow-through.
7. Invert Amicon filter onto a new Amicon microcentrifuge tube (*see* **Note 5**).
8. Spin at 4 °C for 2 min at 1000 g. Do not discard flow-through (*see* **Note 6**).
9. Wash filter 5× per side of filter using 30 μL conjugation buffer (*see* **Note 7**).
10. Invert Amicon filter back into microcentrifuge tube. Spin at 4 °C for 2 min at 1000 g. Do not discard flow-through.
11. Measure the concentration of antibody and adjust the concentration to 0.25 mg/mL with Simoa Planar Array Diluent A. Vortex to combine and store at 4 °C until ready to use (*see* **Note 12**).

3.6 Ultrasensitive Planar Array p24 Gag ELISA

1. Turn on automatic plate washer and run start of day wash protocol (*see* **Note 13**).
2. Calculate the volume of capture antibody needed for 1 μg/mL concentration and dilute in Simoa Planar Array Diluent A (*see* **Note 14**).
3. Prepare standard curve dilutions of the p24 antigen in matrix matching the samples (*see* **Note 15**).

4. Pipet standard curve dilutions and samples into 96-well round bottom preparation plate (*see* **Note 16**).
5. Open plate pouch and wash using SPX wash 2.0. Flick away excess wash buffer from wells (*see* **Note 17**).
6. Fill MicroClime lid with diH_2O (*see* **Note 18**).
7. Using a multichannel pipette, add 50 μL of diluted capture antibody per well.
8. Cover the plate with MicroClime lid and incubate at room temperature for 30 min on a plate shaker at 515 rpm.
9. Wash the plate using SPX wash 2.0 protocol on plate washer. Flick away excess wash buffer from wells.
10. Using a multichannel pipette, add 50 μL of standard curve dilutions in triplicates and sample from prep plate per well.
11. Cover the plate with MicroClime lid and incubate at room temperature for 2 h on a plate shaker at 515 rpm.
12. Wash the plate using SPX wash 2.0 protocol on plate washer. Flick away excess wash buffer from wells.
13. Prepare blocking buffer and use a multichannel to add 50 μL per well (*see* **Note 19**).
14. Cover the plate with MicroClime lid and incubate at room temperature for 30 min on a plate shaker at 515 rpm.
15. Wash the plate using SPX wash 2.0 protocol on plate washer. Flick away excess wash buffer from wells.
16. Calculate the volume of detection antibody needed for 1 μg/mL concentration and dilute in Simoa Planar Array Diluent A (*see* **Note 15**).
17. Cover the plate with MicroClime lid and incubate at room temperature for 30 min on a plate shaker at 515 rpm.
18. Wash the plate using SPX wash 2.0 protocol on plate washer. Flick away excess wash buffer from wells.
19. Using a multichannel pipette, add 50 μL of streptavidin-HRP per well (*see* **Note 20**).
20. Cover the plate with MicroClime lid and incubate at room temperature for 30 min on a plate shaker at 515 rpm.
21. Wash the plate using SPX post-HRP wash 2.0 protocol on plate washer. Flick away excess wash buffer from wells (*see* **Note 21**).
22. Prepare chemiluminescent substrate by mixing 3 mL of SuperSignal Luminol and 3 mL SuperSignal peroxide (*see* **Note 22**). Vortex briefly.
23. Using a multichannel pipette, add 50 μL of chemiluminescent substrate per well.

24. Read the plate immediately on Quanterix SPX imaging system.
25. The end of day wash 2.0 should be run following the conclusion of the assay (*see* **Note 23**).

3.7 Running Plate on Quanterix SPX Imager System

1. Enter field study ID (*see* **Note 24**).
2. Scan the barcode on the side of the p24 homebrew plate. Barcode will populate into imager software.
3. Enter the kit number, 100-0461, for the homebrew kit under kit barcode section of imager system.
4. Set the reference wells to the highest calibrator on the standard curve (*see* **Note 25**).
5. Confirm the plate processing is set to standard.
6. Press the measure button to initiate first scan of the plate (*see* **Note 26**).
7. If the target intensity has changed after initial scan, set it to 90%.
8. If the target exposure time is >300 s, or "inf," manually enter 300 s for target exposure time (*see* **Note 27**).
9. Press capture button and let imager read the plate. Export the . kit run file to be used in the analysis software.

3.8 Analyzing Results Using Quanterix SPX Software

1. Open the .kit run file generated by the imager system into the analysis software.
2. Open the template file matching the layout of the plate (*see* **Note 28**).
3. Press run plate option.
4. After the software analyzes the .kit run file; the data can be exported to an excel spreadsheet for further analysis (*see* **Note 29**).

3.9 Concluding Remarks

We have optimized a planar array p24 ELISA that allows for the detection of HIV-1 p24 in the fg/ml in different biological fluids and cell lysates. This assay can be used to evaluate translation-competent HIV reservoirs in different cell types and different anatomical compartments. As such, this assay could be used to evaluate the efficacy of HIV cure clinical trials deemed to reduce or eliminate latent reservoirs. Furthermore, another advantage of this platform is that it allows optimization with other antibodies as well as for other HIV antigens. In conclusion, we present here an additional tool to understand HIV persistence and to evaluate the efficacy of current and future clinical trials aimed at reducing translation-competent HIV reservoirs in PWH.

4 Notes

1. Magnetic stir bar/plate can be used to help dissolve the powdered PBS if clumps remain after initial mixing.
2. Sample matrix is the biological fluid of the sample such as assay diluent, tissue culture media, CSF, breast milk, or other fluid.
3. Remove Amicon filter by lifting up on rim. Discard flow-through in the microcentrifuge into waste container.
4. Calculations can be done using $C_1 V_1 = C_2 V_2$ with stock concentration of antibody. Subtract value from 500 to determine amount of conjugation buffer needed to dilute to 100 μg. Capture and detection antibodies can be prepared at the same time in different tubes.
5. Lift up on Amicon filter by the rim and place upside down into new Amicon microcentrifuge tube. The lid of the microcentrifuge tube will not close so place facing inward into centrifuge. Dispose of previous microcentrifuge tube.
6. The flow-through produced from this step will be the antibody exchanged into conjugation buffer. Do not discard.
7. Lift Amicon filter out of microcentrifuge and carefully hold the sides of the filter (avoid touching the flat filter part). Dispense the conjugation buffer over the filter and collect from the bottom of the filter, without touching the filter itself, to continue washing step.
8. Add 500 μL of diH_2O to the Sulfo-SMCC vial first and vortex to mix. Transfer to microcentrifuge tube and add an additional 500 μL to microcentrifuge tube. Vortex. Total volume should be 1 mL of diH_2O plus Sulfo-SMCC.
9. Resuspending the Sulfo-SMCC in 1 mL diH_2O creates a 30× molar excess of link to antibody. Can be stored at 4 °C if excess will be used in the same day. Discard any unused Sulfo-SMCC.
10. The process of detection antibody biotinylation can be started during **step 6** of the capture antibody tag conjugation step for joint 30-min incubation periods.
11. Resuspending the NHS-PEG4-Biotin in 170 μL diH_2O creates a 50× molar excess of biotin to antibody. Can be stored at 4 °C if excess will be used in the same day. Discard any unused NHS-PEG4-Biotin.
12. Use parafilm to seal the cap of the Amicon microcentrifuge tube to avoid the lid popping open.
13. The start of day wash protocol should be run at the beginning of the day each time an assay will be run. Make sure the buffer and di water containers are full before starting the wash protocol.

14. The total volume needed should be calculated based on how many wells will be used for each plate. Best to add in extra dead volume to account for pipetting errors.
15. The matrix used for the standard curve should include 1% Triton X-100 to match the samples if they have it. Mix each dilution thoroughly to evenly distribute the p24 antigen before making each dilution.
16. A preparation plate is used so that samples and standard curve can be pipetted onto p24 homebrew plate using a multichannel pipette. This avoids individually pipetting each sample into the well and avoids uneven timing of samples or standards sitting on the plate.
17. Flick plate firmly 4–5 times per side on paper towels to get rid of excess buffer in wells. Do not allow wells to dry completely.
18. The MicroClime lid has open troughs along all four sides on the lid. Place the pipette tip inside the trough and dispense the diH_2O on to the absorbent pad. Only the outer edges of all four sides of the absorbent pad needs to be wet (the middle of the pad does not need to be wet).
19. Ensure that the blocking buffer is mixed thoroughly, and no chunks of milk powder remain.
20. The streptavidin-HRP does not need to be diluted. It can be used straight from the bottle and added to the plates.
21. The post-HRP wash goes through five cycles of three washes each. After every cycle, the excess wash buffer needs to be flicked away before resuming the next cycle.
22. The chemiluminescent substrate needs to be prepared right before adding it to plate. Carefully pipette it onto plate to avoid creating bubbles. If bubbles arise, gently pop them.
23. The end of day wash uses di water only so ensure that bottle is filled, and the tubes are connected properly to that bottle. Turn off plate washer after wash protocol is finished.
24. The field study ID is the name of the experiment running. This makes identifying the plate file later easier if running multiple plates.
25. The imaging software scans the highest calibrator wells first to determine the target intensity and the time it will take to scan the whole plate. Important to always double check the highest calibrator is selected each time.
26. The first scan takes an initial 6 s scan of the plate based on the highest calibrator. This scan will give you the estimated measure of intensity and the time for the primary scan.

27. It is ideal to not let the plate scan for longer than 300 s. The longer the plate scans, the more the chemiluminescent substrate will wear down. A secondary scan can be added to lessen background noise if the initial target exposure time is low. A secondary scan is not necessary for every plate.
28. A plate layout template can be created in advance of running the plate so that it can be seamlessly uploaded and applied to the plate. On the layout template, the kit number will be added so that the software recognizes the area of the well that has been pre-spotted. In the layout, the concentrations of the standard curve can also be added as well as dilution factor for the samples if used.
29. The standard curve can be optimized manually in the software if there are any wells in the run that our outliers.

Acknowledgments

The work in A.B.'s lab is currently supported by the National Institute of Allergy and Infectious Diseases and National Institute of Health grants R01-AI154518, R01-AI147812, UM1-AI164565, T32-AI158105, R21-AI172042, R61-AI176582, and P30-AI117970

References

1. Abdel-Mohsen M, Richman D, Siliciano RF, Nussenzweig MC, Howell BJ, Martinez-Picado J, Chomont N, Bar KJ, Yu XG, Lichterfeld M, Alcami J, Hazuda D, Bushman F, Siliciano JD, Betts MR, Spivak AM, Planelles V, Hahn BH, Smith DM, Ho YC, Buzon MJ, Gaebler C, Paiardini M, Li Q, Estes JD, Hope TJ, Kostman J, Mounzer K, Caskey M, Fox L, Frank I, Riley JL, Tebas P, Montaner LJ, infection B-HDCtCH (2020) Recommendations for measuring HIV reservoir size in cure-directed clinical trials. Nat Med 26(9):1339–1350. https://doi.org/10.1038/s41591-020-1022-1
2. Procopio FA, Fromentin R, Kulpa DA, Brehm JH, Bebin AG, Strain MC, Richman DD, O'Doherty U, Palmer S, Hecht FM, Hoh R, Barnard RJ, Miller MD, Hazuda DJ, Deeks SG, Sekaly RP, Chomont N (2015) A novel assay to measure the magnitude of the inducible viral reservoir in HIV-infected individuals. EBioMedicine 2(8):874–883. https://doi.org/10.1016/j.ebiom.2015.06.019
3. Butler SL, Hansen MS, Bushman FD (2001) A quantitative assay for HIV DNA integration in vivo. Nat Med 7(5):631–634. https://doi.org/10.1038/87979
4. O'Doherty U, Swiggard WJ, Jeyakumar D, McGain D, Malim MH (2002) A sensitive, quantitative assay for human immunodeficiency virus type 1 integration. J Virol 76(21):10942–10950
5. Bruner KM, Wang Z, Simonetti FR, Bender AM, Kwon KJ, Sengupta S, Fray EJ, Beg SA, Antar AAR, Jenike KM, Bertagnolli LN, Capoferri AA, Kufera JT, Timmons A, Nobles C, Gregg J, Wada N, Ho YC, Zhang H, Margolick JB, Blankson JN, Deeks SG, Bushman FD, Siliciano JD, Laird GM, Siliciano RF (2019) A quantitative approach for measuring the reservoir of latent HIV-1 proviruses. Nature 566(7742):120–125. https://doi.org/10.1038/s41586-019-0898-8
6. Yukl SA, Kaiser P, Kim P, Telwatte S, Joshi SK, Vu M, Lampiris H, Wong JK (2018) HIV latency in isolated patient CD4(+) T cells may be due to blocks in HIV transcriptional elongation, completion, and splicing. Sci Transl Med 10(430). https://doi.org/10.1126/scitranslmed.aap9927

7. Pardons M, Baxter AE, Massanella M, Pagliuzza A, Fromentin R, Dufour C, Leyre L, Routy JP, Kaufmann DE, Chomont N (2019) Single-cell characterization and quantification of translation-competent viral reservoirs in treated and untreated HIV infection. PLoS Pathog 15(2):e1007619. https://doi.org/10.1371/journal.ppat.1007619
8. Grau-Exposito J, Luque-Ballesteros L, Navarro J, Curran A, Burgos J, Ribera E, Torrella A, Planas B, Badia R, Martin-Castillo-M, Fernandez-Sojo J, Genesca M, Falco V, Buzon MJ (2019) Latency reversal agents affect differently the latent reservoir present in distinct CD4+ T subpopulations. PLoS Pathog 15(8):e1007991. https://doi.org/10.1371/journal.ppat.1007991
9. Cabrera C, Chang L, Stone M, Busch M, Wilson DH (2015) Rapid, fully automated digital immunoassay for p24 protein with the sensitivity of nucleic acid amplification for detecting acute HIV infection. Clin Chem 61(11): 1372–1380. https://doi.org/10.1373/clinchem.2015.243287
10. Chang L, Song L, Fournier DR, Kan CW, Patel PP, Ferrell EP, Pink BA, Minnehan KA, Hanlon DW, Duffy DC, Wilson DH (2013) Simple diffusion-constrained immunoassay for p24 protein with the sensitivity of nucleic acid amplification for detecting acute HIV infection. J Virol Methods 188(1–2):153–160. https://doi.org/10.1016/j.jviromet.2012.08.017
11. Passaes C, Delagreverie HM, Avettand-Fenoel-V, David A, Monceaux V, Essat A, Muller-Trutwin M, Duffy D, De Castro N, Wittkop L, Rouzioux C, Molina JM, Meyer L, Delaugerre C, Saez-Cirion A (2021) Ultrasensitive detection of p24 in plasma samples from people with primary and chronic HIV-1 infection. J Virol. https://doi.org/10.1128/JVI.00016-21
12. Wu G, Cheney C, Huang Q, Hazuda DJ, Howell BJ, Zuck P (2021) Improved detection of HIV gag p24 protein using a combined immunoprecipitation and digital ELISA method. Front Microbiol 12:636703. https://doi.org/10.3389/fmicb.2021.636703
13. Stuelke EL, James KS, Kirchherr JL, Allard B, Baker C, Kuruc JD, Gay CL, Margolis DM, Archin NM (2020) Measuring the inducible, replication-competent HIV reservoir using an ultra-sensitive p24 readout, the digital ELISA viral outgrowth assay. Front Immunol 11: 1971. https://doi.org/10.3389/fimmu.2020.01971
14. Bosque A, Nilson KA, Macedo AB, Spivak AM, Archin NM, Van Wagoner RM, Martins LJ, Novis CL, Szaniawski MA, Ireland CM, Margolis DM, Price DH, Planelles V (2017) Benzotriazoles reactivate latent HIV-1 through inactivation of STAT5 SUMOylation. Cell Rep 18(5):1324–1334. https://doi.org/10.1016/j.celrep.2017.01.022
15. Macedo AB, Novis CL, De Assis CM, Sorensen ES, Moszczynski P, Huang SH, Ren Y, Spivak AM, Jones RB, Planelles V, Bosque A (2018) Dual TLR2 and TLR7 agonists as HIV latency-reversing agents. JCI Insight 3(19). https://doi.org/10.1172/jci.insight.122673
16. Puertas MC, Bayon-Gil A, Garcia-Guerrero MC, Salgado M, Urrea V, Moron-Lopez S, Pena R, Jimenez-Moyano E, Clotet B, Prado JG, Martinez-Picado J (2021) Erratum for Puertas et al., "VIP-SPOT: an innovative assay to quantify the productive HIV-1 reservoir in the monitoring of cure strategies". mBio 12(5):e0253521. https://doi.org/10.1128/mBio.02535-21
17. Lambert J, Mendes M, Yan Y-X, Ball A (2019) A universal planar assay format for high sensitivity cytokine quantification in human serum and plasma. J Immunol 202(1 Supplement):131.111
18. Cheng J, Myers TG, Levinger C, Kumar P, Kumar J, Goshu BA, Bosque A, Catalfamo M (2022) IL-27 induces IFN/STAT1-dependent genes and enhances function of TIGIT(+) HIVGag-specific T cells. iScience 25(1): 103588. https://doi.org/10.1016/j.isci.2021.103588
19. Innis EA, Levinger C, Szaniawski MA, Williams E, Alcami J, Bosque A, Schiffer JT, Coiras M, Spivak AM, Planelles V (2021) Pharmacologic control of homeostatic and antigen-driven proliferation to target HIV-1 persistence. Biochem Pharmacol 194:114816. https://doi.org/10.1016/j.bcp.2021.114816
20. Levinger C, Howard J, Cheng J, Tang P, Joshi A, Catalfamo M, Bosque A (2021) An ultrasensitive planar array p24 Gag ELISA to detect HIV-1 in diverse biological matrixes. Sci Rep 11(1):23682. https://doi.org/10.1038/s41598-021-03072-7

INDEX

Vinayaka R. Prasad and Ganjam V. Kalpana (eds.), *HIV Protocols*, Methods in Molecular Biology, vol. 2807, https://doi.org/10.1007/978-1-0716-3862-0,